THE CONTINGENT NATURE OF LIFE

INTERNATIONAL LIBRARY OF ETHICS, LAW, AND THE NEW MEDICINE

VOLUME 39

The titles published in this series are listed at the end of this volume.

The Contingent Nature of Life

Bioethics and Limits of Human Existence

Edited by

MARCUS DÜWELL
Universiteit Utrecht, The Netherlands

CHRISTOPH REHMANN-SUTTER
Universität Basel, Switzerland

and

DIETMAR MIETH
Universität Tübingen, Germany

Springer

Marcus Düwell
Universiteit Utrecht
The Netherlands

Christoph Rehmann-Sutter
Universität Basel
Switzerland

Dietmar Mieth
Universität Tübingen
Germany

ISBN: 978-90-481-7717-2 e-ISBN: 978-1-4020-6764-8

Library of Congress Control Number: 2007938399

Printed on acid-free paper

9 8 7 6 5 4 3 2 1

springer.com

Contents

Author Biographies

Ahmet Hadi Adanali currently teaches at the Divinity School of Ankara University. He holds an MA in analytical philosophy from the Middle East Technical University at Ankara and a Ph.D. in Islamic Philosophy from the University of Chicago. He has taught at various universities including Gregorian University in Rome and has published in scholarly journals. In the years 2005–2006, he was involved as the Turkish coordinator in the UN Project of the Alliance of Civilizations. His research focuses on questions that relates to borderline topics between philosophy and theology, and science and religion.

Brenda Almond is Emeritus Professor of Moral and Social Philosophy at the University of Hull and is a Member of the Human Genetics Commission in the UK. She is the author of a number of books including The Fragmenting Family (Oxford University Press, 2006) and Ethiek: Reis naar het land van goed en kwaad (Wereldbibliotheek, Amsterdam, 2001).

Joel Anderson was educated at Princeton, Northwestern, and Frankfurt Universities and taught at Washington University in St. Louis for 9 years before joining the Philosophy Department of Utrecht University. His research focuses on philosophical anthropology (esp. links between autonomy, agency, and normativity), ethics (esp. discourse ethics and neuro-ethics), and social theory (esp. "recognition theory" and the conditions for developing autonomy skills). He edited Free Will as Part of Nature: Habermas and His Critics (special issue of Philosophical Explorations, March 2007) and Autonomy and the Challenges to Liberalism (with John Christman, Cambridge UP, 2005). His current book project is entitled "Scaffolded Autonomy: The Construction, Impairment, and Enhancement of Human Agency".

Bernard Baertschi is Maître d'enseignement et de recherche (MER) at the Institute of Biomedical Ethics in the University of Geneva (Switzerland). He graduated from the University of Fribourg and obtained his doctoral degree in Philosophy at the University of Geneva in 1979 (with a study on a French philosopher of the early nineteenth century: Maine de Biran). After having published several papers and books on this period, his philosophical interests shifted to ethics and bioethics. On those topics, his two main books are La valeur de la vie humaine et l'intégrité de la personne (Paris, Vrin, 1995) and Enquête philosophique sur la dignité. Anthropologie et éthique des biotechnologies (Genève, Labor and Fides, 2005).

Gerhold K. Becker, currently Visiting Professor at the Graduate School of Philosophy, Assumption University, Bangkok, retired in 2004 as Chair Professor of Philosophy and Religion and Founding Director of the Centre for Applied Ethics after more than 18 years of teaching and research at Hong Kong Baptist University. From 1996–2004, he was a member of the Hong Kong Government's Council on Human Reproductive Technology and from 2000–2004 he served as chairman of the Council's Ethics Committee. He is a Research Fellow of the Centre for Business Ethics of the Shanghai Academy of Social Sciences, associate editor of Rodopi's Value Inquiry Book Series (Amsterdam: New York), and the editor of its special series Studies in Applied Ethics.

Deryck Beyleveld is Professor of Law and Bioethics at Durham before which he was Professor of Jurisprudence at the University of Sheffield (where he founded and directed the Sheffield Institute of Biotechnological Law and Ethics (SIBLE)). In 2006 he held the Belle van Zuylen Visiting Chair (in Human Rights and Bioethics) at the University of Utrecht. His publications cover many areas of law, legal philosophy, moral philosophy, and applied ethics (especially bioethics), and include The Dialectical Necessity of Morality (1991) and (with Roger Brownsword) Law as a Moral Judgment (1986), Mice, Morality and Patents (1993), Human Dignity in Bioethics and Biolaw (2001), and Consent in the Law (2007).

Rosi Braidotti is Distinguished Professor in the Humanities at Utrecht University in the Netherlands. She has published extensively in feminist philosophy, epistemology, poststructuralism and psychoanalysis. Her books include Patterns of Dissonance. Cambridge, Polity Press, 1991; Nomadic Subjects: Embodiment and Sexual Difference in Contemporary Feminist Theory. New York: Columbia University Press, 1994; Metamorphoses: Towards a Materialist Theory of Becoming Polity Press, 2002. She has co-edited the following: Women, the Enviromment, and Sustainable Development. Towards a Theoretical Synthesis (with E. Charkiewicz, S. Hausler and S. Wieringa) Zed Books, 1994; Between Monsters, Goddesses and Cyborgs (with Nina Lykke). London: Zed Books, 1996; Thinking Differently. A European Women's Studies Reader (with Gabriele Griffin) Zed Books, 2002; and numerous essays and chapters in various texts. Her latest book is : Transpositions. On Nomadic Ethics, Polity Press, 2006.

Daniel Callahan was a cofounder of The Hastings Center in 1969 and served as its President from 1969–1996. He is a Senior Fellow at the Harvard Medical School and a Senior Fellow at Yale and also an Honorary Faculty Member of the Charles University Medical School in Prague, the Czech Republic. He won the 1996 Freedom and Scientific Responsibility Award of the American Association for the Advancement of Science. His research interests have ranged from the beginning to the end of life, health policy and research policy. His project on medicine and the market is examining the impact of market theory, thinking, and practice on health care systems. A related interest is that of globalization and it impact on health status in different parts of the world. He is the author or editor of 39 books.

Norbert Campagna, born 1963, Dr. Phil., Associate Assistant-Professor at the Philosophy Departement of the Université du Luxembourg. He has published 16 books in the domains of political philosophy, philosophy of law and sexual ethics. Last publications: Le droit, la nature et la volonté (Paris 2006), Strafrecht und unbestrafte Straftaten (Stuttgart 2007). Email-Address: norbertcampagna@hotmail.com

Ole Döring holds an MA Phil from Göttingen University and a Ph.D. from Bochum University. He studied Philosophy and Sinology as majors, in Tübingen and Göttingen. Since 1996, he has conducted several research projects on the state of East Asia-related sciences, and on bioethics in China and East Asia. He has been a senior research fellow with the Faculty for East Asian Studies at Bochum University since 2001, with a research project on Culture-Transcending Bioethics (KBE). He has been teaching at the universities of Bochum, Bremen, Hamburg and Lüneburg, and in various universities in the PR China and Taiwan. In 2004, he accepted a position of a Visiting Professor at Hunan Normal University, Changsha (PR China). He has organised several international bioethics conferences, and has conducted extended field research and training courses in China. Since 2006, he has been a core group member of the European-Chinese research consortium BIONET under the EC FP6, at the GIGA-Institute of Asian Studies (Hamburg, Germany). His recent book is entitled Chinas Bioethik verstehen (Understanding China's Bioethics), Hamburg 2004.

Marcus Düwell (born in 1962) holds a chair for philosophical ethics at the Department for Philosophy at Utrecht University. He is research director of the Ethics Institute of Utrecht University, director of the Netherlands Research School for Practical Philosophy and director of the Leiden-Utrecht Research Institute ZENO. From 1993–2001 he was academic coordinator of the Interdepartmental Center for Ethics in the Sciences and Humanities at the University of Tübingen. His research interests include bioethics (especially ethics of genetics, environmental ethics) and basic questions of moral philosophy (foundations of individual rights, human dignity) and the relation between ethics and aesthetics. He is editor-in-Chief of the book series "Ethics and Applied Philosophy" (Springer publisher).

Sigrid Graumann has studied Biology and Philosophy at the University of Tuebingen. She was scholar of the Postgraduate College, "Ethics in the Sciences and Humanities" at the University of Tuebingen. She has got her Ph.D. with the thesis: "Somatic Gene Therapy for Monogenetic Diseases – Conceptual and Ethical Questions". Since April 2002 she is Senior Researcher at the Institute Man, Ethics and Science, Berlin. She was member of the Enquete-Commission "Ethics and Law of Modern Medicine" of the German Bundestag 2000–2002 and 2003–2005, and she is member of the Central Ethics Commission of the German Medical Association and board-member of the Academy of Ethics in Medicine (Germany). Her research interests are ethical questions of biomedical research, feminist and disability ethics.

Hille Haker is Professor of Theological Ethics at Goethe-University, Frankfurt, and since 2005 a member of the European Group on Ethics in Sciences and New Technologies (EGE). From 2003–2005, she was Associate Professor of Christian Ethics at Harvard University. From 1989–2003 she taught at the University of Tuebingen and directed EU-projects in bioethics at the Center for Ethics. Her books include Ethik der genetischen Frühdiagnostik (2002) Moralische Identität (1999), and the co-edited volumes: Ethik-Geschlecht-Wissenschaften (2006) and The Ethics of Genetics in Human Procreation (2000). She is director of a international critical Catholic journal "Concilium", and in this function co-edited several issues, the latest on Women's voices and HIV-AIDS.

Prof. Dr. Dr.h.c. Ludger Honnefelder is Professor em. of Philosophy at the University of Bonn. Since 2005 Guardini Professor at the Humboldt-University of Berlin. He is a member of the Northrhine-Westphalian Academy of Sciences, the Standing Committee "Science & Ethics" of All European Academies (ALLEA), and the Steering Committee for Bioethics (CDBI) of the Council of Europe. Professor Honnefelder was director of the Institute for Science and Ethics at the University of Bonn and the German Reference Centre for Ethics in the Life Sciences (Bonn) and a member of the Enquete Commission "Law and Ethics of Modern Medicine" of the German Bundestag (from 2000–2002). His main scientific interests are Ethics, Applied Ethics/Bioethics, Metaphysics, History of Medieval Philosophy.

Matthias Kettner is dean of the faculty of humanities and liberal arts at Witten/Herdecke university (www.uni-wh.de) in Germany. He was assistant professor from 1994–2000 at the Institute for Advanced Studies in the Humanities at Essen (www.kwi-nrw.de) where his research focused on clinical ethics committees in Germany, communicative ethics, and the relationship of applied ethics, biopolitics and democracy. Prior to that, he collaborated with Karl-Otto Apel and Jürgen Habermas in research on discourse ethics at Frankfurt University where he had earned a Ph.D. in philosophy and a diploma in psychology. Recently edited books: "Biomedizin und Menschenwürde" (2004) "Angewandte Ethik als Politikum" (2001), both by Suhrkamp Verlag, Frankfurt. contact: kettner@uni-wh.de

Michiel Korthals is Professor of Applied Philosophy at Wageningen University. He studied Philosophy, Sociology, German and Anthropology at the University of Amsterdam and the Karl Ruprecht Universität in Heidelberg. His academic interests include bioethics and ethical problems concerning food production and environmental issues, deliberative theories, and American Pragmatism. Main publications: Filosofie en intersubjectiviteit, Alphen a/d Rijn, 1983; Duurzaamheid en democratie (Sustainability and democracy, Boom, 1995); Philosophy of Development (Kluwer, 1996 with Wouter van Haaften and Thomas Wren), Tussen voeding en medicijn (Between Food and Medicine), Utrecht 2001, Pragmatist Ethics for a Technological Culture (with Keulartz et al.; Kluwer 2002), Ethics for Life Sciences (Springer, 2005), Before Dinner. Philosophy and Ethics of Food (Springer 2004), and in 2006, Pépé Grégoire, Een filosofische duiding van zijn beelden/A Philosophical Interpretation of his Sculptures, Zwolle: Waanders.

Professor Sheila A.M. McLean LL.B., M.Litt., Ph.D., LL.D. (Edin), LL.D. (Abertay, Dundee), F.R.S.E., F.R.C.G.P., F.Med.Sci., F.R.C.P.E. (Edin), F.R.S.A. Professor McLean is the first holder of the International Bar Association Chair of Law and Ethics in Medicine at Glasgow University and is Director of the Institute of Law and Ethics in Medicine at Glasgow University. She has acted as a consultant to the World Health Organisation, the Council of Europe, and a number of individual states. She has acted as an expert reviewer for many of the major grant awarding bodies and similar organisations outwith the United Kingdom. She has published extensively in the area of medical law, is on the Editorial Board of a number of national and international Journals and is regularly consulted by the media on matters of medical law and ethics. In 2005 she was awarded the first ever Lifetime Achievement Award by the Scottish Legal Awards.

Dietmar Mieth (*1940) has a Chair for Theological Ethics at the Catholic Faculty of Theology at the University of Tübingen. From 1990–2001 he was founding director of the Inter-Departemental Center for Ethics in the Sciences and Humanities at the same university. He was Member of the ethical advisory groups of the European Commission (European Group on Ethics in the Sciences and New Technologies), the German Federal Ministry of Health and the Enquete-Committee of the German Parliament "Ethics and Law in Modern Medicine". His research interests include social ethics, bioethics, narrative ethics, religious experience and medieval mysticism.

Maurizio Mori is professor of bioethics at the University of Turin, Italy. Since 1985 he is coordinator of the bioethics branch of the Center "Politeia" in Milan; and since 2006 he is President of the Consulta di Bioetica, a national Association founded in 1989 to promote bioethics in a pluralistic attitude. He is editor in chief of Bioetica. Rivista interdisciplinare, the only bioethical journal open to ethical pluralism in Italy. He has written 5 books and over 250 papers published in Italian and in international journals. His major interests are in reproductive issues and end of life; stem cells; history and nature of bioethics, truth telling, Ethics Committees, resource allocation. He contributed also to ethical issues on environment and non-human animals, as well as to business ethics.

Albert W. Musschenga (1950) is Director of the Blaise Pascal Institute and Professor of Ethics at the Vrije Universiteit, Amsterdam. He published books on Quality of Life (1987), Integrity (2004), (co-)edited a number of volumes about a wide range of subjects. He also published a great number of articles on practical and theoretical ethics. His current research focuses on the relation between empirical sciences and ethical theory.

Christoph Rehmann-Sutter, born 1959, is currently directing the Unit for Ethics in the Biosciences at the University of Basel, Switzerland. He studied first molecular and subsequently philosophy and sociology. Ph.D. in philosophy of biology and bioethics in 1995 at Technical University of Darmstadt. 2000 venia docendi in Philosophy at University of Basel. 1997–1998 he was research fellow at the University of California Berkeley. Since 2001 President of the Swiss National Advisory Commission on Biomedical Ethics (NEK-CNE), which gives recommendations for the Swiss government and parliament. Since 2008 also Visiting Professor at the London School of Economics and Political Science (LSE). Research and writing interests: philosophical foundations of bioethics, sociological methods in bioethics and medical ethics, ethical issues of gene therapy, genetic testing, embryo research, and technical risks, genome theories, ethics of counselling.

Carlos M. Romeo-Casabona is Professor of Criminal Law at The University of the Basque Country and Director of the Inter-University Chair BBVA Foundation-Provincial Government of Biscay in Law and the Human Genome, at the UD and of

UPV/EHU. He has a Doctorate in Law and in Medicine. He is author, co-author and editor of several books and numerous articles, published in seven languages. He is a member of the Spanish National Commission on Assisted Human Reproduction, of the steering Bioethics Committee of the Council of Europe, of the Ethics Committee of HUGO, Chairman of the Ethics Committee on Clinical Research of the Autonomous Community of the Basque Country. He is an Adviser of the Spanish Minister of Health and of the Ministry of Justice. He has been awarded five Honorary Degrees "Doctor Honoris Causa" from several Latin-American and Spanish Universities.

Tom Shakespeare was trained in sociology at the University of Cambridge, and has researched and taught at the Universities of Sunderland, Leeds and Newcastle, where he is currently research fellow at the Policy, Ethics and Life Sciences Institute. His books include The Sexual Politics of Disability and most recently Disability Rights and Wrongs.

Dr. Carmel Shalev is a human rights lawyer and ethicist, who specializes in reproduction, health and biotechnology. She has served as legal advisor to the Ministry of Health and on numerous public committees in the area of health ethics in Israel. She also served as an expert member of the United Nations Committee on the Elimination of All Forms of Discrimination Against Women (CEDAW), where she specialized in women and health. Between 1998 and 2004 Dr. Shalev established and directed the Unit of Health Rights and Ethics at the Gertner Institute for Health Policy Research, Tel Hashomer. She now teaches at the Tel Aviv University Law Faculty, and other academic institutions in Israel, and works as a local and international consultant in the field of biomedicine and research ethics.

Ludwig Siep teaches philosophy at the university of Münster/Germany. He is chairman of the Central Ethics Committee of Stem-Cell Research (Berlin), head of the Center for bioethics at the University of Münster, member of the Northrhine-Westphalian and corresponding member of the Bavarian Academy of Sciences. His fields of research are ethics, applied ethics and history of philosophy, especially German Idealism. Among his books: Hegels Fichtekritik und die Wissenschaftslehre von 1804 (1970, Japan Transl. 2001); Anerkennung als Prinzip der praktischen Philosophie (1979, Italian Transl. 2007); Praktische Philosophie im Deutschen Idealismus (1992); Der Weg der Phänomenologie des Geistes (2001); Konkrete Ethik (2004, Japan Transl. 2006); (Ed.), Hegels Grundlinien der Philosophie des Rechts (2005).

Theo van Willigenburg (b. 1960, Ph.D. 1991 Utrecht University) was Socrates professor of medical ethics at the Amsterdam Medical School, professor of practical philosophy in the faculty of Philosophy Erasmus University Rotterdam and professor of business ethics at the Rotterdam School of Management. He is currently affiliated to the University of Amsterdam and is director of the Kant Academy Utrecht. His main research interest are in metaethics (sources of normativity and motivation), neurophilosophy and the philosophy of emotions. He has written several textbooks and has published in journals like Philosophical Investigations, Journal of Applied Philosophy, The Journal of Medicine and Philosophy, Utilitas, Philosophical

Explorations, Ethical Theory and Moral Practice, The Journal of Value Inquiry, and The Journal of Medicine and Philosophy.

Boris Yudin, born in 1943 in Moscow, Russia, philosopher, professor of the Lomonosov's Moscow State University, Dr. of Sciences (philosophy), Corresponding Member of the Russian Academy of Sciences (RAS). Main fields of research interests – ethics of science and technology, bioethics (ethics of biomedical research, ethical issues of the beginning and the end of human life, human enhancement). Head of department of Comprehensive Problems of Human Studies, Institute of Philosophy of the RAS. Editor-in-Chief of the scientific and popular journal "Chelovek" (the human being). Vice-Chairman of the Russian Committee of Bioethics. Representative of Russia in the Steering Committee on Bioethics of Council of Europe. In 1998–2007 – Member of Board of Directors, International Association of Bioethics.

Laurie Zoloth is Director of the Center for Bioethics, Science and Society and Professor of Medical Ethics and Humanities at Northwestern University, Feinberg School of Medicine, and Professor of Religion and a member of the Jewish Studies faculty at Northwestern University, Weinberg College of Arts and Science. In 2005, after two terms of service on NASA's National Advisory Council, she received the NASA Public Service Medal. In 2001, she was President of the American Society for Bioethics and Humanities. Professor Zoloth is the chair of the HHMI Bioethics Committee and is a founding board member of the International Society for Stem Cell Research, and of the Society for Neuroethics, on NASA's first Planetary Protection Advisory Committee, and on the boards of several national organizations. In 2007 she was nominated to the Institute of Medicine. She is the author of The Ethics of Encounter: A Jewish Perspective on Health Care Justice, and is co-editor of two books on religion, science, and ethics. In 2000, she was co-PI for an NIH ELSI grant on identity, ethnicity, and citizenship after the Human Genome Project. She has published extensively in the areas of ethics, family, feminist theory, religion and science, Jewish Studies, and social policy. Her current research projects include work on the emerging issues in medical and research genetics and on the ethical issues in stem cell research, and her research interest in distributive justice in health care continues.

Introduction

The development of bioethics has presented us with an ever increasing number of very different discussions over the last four decades. Bioethicists were initially concerned about questions of reproduction, end of life, organ transplantation, and a broad range of moral problems raised by the forward march of the life sciences. Meanwhile these sciences grew to be a major influence in nearly all areas of our lives. Biotechnology has brought about considerable changes in agriculture, plant breeding, pharmacy, veterinary medicine and medicine in general. These scientific and technological changes in turn are having a profound influence on economy, law, politics and culture. The life sciences are now certain to change our world in important ways.

Because of their potentially all-pervasive and highly diverse impact, bioethical discussions concerning the life sciences are no longer simply about ethical guidelines or legal regulation of concrete technologies. Certainly, the on-going debates concerning rules and regulations are complicated – and becoming more so. Nevertheless, bioethics cannot be restricted to these topics – they cover but a fraction of the social and personal consequences of bio-technological change. The life sciences drive us to rethink long-time-honoured concepts of humanness, of personhood, of nature. Bioethics therefore needs to develop an understanding of the impact those changes have on the conceptualization of the ethical dimension of the life sciences.

The normative framework we might use for the evaluation of the life sciences is itself a matter of dispute. Not only are we confronted with a variety of ethical theories – a challenge for ethics in general – but also with very specific conceptual issues arising in the more specialized area of bioethics. It seems unavoidable therefore to choose a much broader perspective for an adequate discussion of the moral dimension of the whole impact of the life sciences.

The focus of this book is the notion of "contingency". Why? Because it seems as if the self-imposed mission of the life sciences amounts to a declaration of war on a specific characteristic of nature in general and of human nature in particular. Key words here are: imperfect, uncontrollable, largely (and perhaps permanently) unknowable, that is to say: contingent. Nature and Life are like deities fond of surprising us. And surely the unpredictable nature of life is what makes it so exiting. But at the same time it sets the limits for regulation and control. The contingency of life is a challenge for medicine and technology. Life sciences seem to broaden

M. Düwell et al. (eds.), *The Contingent Nature of Life*,

the possibilities of control to an extent that the contingency of life and nature is no longer self-evident. Today's very broad diagnoses raises a lot of serious questions. Are they valid diagnoses? Are the life sciences really defying the contingency of our existence? Or are they only manipulating us with utopian promises? And if contingency is really being challenged, why should we worry about it? After all, contingency is just a disturbing factor in our worldview, is it not? Or should we say that the contingency of our natural existence provides us with important sources of meaning and motivation? Is contingency essential for a meaningful life and way of life? To focus on contingency is to explore a notion that is of crucial importance in many cultures and religions and, simultaneously, a driving force in the life sciences.

This volume presents several perspectives on current debates in bioethics. It is part of a series of research activities that were discussed at a number of conferences supported by the European Science Foundation. The first volume of this series was published in 2006 under the title *Bioethics in Cultural Contexts*. That book had a methodological focus and was a collection of papers about different approaches in bioethics. The present volume concentrates on some fundamental philosophical concepts crucial to bioethics. The title, *The contingent nature of life. Bioethics and the limits of human existence*, refers to some of the ethically most challenging theoretical ideas touched by the life sciences.

The first section, "Contingency of life and the ethical", explores the different dimensions of how the contingency of life, and especially human life, is relevant to ethical discussions. The aspiration of the life sciences as a global enterprise is *knowledge* about life and nature and, concurrently, the development of methods for *intervention* in life and nature. These sciences challenge the contingent aspects of the natural environment and of the nature of humans. Life sciences are driven by the idea that we are about to achieve a far more powerful and specific influence on natural processes than ever before. However, biologists are very careful about promising control over the biological basis of human beings and their life conditions. The genetic determinism that seems to be a necessary presupposition for the project of the life sciences is highly controversial. In the discussions, the notion of "contingency" is on the agenda again and is an index of the highly complex relationship between life sciences and the philosophical self-interpretation of human beings. The main goal of this section is to identify the new dimensions of our philosophical concepts of nature, life and contingency in the context of the life sciences, and to explore to what extent this influences ethical debates. Changes in our understanding of nature and life will change the limits and scope of human existence. Therefore this section is also closely linked to the issue of the first conference at which the notion of "finitude" was very prominent.

In the second section, "Ethical theories and the limits of life sciences", several papers deal with the challenge issued by the life sciences to our normative frameworks. It is the task of ethical analysis to provide us with a justification of the principles by which we morally evaluate the life sciences, and to determine moral limits in a transparent and non-arbitrary manner. This task of ethical reflection is, however, challenged by the life sciences in several respects. The impact of the life sciences on our concepts of personhood, human nature, vulnerability and the like is not only important for the self-interpretation of human beings from an anthropological and hermeneutical perspective, but is also significantly influential for normative concepts. On the one hand the nature of ethical reflection and moral judgment are much debated. Bioethical discussions have forced us to reflect more deeply on how ethical evaluations are made, how to combine empirical and philosophical reflection, and how to come to concrete but philosophically defensible judgments. On the other hand, concepts of moral protection are questioned in view of new fields of activity such as intervention in human procreation and reproduction. In this context the issue came to be: what needs to be morally protected after all? Debates on moral protection and moral rights refer to aspects of human existence that may deserve protection. Moral protection presupposes vulnerability, need, capacities and desires as possible objects of protection. Therefore this section contains conceptual and philosophical reflections on these notions.

The contributors to this volume did not want to remain on a purely philosophical level. The aim was to effect a linkage, a combination of fundamental, conceptual reflections and concrete bioethical debates. In the section "Cases of limits" such interfaces with concrete debates are explicitly made, and the majority of papers deal with issues concerning human reproduction. Human reproduction seems to be an area in which the developments in the life sciences touch the most private and intimate areas of human existence. In this context, but also in other areas of bioethics, we meet with sometimes acrimonious discussions of the meaning of freedom, autonomy and informed consent. Although in this section special attention is paid to human reproduction, the implications discussed are much broader in scope.

Particularly important for the discussion about the limits of human existence is the impact of the life sciences on people with disabilities. Accordingly, the section "Abilities and disabilities" addresses this topic. The life sciences are exploring possibilities to ease disabled people's burdens, to enhance their lives, ultimately to get rid of disabilities altogether. Throughout the world, however, there are disability movements that in many respects consider the new developments as constituting a challenge. The impact of these developments on the identity of people with disability, their need for social recognition, the extent to which societies owe them justice and respect have all received too little attention in bioethical debates. For future bioethical work the implications of the life sciences for thinking about and living with disabilities should be a central topic. This section aims to identify some of the issues that need to be put on the research agenda.

In the last section, "Others' views: Intercultural perspectives", several scholars offer insights into how different cultures may perhaps converge in our bioethical debates. There is no doubt that cultural traditions, whether from Asia (especially China), Israel, Russia or other parts of the world, are putting their indelible stamp on bioethics. The different roles of the family in different cultures, different concepts of individuality or concepts of nature are each in their own way framing the debates in nearly all areas of bioethics. However, it is only very recently that the importance of an intercultural perspective has been acknowledged. The papers in this volume present a variety of interesting perspectives to open the philosophical discussion on the intercultural dimension of bioethics.

It was the goal of the editors to offer a variety of perspectives and a diversity of approaches. We are convinced that for a discussion of the ethical and philosophical dimensions of the life sciences an interdisciplinary debate involving a broad range of approaches is urgently needed. We hope that this volume may contribute to a more intense debate on the cultural importance of the life sciences.

We want to thank several people. First of all the participants of the two conferences in Davos (2001) and Doorn (2005) who made the discussion about the philosophical perspectives of the life sciences a really exciting experience. We would also like to thank the European Science Foundation for its financial support of those conferences.

For the conference in Doorn we also want to offer our sincere thanks to those who provided additional funding: Utrecht University, the Utrecht-Leiden research institute for philosophy ZENO, the Netherlands School for Research in Practical Philosophy, the Netherlands Organization for Scientific Research, the Royal Netherlands Academy of Arts and Sciences, and the International Association for the Promotion of Cooperation With Scientists from the New Independent States of the Former Soviet Union (INTAS). Our special thanks go to all those who participated in the organization of the conference: Anna van Dijk and Lenka Chludova, and especially Stephanie Roels who for months did a tremendous job organizing the conference. During the preparation of the book, Frederike Kaldewaij and Gerhard Bos have spent a lot of time and effort supervising the editing of the manuscripts. We would also like to thank Fritz Schmuhl of Springer Publishing for his support. Finally we would like to thank two anonymous reviewers of the manuscript; their suggestions certainly helped to improve the quality of the texts.

Part I
Contingency of Life and the Ethical

The Value of Natural Contingency

Ludwig Siep

1 Contingency as Imperfectness

Despite the recent increase in the occurrence of natural catastrophes, of which only a part is caused by human activities and forms of life, human control of natural processes is increasing in many spheres of the natural world, especially in the realm of living organisms. From the time of Francis Bacon and the New Science through the Enlightenment and the Rationalization Processes of the nineteenth and twentieth centuries there has been little doubt concerning the value of human control of natural processes. Just as in the metaphysical tradition, natural contingency was considered as something imperfect and disadvantageous for human beings.

This view remains unshaken as it relates to some areas of natural processes and events, like the mentioned catastrophes, and also to human health. Today, the attempt to eliminate contingency has reached the threshold to an improvement of natural processes including human reproduction and individual genetic outfit. However, these possibilities begin to cast doubt on the traditional "contempt" of contingency and the aim of complete control of natural processes. Even concerning *animal* breeding the attempt to secure results by cloning the optimal stock animal is not beyond doubt and criticism (cf. Siep 1998: 191–198). There are three main reasons for those doubts and for a possible re-evaluation of natural contingency: *Firstly*, control implies responsibility and this may generate severe problems in interpersonal and intergenerational relations. *Secondly*, technical forms of breeding may reduce biodiversity. *Thirdly*, the ideal of a completely foreseeable process of natural events may deprive human beings of valuable life experiences with events and circumstances that are independent of their wishes and expectations.

In the following, I will discuss these problems within the broader framework of the philosophical conditions of an evaluative view of nature (for the following cf. Siep 2004b). Such a view existed in pre-modern metaphysics and survives in the common sense of present times. But the "scientific image" regards nature as completely value-free. Values seem to be generated by individual wishes and private

L. Siep
e-mail: siep@uni-muenster.de

M. Düwell et al. (eds.), *The Contingent Nature of Life*,

evaluations. The pluralism of modern societies seems to be based on the privacy, subjectivity and irrationality of values. This, however, may be a view which cannot account for common value experiences and the possibility of rational deliberations guiding public decisions on technical options concerning natural processes and events.

Before I turn to the questions of subjective and objective values, however, I will try to clarify the concept of "natural contingency" to some extent.

2 The Meaning of Natural Contingency

Natural contingencies may have a broad range. The seaquake which devastated the brim of the Indian Ocean was unforeseeable and therefore contingent as to human knowledge and expectations. This does not mean, of course, that it had no causes or did not follow laws of nature. But its exact location in time and space and the size of its effects could not be predicted. In the following, I will not discuss the logical, epistemological and ontological questions of necessity and contingency. By contingency of natural events and processes I understand the quality of such entities to be uncontrolled and unpredicted by human behaviour and knowledge. Earth- and seaquakes are still contingent in this sense, since their exact occurrence in time, the location of their focus, their size and effects are beyond control and precise prediction. Since this use of "contingency" is related to possible human prediction or control, the difference between stochastic processes and those explainable by "classical" Newtonian physics is not of crucial importance here. Not all processes explainable by classical physics are predictable or controllable. Therefore human beings cannot be held responsible for the failure or lack of action to predict, prevent or control them.

The same is true for genetic mutations and for the results of gene-mixing within the process of natural sexual reproduction. However, these particular processes can be influenced by human action – perhaps they could one day be completely replaced by artificial combination of "parental" genes, by cell nucleus transfer or by other yet unknown technical procedures. Thus we have to distinguish between processes beyond human control and others which may either be left uncontrolled and contingent or be replaced by technical procedures.

To a certain degree, the concept of nature and the concept of contingency depend on each other. The very meaning of "nature" seems to be "independent of human will and wishes" – even if for most natural processes and objects there is not a great deal of naturalness left over in modern times. These processes and objects are influenced by human behaviour and since this behaviour depends to a great extent on wishes and voluntary decisions, the whole distinction between nature and culture, artificial products and natural beings seems to break down. However, there is still an important difference between processes and beings *directed* by the human will and those who are – unintentionally and sometimes by many detours and side effects – merely *influenced* by human behaviour. This difference pertains even to the human body and its completely or partially uncontrollable processes compared with those we can direct voluntarily.

If "natural" means at least partly uncontrollable, it may follow that nature as a whole is contingent. But according to the common use of the concept it is certainly not contingent that the sun rises at a time which we can predict exactly and that the earth repeats its movement with long-enduring regularity. Even if this, as Hume pointed out, is not a logical or conceptual necessity, and even if it may be changed by some cosmic catastrophe, it does constitute a series of events and processes which are explicable, predictable and reliable. In contrast, by "contingent" we understand only those processes and their outcomes which are not only independent from our will, but also unpredictable by our best available knowledge – leaving open the question whether some of them may be in principle beyond human knowledge and control.

According to most of the *traditional* views, natural contingency had no ultimate reality. In the teleological conception (which despite the rise of mechanical physics lasted up to the first decades of the nineteenth century), nothing was without purpose in nature. If the purpose was unknown or not yet understandable, this was due to a limit in human understanding, either temporary or in principle. Since the very concept of an organized living being was not possible without the presupposition of its intentional production – as Kant maintained in his third critique – an infinite understanding and will had to be presupposed (cf. Kant 1968: 398ff.). In the Aristotelian conception, there was at least room for a sphere of irregular and unpredictable processes in the sub-lunar sphere, but this was a sphere of imperfection and futile striving for the perfection of the supra-lunar movements and bodies.

Starting with the new science and philosophy of the sixteenth and seventeenth centuries, it was argued that this imperfection had to be improved by human science and technology. For thinkers in the Calvinist tradition such as Bacon or Locke this was even a divine command whose fulfilment would be assessed at the Day of Judgement (cf. Locke 1959: 360–362). Contingency of natural processes remained something negative and to be overcome throughout the age of enlightenment and the technical and industrial era. Fichte referred to the "last convulsions" of a nature not yet conquered by the autonomous human will (Fichte 1968: 268f.).

However, with the theory of evolution, with classical and modern genetics, with the evolutionary view of the cosmos, perhaps also with the theory of chaotic natural and artificial processes, the realm of contingency seems to grow. On the other hand, the techniques for manipulating and controlling natural processes are improving with increasing speed.

It is a well-known cultural and historical phenomenon that the value of goods and processes becomes apparent just at the time when the quantity and accessibility of them is reduced dramatically. Rousseauism, Romanticism and other pro nature movements started with the peak of the Enlightenment and the early phase of industrialisation. And the twentieth century ecological movement was fuelled by the growth of mega-cities, industrial (and therefore predominantly mono-cultural) agriculture and the serial mass-production of technological goods. It seems that the new biotechnical possibilities raise awareness of potential losses in biodiversity and spontaneity. Correspondingly the discovery of the value of cultural diversity may be a reaction to the rationalization processes of the twentieth century, concerning urban planning, administration reforms, global business, entertainment and mobility.

Before the possibility of technically optimizing life processes existed, contingency, spontaneity, non-repeatability, unpredictable variation etc. belonged to their distinctive features. Of course, some of these processes resulted in deformations causing suffering to organisms and their environment. Many of the "copy mistakes" in the process of gene-directed development of animal and human organisms lead to severe illnesses or disabilities. But there are other processes and their outcomes which many human beings regard as valuable, such as diversification, individualisation and also the contingent composition of individual genomes which no one has to take responsibility for.

Before discussing further the value of such aspects, however, I will briefly turn to some general questions concerning values, particularly the value-character of natural properties.

3 Nature and the Status of Values

Like modal logic, epistemology and ontology, value theory is a broad field which I can only touch on here. There is a longstanding and ongoing discussion in value theory and meta-ethics concerning ontological and epistemological options. The three main options in value ontology are realism, projectivism and versions of a relational theory of value. In epistemology different sorts of value-subjectivism and -objectivism are defended. I have discussed some of these options elsewhere (see Siep 2004b: Ch. 3; cf. also Quante 2003: 95ff.). Here I confine myself to some sketchy arguments for a relational theory of values which can lay claim to intersubjective and intercultural approvability. I start with three examples of commonly accepted values: justice, health and biodiversity.

1. It is certainly not easy to defend the value of justice as objective against the positions of projectivism and subjectivism. It is a standard argument of nineteenth and twentieth century historicism and legal positivism that there have been almost as many conceptions of justice as there are political constitutions and systems of positive law. But the value of justice is not simply a question of convention. It is tied to social and natural facts and to basic conceptions of culture or community, even if they are controversial. Justice, as Aristotle observes, has to do with special sorts of equality (cf. *Ethica Nicomachea* 1131 a 10ff.). There must be some equality between deed and punishment or burden and benefit in joint action. If criminal justice is a cultural successor of injury and revenge, as for instance Mill suggests, there must be a coherent relation between natural events, cultural criteria and their narrative or rational legitimization (Mill 1993: 53f.).

   One could extend this argument regarding virtues in general (regarding the following cf. Siep 2005: 83–98): The origins of virtues are very likely narratives about praiseworthy and exemplary deeds and characters – such as the Homeric or Babylonian epic poems. They contain rather precise descriptions of "virtuous" deeds. Instead of being derived from ideas or fixed norms, virtues are discovered by describing particular actions and reactions regarded as appropriate, as successful and exemplary solutions of social problems or as astonishing manifestations of human faculties. In the cultural process, virtues gain ever more stability and independence from the "mental states" of their "executors" as well as their observers. Virtuous actions are "incorporated" in the behaviour and physiological states of a body. At the same time, they belong to a social world of public rules and interactions. The reactions of a virtuous or non-virtuous person can be to a certain degree foreseen and relied on by other persons – in analogy to natural events following laws of nature. The social world can be regarded as a second nature with its own laws and in many respects connected with the order of the "first" nature.[1]

2. Turning to the second example and to a more "natural" value, there is little question, that "health" is a personal and social value of almost universal acceptance. Of course, human beings can abstain from fulfilling their bodily needs, but that does not diminish the general value of many natural properties of the body and its environment for typical human beings. The same is true regarding illness as a means to moral or artistic virtues: What may favour valuable dispositions in some people does not represent a value in itself and is usually an obstacle for the achievement of many normal values and expectations.

   To achieve and preserve health is a social aim of many communities pursued in institutions such as hospitals, medical professions, health care systems etc. Of course these institutions can have very different shapes and degrees of complexity, specialization, professionalism, financing, technology, and so on. But health is a common goal and a firmly established value in most societies. It has a natural basis in conditions and functions of the human body. To be sure there are different cultural and theoretical concepts of health and different social standards of expectations concerning the normal faculties of a healthy human being. But they have to be related to physiological states and functions. One may call health a supervenient value-relation with a basis in physiological, mental and social facts. It is not simply a social fashion or a private wish which could be projected on every possible state of a value-free social or natural "reality".

3. My third example is the value of biodiversity. This will take us closer to the value of natural contingency. Skipping over the problem of how to quantify and measure biodiversity (cf. Gutmann and Janich 2001: 3–27; Kim et al. 2001), I understand it simply as a variety of forms of life generated by the process of evolution. In recent discussions a common distinction is made between natural varieties in species, genes and ecosystems (cf. Janich 2001). Now these varieties are valuable at least in the following respects: *Firstly*, valuable for non-human living beings regarding their survival and fitness, *secondly* valuable for ecosystems, and *thirdly* for human beings in different respects – among them medical, agricultural and aesthetical.

A projectivist will argue that all three respects depend on human wishes and valuations. Against this it may be argued, *firstly*, that whether something is good for a non-human living being or an ecosystem, meeting human needs or improving fitness depends on many biological properties "in the world". A considerable extent of biodiversity seems to be among these properties.

Secondly, the response that the existence of something in nature is only valuable if human beings like, want or esteem it, presupposes a strong anthropocentrism. In my view this is hard to match with an evolutionary perspective and with the semantic of "good" in the ethical sense. Of course, conscious judgement and verbal expression is only possible for human beings. But other beings have their own way of valuing or realising valuable relations. To argue that only human beings "create" goodness by their valuation sounds like a secular version of the view that the world exists only for the sake of human or moral beings.[2] And the ethical meaning of "good" is certainly not, "to be valued or preferred by human wishes", but "worthy of being esteemed, approved, striven for etc.".

To be sure, biodiversity may be judged as valuable for human beings because it serves human needs and therefore possesses value in a more than private sense (for the following cf. Siep 2004a: 17–24). But to what degree biodiversity is good for human beings in this respect is a question much discussed in recent ecology. It may be argued that human needs could be met with much less diversity than that which recent international conventions seek to protect. Similar doubts have been raised as to the degree of economic profit to be drawn from biodiversity for the pharmaceutical industry. And even tourism may continue with much less than the current level of natural biodiversity.

Nevertheless, as some recent comments even from law experts have pointed out, since the conferences of Stockholm and Rio the international documents show a trend towards "protecting nature for its own sake" (Wolfrum 2001). Perhaps this tendency shows that the value many religions and cultural traditions have seen in the natural diversity of the cosmos is about to be rediscovered by humanity. This would be another example of becoming aware of a value in the very historical moment in which it is endangered by human activity as never before.

In order to defend such a position one has to argue that the value-free conception of nature in the sciences is not the original and not the only objective perspective on nature. Rather, it is only one type from the manifold perspectives human beings can develop. In all of these perspectives evaluative moments are included, even in the sciences they cannot be completely eliminated. These implicit evaluations may either be developed to a conscious and rationally defensible form, as in aesthetics, ethics and to some extent in politics. Or they can be diminished for the sake of neutral methods of research and experimentation. But these research methods themselves are tied up with values such as truth, sincerity, fairness etc. And they serve values like the prediction of events, technical applicability, the improvement of medicine etc. They presuppose discoveries of what is valuable for human beings and of course also settlements about which individual wishes should be allowed or forbidden.

Such discoveries and settlements can be the result of common experiences in the process of the cultural formation of values and norms. The structure of experiences supporting or changing common values is another subject which I can only touch on here. Collective cultural experiences are of a character quite different from methodical experiments or observations in the sciences. They involve shared feelings of suffering or relief, commonly accepted interpretations of historical events, convincing theoretical reflection, critique or justification of values – and, of course, processes of the enactment and change of laws or law-like conventions, sometimes of a global range.

4 Valuable Aspects of Natural Contingency

Let me draw some consequences from these reflections regarding the value of natural contingency. Human evaluations of natural objects and processes may depend on a broad range of perspectives. Such objects may serve basic human needs or may derive their value from cultural perspectives in art, religion etc. But, in a long-standing intercultural perspective, they may also have been regarded as belonging to a good shape of nature to be maintained and protected for itself. One may argue that the very concept of "good" in its ethical sense has been developed in view of a shape of nature as "cosmos" in the evaluative sense. That is, as a state of the world which could be universally affirmed and regarded as worth to be attained.

Now natural contingency is certainly a relationship between things and processes on the one hand and human cognition and will on the other. It is characterized, as we have seen, by a special kind of independency regarding these human faculties. This contingency and its results may be considered good or bad. As recent experiences show, it does not seem rational to regard natural contingency as intrinsically bad in the way many metaphysical and scientific positions have done in the past. Some natural processes and events are good for human beings *just because* they are uncontrollable and unpredictable. But others are bad due to the same property and the resulting impossibility to prevent them or protect oneself against them. It is possible that the contingency of processes like bisexual reproduction in mammals may have positive and negative value regarding different aspects of their outcomes. Thus what is needed is a perspective which can distinguish between valuable and negative aspects of such processes. In other words: an evaluative perspective on nature beyond private wishes or tastes.

There seem to be three principal positive effects of the contingency of reproduction and evolution: The *first* is the social effect of easing the burden of responsibility for the natural equipment of human beings. If this were to be replaced by voluntary action and technological construction, individuals or societies would have to accept responsibility for the outcome – and eventually compensate those which suffer from them. This would put a heavy burden on the relation between parents or other people responsible for the designed "gene-mix" and the offspring endowed with it. But it would also affect the relation of society to those of its members which are less fortunate in gaining the means for securing an advantageous physical outfit. Beyond the compensation of the effects in everyday life, society could be made responsible for the distribution of natural properties itself. To leave this burden to the contingent processes of genetically uncontrolled reproduction represents a positive value for social relations.

The *second* positive effect is the genetic variety between species and between individuals which so far have resulted from the contingent processes of reproduction, species formation and evolution in general. I have discussed the different value-aspects of biodiversity in the preceding section of this paper.

The *third* positive value-aspect of natural contingency may be the general independence of natural processes from human will and foresight. It is not limited to such seemingly subjective and marginal benefits as surprise, the possibility for discoveries, the avoidance of repetition and boredom etc. Natural contingency could also be valued in a more fundamental way regarding the existence of an independent "partner" of humanity, something which at least in some respects remains unconquered, incalculable etc.

If natural contingency can be regarded as a value, we have to deal with a rather complex relationship: Contingency is a relationship between the world and human faculties – and values are themselves relations between these poles. As in other value-relationships, the value of natural contingency supervenes on a relationship between natural processes and human dispositions. In some cases – as in leaving the burden of genetic distribution to nature – it is contingency *as such* which is valuable. In other cases, such as positive surprises and encounters, contingency is rather *part* of the value. And in still other cases, particular results and aspects of contingent natural processes are the principal bearers of value.

Certainly more analysis is needed to determine the exact properties of this value relationship. For philosophical value theory it is important to realise that on all levels of the relationship the *relata* are to a considerable degree independent of each other. Due to this partial independence in relation to individual wishes and mental states in general, what is valuable belongs to the fabric of the world and is not simply projected on it. In my view, modern bioethics is as much in need of a sort of value realism – if a relational one – as it supports the plausibility of such positions in value theory and meta-ethics. Questions of the value of natural properties, including those of the human body or human reproduction, can only be treated sufficiently by means of such a theory.

Notes

[1] For a renewal of the concept of "second nature" cf. McDowell: 167–197.

[2] That the only value and meaning of the existence of the world ("Dasein einer Welt") lies in its final aim (Endzweck), the moral human being, is Kants view in the Critique of Judgement. Cf. Kant 1968: 434 f., 449.

References

Birnbacher, D., Natürlichkeit. Berlin/New York: de Gruyter 2006.

Fichte, J.G. "Die Bestimmung des Menschen." In *Gesamtausgabe der Bayerischen Akademie der Wissenschaften*. Vol. 6. R. Lauth and H. Jacob (eds.). Stuttgart: Frommann-Holzboog, 1968.

Gutmann, M. and Janich, P. "Überblick zu methodischen Problemen der Biodiversität." In *Biodiversität – Wissenschaftliche Grundlagen und gesetzliche Relevanz*. P. Janich; M. Gutmann and K. Prieß (eds.). Berlin: Springer Verlag, 3–27, 2001.

Janich, P. "Zusammenfassung." In *Biodiversität – Wissenschaftliche Grundlagen und gesetzliche Relevanz*. P. Janich; M. Gutmann and K. Prieß (eds.). Berlin: Springer Verlag, XXI, 2001.

Kant, I. *Kritik der Urteilskraft*. Bd. V. Akademie-Textausgabe. Berlin: de Gruyter, 1968.

Kim, J.T.; Schwöbbermeyer, H.; Theißen, G. and Saedler, H. "Biodiversitätsmessung anhand molekularer Daten." In *Biodiversität – Wissenschaftliche Grundlagen und gesetzliche Relevanz*. P. Janich; M. Gutmann and K. Prieß (eds.). Berlin: Springer Verlag, 181–234, 2001.

Locke, J. *An Essay Concerning Human Understanding*. Vol. II. A.C. Fraser (ed.). New York: Dover, 1959.

McDowell, J. "Two Sorts of Naturalism." In *Mind, Value and Reality*. Cambridge (Mass.), London: Harvard University Press, 167–197, 1998.

Mill, J.S. "Utilitarianism." In *Utilitarianism. On Liberty. On Representative Government. Remarks on Bentham's Philosophy*. G. Williams (ed.). London: Everyman's Library, 1993.

Quante, M. *Einführung in die allgemeine Ethik*. Darmstadt, Wissenschaftliche Buchgesell schaft, 2003.

Rolston, H. "Value in Nature and the Nature of Value". In: *Philosophy and the Natural Environment*. R. Atfield and A. Belsey (eds.). Cambridge: Cambridge University Press, 13–30, 1994.

Rüdiger Wolfrum, "Biodiversität – juristische, insbesondere völkerrechtliche Aspekte ihres Schutzes" In: Peter Janich, Mathias Gutmann, Kathrin Prieß (Hrsg.), Biodiversität. Wissenschaftliche Grundlagen und gesellschaftliche Relevanzen. Berlin/Heidelberg/New York (Springer), 2001.

Siep, L. "'Dolly' oder Die Optimierung der Natur." In *Hello Dolly? Über das Klonen*. J.S. Ach; G. Brudermüller and Ch. Runtenberg (eds.). Frankfurt/M, 191–198, 1998.

Siep, L. "Erhaltung der Biodiversität – Nur zum Nutzen des Menschen? In *Biologische Vielfalt und Schutzgebiete – Eine Bilanz 2004*. Vorträge und Studien, Heft 15. B. Hiller and M.A. Lange (eds.). Münster: Zentrum für Umweltforschung, 17–24, 2004a.

Siep, L. *Konkrete Ethik*. Frankfurt/M.: Suhrkamp, 2004b.

Siep, L. "Virtues, Values, and Moral Objectivity." In: *Virtue, Norms, and Objectivity*. Ch. Gill (ed.). Oxford: Oxford University Press, 83–98, 2005.

Between Natural Necessity and Ethical Contingency

Ahmet Hadi Adanali

1 Concepts of Necessity and Contingency

It is a well-known fact that the human mind is good at performing some theoretical activities such as logical and mathematical thinking, and also good at practical ones such as building airplanes and designing computers. The human mind is not very good or rather weak at some speculative activities such as metaphysical and moral thinking. It is not a logical contradiction to assume a world in which the human mind is good at metaphysics and ethics, and not so good at logic and mathematics or even to assume a world in which the human mind marvels in both areas. That the human mind cannot have it both ways is a contingent fact; or is it? To answer this question, we need to clarify the concepts of contingency, necessity, impossibility, and we also need to have a proper concept of mind; neither need could be satisfied easily. The question in its bare essentials relates to who we are and what we can know.

Modal concepts are usually grouped into the following three modalities: necessary, possible (contingent) and impossible.[1] Since these modal concepts are semantically related, they are usually defined in relation to each other; in other words, if we can understand one of them "intuitively", we can easily understand the other two. Necessity, contingency and impossibility are also defined in relation to three different disciplines of philosophy: logic, ontology and ethics. From a logical point of view, these concepts are seen as the property of propositions. A proposition is necessary if it is always true, or if its negation is a logical contradiction; and conversely a proposition is possible if is not always true, or if its negation is not a logical contradiction. Finally, a proposition is impossible, if it is never true (or always false). For example, it is always true and thus logically necessary that circles are round; it is not always true and thus logically possible that a certain circle has five centimeter diameter; and it is never true and thus logically impossible that circles have three angles.[2]

Ontological necessity, on the other hand, is a matter of natural or physical laws. Something is ontologically necessary, if a physical law makes it exist the way it does and not any other way. For example, laws of motion make it necessary that the

A.H. Adanali
e-mail: hadiadanali@yahoo.com

M. Düwell et al. (eds.), *The Contingent Nature of Life*,

world follows an elliptical orbit around the Sun, and not a full circle (Will 1989). Ontological contingency, on the other hand, refers to the case in which something may or may not exist. It is a contingent fact that the solar system has the number of planets that it has rather than more or less.[3] To imagine that the solar system in which the sun had only five planets or the world had more than one moon is not to think something against a physical law. Furthermore, something is ontologically impossible, given the laws of physics, if it can never exist. Given what we know about the surface and atmospheric conditions about Mars, it is physically impossible that life exists in it.

Finally, ethical necessity is what accords with an ethical rule or law that has a normative moral value. Something is ethically necessary if it is always true; for example, it is always true that one should not harm innocent persons. Conversely, something is ethically contingent, if its rejection does not violate an ethical law. One may pursue a career in philosophy or in politics, and neither preference is ethically binding, there is no moral duty or obligation to prefer one to the other. Ethical impossibility is rarely discussed in the literature, thus it is hard to find a definition for it. The closest example to ethical impossibility, I think, is to hold babies responsible for their actions.

The relationship among these three kinds of necessity is a matter of controversy. Some philosophers believe that the distinctions among them are definite, and that any attempt to reduce one to the other will necessarily fail. Others who see parallels between ethical norms and logical necessity want to establish ethical norms on a transcendental (a priori) base. There are also those who want to reduce ethical values to the statements of science, i.e., psychology or biology (Will 1989).

Since modality is primarily a subject matter of logic, our theory of logic has bearing on how we evaluate and apply these modal terms. The concept of logic has undergone substantial changes in the last couple of centuries. Traditional Aristotelian logic was criticized as being unfit for science because science is (or has to be) empirical in principle. There is no need here to go into the reasons and explanations why Francis Bacon, David Hume, John Stuart Mill and others replaced the deductive methods of classical sciences with inductive logic. Furthermore, modern logic that was initiated through the pioneering works of Giuseppe Peano, Gottlob Frege and Bertrand Russell rejected certain assumptions on which the traditional Aristotelian logic was built.

Russell, for example, criticized the traditional logic on a number of points: for its negligence of empirical knowledge and observation, for its attempt to deduce all facts theoretically, and more importantly, for its fixation on the concept of necessity. The traditional logic, according to him, attempted to construct the world through the method of negation without or with little appeal to experience. Traditional logic first considered a set of seemingly possible alternatives about the world, then negated all but one, and claimed that this must be the actual world. Traditional logic looked for alternatives that were impossible. Its main concern was to find out how the world cannot be, instead of how it can be. The true function of logic, Russell claims, is exactly the opposite; to show the possibility of alternatives that are previously considered necessary or impossible. In this way, the new logic "liberates imagination as to what the world *may* be, it refuses to legislate as to what the world *is*" (Russell 1960: 15). The new logic refuses to set limits on the extent and nature of what can be known. In this matter as well, the new logic shows what may happen, not what must happen.

Russell's evaluation of traditional logic seems a bit unfair for several reasons. First, it is not true that traditional logic, in its widest sense, ignored possibilities and tried to find only what is necessary or impossible. Second, modern logic, as least some interpretation of it, considers the same kind of necessity between premises and conclusions, without which the inference would have been impossible. More importantly, there had been philosophers of Islamic tradition who accepted Aristotelian logic, but rejected necessary entailment between the logical principles and metaphysical theories. The famous theologian Ghazâlî, for example, was a firm supporter of Aristotelian logic, but he was critical of linking logical necessity to metaphysical or physical theories. He criticized the peripatetic philosophers and accused them of "imposing conclusions" and "dictating what is necessary, impossible, or improbable" without giving any proofs for them. He took "possibility" and "probability" as his guide in natural events and argued whatever is not proved to be impossible, cannot be rejected just because it is improbable. He even suggested that if universals are mental constructions, it is plausible that the concepts of possibility, impossibility, and necessity are so as well.

2 Necessity and Contingency as Natural Concepts

Ghazâlî's critique of causal necessity in nature is not unprecedented and it was motivated by some theological concerns, mainly by the free will and omnipotence of God. The idea that natural events are orderly and sequential is accepted commonly by all Muslim theologians and philosophers. The philosophers differ from the theologians in their explanation of the source and cause of this natural order. Muslim philosophers, or at least majority of them, tend to see a necessary causal nexus in the world that maintains the cosmic order.[4] This order, according to them, ends in the First Uncaused Cause.

Muslim Neo-Platonists, such as al-Fârâbî and Avicenna, believed that the universe emanated from the Necessary Being spontaneously and causally, and this emanation continued until it reached to the lower levels of the existence, namely, to our world of generation and corruption. Avicenna made a distinction between essence and existence. According to him, we cannot derive the existence of something solely from its essence. Nothing among the essential properties of a thing necessitates its existence. The universal concept of man, for example, does not entail that there are or there should be actual men. Something else other than the essence makes a specific thing exist, which is its necessitating cause. This cause itself may be necessitated by another cause, and so it ends in ad infinitum or it ends in the Uncaused Cause of Everything who is also the Necessary Being. Necessary, for Avicenna, is what exists all the time; the Necessary Being is the one whose existence does not depend on anything other than itself. The possible or contingent, on the other, is the thing whose existence depends on something other than itself. Impossible for Avicenna is what can never exist (see Marmura and Michael 1972).

In addition to these three modes of being, Avicenna added a new one or rather he combined two in an unaccustomed way: a thing that is both necessary and contingent. This new concept of a necessary-contingent being, though bordering the logical contradiction, has these two modes in a relational way. According to Avicenna's cosmological theory, the First Intellect (the Being that emanated from God first) is both necessary and contingent: It is necessary with regard to God, because it is caused by God. It is also contingent, because its existence depends on other than itself, namely on God. Thus, the same being, according to Avicenna, can become both necessary and contingent.

As for the orthodox theologians (Ash'arites), majority of them understand necessity, impossibility, and contingency as logical concepts and take logical contradiction as the criterion in defining necessity and contingency. For them, whatever is not a logical contradiction is contingent. Ghazâlî, like other orthodox theologians, defines the modal concepts as logical judgments. He says:

> The concept of contingency that you [philosophers] set forth is a judgment of mind. What reason conceives its existence and what is not considered impossible by reason is contingent, according to us; and if it is not possible for the reason to think of something as existent, it is impossible, and if it is not possible for reason to think of something as non-existent, it is necessary. These are the judgments of reason and they do not need material things in order that these judgments be attributed to them
>
> (Averroes 1987: 60; translation is mine).

For Ghazâlî, necessity, contingency and impossibility are rational or logical concepts, and they can be understood without any reference to the existent things. Unlike philosophers who claimed that there has to be something prior (i.e., matter) for the predicate contingency to apply, Ghazâlî argued that contingency, being logical concept, do not need any prior existing thing. If this were true, he claimed, we could not have understood impossibility, i.e., something that can never exist. Since we can conceive impossibility without reference to an existent thing, we can also conceive necessity and contingency similarly. Furthermore, if there is no logical entailment between two things, than the existence of one thing does not necessitate the existence of another. Thus, whatever exists in the natural world is contingent, not physically necessary. This led him to reject the causal necessity in nature:

> According to us the connection between what is usually believed to be a cause and what is believed to be an effect is not a necessary connection; neither the affirmation nor the negation, neither the existence nor the non-existence of the one is implied in the affirmation, negation, existence and non-existence of the other—e.g. the satisfaction of thirst does not imply drinking, nor satiety eating, nor burning contact with fire, nor light sunrise, nor decapitation death, nor recovery the drinking of medicine ... and so on for all the empirical connections existing in medicine, astronomy, the sciences, and the crafts
>
> (Averroes 1987: 316).

Ghazâlî claimed that the connection between the cause and the effect is not necessary but contingent. The effect exists *with* the cause, not *due to* it. When cotton is brought with fire, all we observe is that burning takes place with fire, but not by or through fire. He further indicated that necessary causal relation between the cause and the effect is not *observable* in the world and neither can it be logically *demonstrated*. To think the cause without the effect or to think the effect with the cause is not a logical contradiction. How is then, we have the impression of continuity and sequential order in the word? Ghazâlî's answer for this question was that God creates this impression within us. How can we be sure that the future would resemble the past? Again God insures us through revelation that His custom of running events would not change. It is important to keep in mind that Ghazâlî does not deny regularities in nature. When fire touches cotton, surely enough, cotton burns; it is not the act of burning that is questioned here, but the *necessary* link between the act of burning and the coexistence of fire and cotton.

According to a recent commentary, Ghazâlî's distinction between conceptual and natural possibilities represents "a genuine breakthrough," since his interpretation of the modal terms allows him to treat counterfactual possibilities freely, "an important feature in any refiguring of the limits of possibility" (Kukkonen 2000a: 480). Ghazâlî's theory of contingency when developed to its logical conclusion entails that for every actual state of affairs, there in an infinite number of possible states that were not realized (ibid.: 481). The actual world is the one that God has chosen, thus it is necessary; each possible world is contingent: "everything could be other than it is" (ibid.: 483).

There are limits, however, to Ghazâlî's theory of contingency, and these limits are the limits of reason and intelligibility. Impossible is, as we have seen, what can never exists, and even God cannot make what is impossible exist. Ghazâlî defines the impossible as something that "consists in the simultaneous affirmation and negation of a thing, or the affirmation of the more particular with the negation of more general, or the affirmation of two things with the negation of one them, and what does not refer to this is not impossible and what is not impossible can be done."[5] Thus, a thing cannot be white and non-white at the same time; similarly, a thing cannot be white and black simultaneously, since to affirm that something is white implies the negation of blackness. Furthermore, a man cannot be at home and outside at the same time; an inorganic thing cannot have knowledge, since it cannot perceive; and Ghazâlî claims that if knowledge is attributed to an inorganic thing, "it would become impossible to call it inorganic in the sense in which this word is understood" (Averroes 1987: 329). All these are logical contradictions according to Ghazâlî, and they are impossible in the logical and thus in the physical sense.

Averroes, the famous Muslim philosopher who has become a controversial figure in the West due to his ideas on the nature of soul and its status after death, heavily criticized Ghazâlî's occasionalist theory of nature and claimed that denial of causality in nature is tantamount to the denial of the knowledge. Denial of causality, for him, leads to the rejection of sense perception and the reliability of cognitive knowledge. Furthermore, denial of causality leads to the denial of nature (or essence if you will) of things. He also claimed that rejection of natural causality would turn everything into one entity (and perhaps not even one entity) by making the things devoid of their essences. In addition, denial of causality is the rejection of the human desire to know things. And finally, denial of cause, according to him, is the denial of God as the Cause of the natural world.

As we have seen, for Ghazâlî, although the current states of affairs are necessitated by the Divine choice, it is nevertheless purely contingent in the sense that it was possible to be otherwise. There is nothing prior to its existence that makes it physically necessary; nothing determines God's choice in one way or the other; He has absolute freedom. A natural conclusion of this is that human mind or nature is also contingent. Given the Divine choice was different, it could have been different as well.

This cannot be accepted by Averroes who believes that it is not possible for the intellect to have properties different from the ones that it has, because it is not contingent. The natural properties including the modal ones (necessary and impossible) are required to have true knowledge. Thus, "the contents of the intellect also cannot be possible (i.e., contingent) but must instead be necessary" (Kukkonen 2000b: 499). According to Averroes, the mind has a necessary structure and human cognition has its fixed ways of perceiving the world. These ways, in turn, is determined by the properties of the natural world.

Going back to the question that was posed at the beginning, whether it could have been possible for the human mind to marvel both at mathematics and metaphysics, Ghazzali's answer would be a definite "yes". He would consider it a contingent fact that human mind is good at mathematics and rather weak at metaphysics. Nothing in nature necessitates its being the way it is. Averroes, on the other hand, as the previous paragraph made it clear, would argue that human mind follows certain natural laws and that it is not possible for it to be different than it is, given that its cognitive structure is closely related to physical features of the world.

Averroes' discussion of the meanings of possibility is also motivated by the theological problems such as the eternal creation of the universe. For him, like many other medieval Aristotelians, what is impossible cannot become possible. He also believes that if possibility always exists, then the possible must also always exist. Thus, the universe as possible has always existed and will always exist (ibid.: 332). The concept of possibility which underlies this argument is the so-called "statistical model of modality." In this sense, the terms "necessity", "possibility", "impossibility" can be translated into the temporal terms "always", "sometimes" and "never" respectively (ibid.: 338–39).

Averroes in his explanation of possibility relies on an Aristotelian principle that a genuine possibility will always become real at one time or another. In other words, it is impossible for a thing to be possible and never exist. To imagine an infinite number of possibilities in the universe that will never become real is an exercise in futility. Averroes also argues that if something is possible and it always exists, then it must really be necessary. He claims that "the possibility in eternal beings is necessary." The logical corollary of this supposition is that both the universe and its eternal states are necessary (determined). Not even God could make the universe cease to exist, nor can He change its essential properties. Finally, Averroes draws upon the notion of Aristotelian concept of potentiality. Contingency in this sense is the contradictory of actual and it is inherent in the nature of things; without this pre-existing potentiality, the possible could not have become real.

Ghazâlî was quick to realize that the statistical as well as potentiality models of modality have deterministic implications for God and for men equally. For this reason, he rejected both models and extended the realm of possibilities beyond the actual events in the world. "For there to be a will and a choice, there have to be real alternatives, and for this, unrealized possibilities have to be regarded as genuine" (ibid.: 348).

3 Contingency and Necessity as Ethical Concepts

The occasionalist theory of the theologians was carried to a moral domain, and the human acts are seen not morally good or bad in themselves but they are so just because God willed them to be good or bad. The theologians claimed that there is no rational ground on which the ethical values can be evaluated; there is no real evidence that these values are rationally justified other than the belief itself that they are. The theologians supported their claim pointing to the fact that even those who base the ethical value on a rational ground disagree among themselves: some believe that killing animals for consumption is bad and others do not consider it so.

Against this ethical voluntarism, the philosophers and rationalistically oriented theologians (i.e., Mu'tazilites) argued that ethical norms, the nature of right or wrong can be determined by reason alone and they are independent of the Divine prescriptions. They defined human actions ethically in relation to its agent. An action is moral if it has a relation to an agent who is rational and capable. Not all actions are moral, only those that have a certain intrinsic quality that can be defined as good or bad, right or wrong (Fakhry and Majid 1991: 32). They rejected the idea of ethical voluntarism and argued that what makes an action good or bad is the intrinsic quality (nature) of these acts, and this intrinsic quality gives the ground for Divine commands or prohibitions. Furthermore, this intrinsic quality is known to us "by necessity" (by intuition). Anyone who questions the distinction between good and bad actions rejects this necessity (ibid.: 33).[6]

The two different concepts of modality that were developed by Muslim philosophers and theologians need to be studied further, particularly in respect to their implications for the contemporary ethical problems. The similarity between occasionalist theory of Ghazâlî and pragmatist approach developed by Richard Rorty is striking. Both would join hands in solidarity for anti-essentialism and even for rejecting a rational ground for ethical values, though they might part company for the justification of their approaches.

Another potential avenue to be pursued is Averroes' approach to the statistical and potentiality theory of contingency particularly in relation to ethical principles: what is always true is necessary, what is never true is impossible. In between lies the realm of contingencies that is sometimes, and perhaps most of the time true, but not always. Ethics, particularly bioethics should try to aim at principles that are true most of the time. Few and seemingly insignificant exceptions that remain outside the general principles are usually the ones that make scientific and cultural development possible and human life worth living.

Notes

[1] Possibility and contingency are sometimes used interchangeably. Unless the difference is indicated, they are also used interchangeably in this article. For the difference between both, see Hamlyn 1972.
[2] Properly speaking, logic is not about specific topics such as circles or triangles (which belong to the domain of geometry) but about any topic in the most general sense.
[3] Some philosophers, following Aristotle, believed that whatever exists necessarily. However, this kind of necessity has to do with cause and effect relation rather than the number of the planets.
[4] In addition to the philosophers, the rationally oriented theologians, known as Mu'tazilites, also accepted necessary causation and a fixed nature of things.
[5] Averroes 1987: 329. Another theologian al-Bâkillânî, defined impossible as something that "had never happened, and can never happen and a logical contradiction." He believed that God can create things, no matter how extraordinary and miraculous they are, as long as they do not entail a logical contradiction such as combination of contraries or something's being at two different places at once. See Macit 1997: 102.
[6] Some Western scholars of Islam claimed that there is no concept of natural law in Islamic tradition. This, however, makes less than justice to the vast corpus of Islamic jurisprudence. For the arguments that clearly show that Muslim theologians and jurists debated themes that were familiar to the themes discussed in natural law literature in the Western tradition, see Emon 2004/2005.

References

Averroes. *The Incoherence of the Incoherence (Tahafut al-tahafut)*. Simon van den Berg (trans.). Cambridge: Cambridge University Press, 1987.
Emon, Anver M. "Natural Law and Natural Rights in Islamic Law." *Journal of Law and Religion*, 20, 2, 351–395, 2004/2005.
Fakhry, Majid. *Ethical Theories in Islam*. Leiden: E. J. Brill, 1991.
Hamlyn, David W. "Contingent and Necessary Statements." In *The Encyclopedia of Philosophy*. Paul Edwards (ed. in chief). New York: Macmillan, 1972.
Kukkonen, Taneli. "Possible Worlds in the 'Tahâfut al-Falâsifa': Al-Gazâlî on Creation and Contingency." *Journal of the History of Philosophy*, 38, 479–502, October 2000a.
Kukkonen, Taneli. "Possible Worlds in the 'Tahâfut al-tahâfut': Averroes on Plenitude and Possibility." *Journal of the History of Philosophy*, 38, 499, July 2000b.
Macit, Muhittin. "İmkân Metafiziği Üzerine: Gazzâli'nin Felsefi Determinizmi Eleştirisi." ("On the Metaphysics of Contingency: The Critique of Ghazâlî's Philosophical Determinism.") *Divan*, 102, 93–140, 1997.
Marmura, Michael E. "Avicenna." In *The Encyclopedia of Philosophy*. Paul Edwards (ed. in chief). New York: Macmillan, 1972.
Russell, Bertrand. *Our Knowledge of the External World*. New York: The New American Library, 1960.
Will, Frederick L. "Necessity." In *Dictionary of Philosophy*. Dagobert D. Runes (ed.). Totowa, NJ: Littlefield, 1989.

Of Poststructuralist Ethics and Nomadic Subjects

Rosi Braidotti

1 Introduction

This chapter rests on a number of assumptions that need to be clarified from the outset. The first point is that I approach the question of ethics from the background of Continental, notably modern French philosophy. It is therefore important to clear the grounds of the on-going polemic regarding French theory in general and post-structuralism in particular. More specifically for the purpose of this collection, I want to dispel from the start any association between poststructuralist ethics and the charges of moral relativism, of a-moral anarchy or romantic radicalism that are often moved against it (Sokal and Bricmont 1998).

These negative charges are allegedly motivated by the emphasis post-structuralism has placed on questioning, deconstructing and de-territorializing the unitary vision of the subject, which postulates the coincidence of the subject with his conscious, rational and reflexive self, in keeping with a humanist idea of the individual. The systematic critique of this implicit or explicit humanist assumption by Foucault, Derrida, Deleuze or Irigaray – to name but a few – has fuelled an over-defensive reaction on the part of those who believe that only a centralized, rationally-based and consciousness-driven notion of the subject – as in the traditional notion of liberal individualism – can guarantee ethical and political agency and a sense of responsibility. One may want to argue that a great deal of this reaction can be read as expressing the fear of loss of cognitive and political mastery on the part of professional philosophers. Such a polemic, however, falls outside the scope of my paper, hence my desire to clear it out from the very start. Rather than falling into reductive simplifications that equate post-structuralism with relativism, I would like to focus on the specific contribution this tradition of thought can make to the debates on ethics in general and bio-ethics in particular.

The charges of moral relativism are incorrect, both historically and conceptually: conceptually, French philosophy does not correspond to postmodernism, but rather refers back to a rich and established tradition of materialism and practical ethics.

R. Braidotti
e-mail: Rosi.Braidotti@let.uu.nl

M. Düwell et al. (eds.), *The Contingent Nature of Life*,

One only has to look across the field of contemporary French thought: Deleuze's ethics of *amor fati* (1992; 1995), Irigaray's ethics of sexual difference (1984), Foucault's search for the ethical relationship (1976; 1977; 1984a,b), Derrida's (2001) and Levinas (1969) emphasis on a non-appropriative ethics of otherness which acknowledges the receding horizons of alterity (Critchley 1992), to realize that one is immersed in ethical concerns. Of great relevance is also the established tradition of Lacanian ethics of psychoanalysis, which defends intersubjectivity, while it also posits a split or process-oriented self and defends a radical form of scepticism towards any foundational notion of truth and unified understanding of the subject.

It is therefore the case that ethics in poststructuralist philosophy is not confined to the realm of Rights, distributive justice, or the Law, but it rather bears close links with the notion of political agency and the management of power and of power-relations. Issues of responsibility are dealt with in terms of alterity or the relationship to others, but are also infused by a commitment to accountability, situatedness and cartographic accuracy (Braidotti 2002). The main thesis of this paper is that a poststructuralist ethical position, far from thinking that a liberal individual definition of the subject is the necessary pre-condition for ethics, argues that such a definition hinders the development of modes of ethical behaviour that respond to the contradictions of our era.

The prejudice against poststructuralist ethics is also inaccurate historically. The historical condition of post-modernity as analyzed by Bauman (1993; 1998) and Deleuze (1972a and b and Deleuze and Guattari 1980) calls for a rigorous assessment of the shortcomings of modernist or humanist ethical values. It is important to stress the point that poststructuralism constitutes a form of critical assessment of high modernism and of the project of modernity. This critical stand has crucial implications for the discussion on ethics; philosophy is about accounting in a cartographic manner for the actual conditions of our present historical location. This embedded form of materialism and respect for history as well as the high level of accountability are all the more relevant in view of historical events such as the Holocaust and other genocides; the devastation caused by development enforced on a colonialist model and the terror introduced in our social and moral universe by the nuclear predicament. Foucault's work on bio-power is of the greatest relevance to this discussion (Foucault 1976; 1977; 1984a; 1984b) as he famously challenges the ideals of the Enlightenment (Foucault 1975; 1997). The question of how to assess their broken promises lies at the heart of poststructuralist ethics.

In other words, post-modernity as an event marks the historical decline of some of the fundamental premises of the Enlightenment, notably the progress of mankind through a self-regulatory and teleological use of reason and of scientific rationality allegedly aimed at the perfectibility of man. According to poststructuralist analysis, we have entered a post-humanistic era as a result of the effects of our own historical development. Post-humanism is a factor of our own historicity (Haraway 1997 and Braidotti 2006) and technology has played a key role in engendering this situation.

My position is therefore pragmatic: we need schemes of thought and figurations that enable us to account accurately for the complexities of our historicity. For instance, most of us already live in emancipated (i.e.: post-feminist), multicultural (i.e.: post-eurocentric), technologically driven (i.e.: post-natural) societies with high degrees of mediation and global interconnections. These are neither simple, nor linear events, but rather multi-layered and internally contradictory phenomena that combine elements of ultra-modernity with splinters of neo-archaism: high tech advances and neo-primitivism at the same time. The simultaneity of radically opposite social effects that defy the logic of excluded middle throws a challenge to critical theory: how to bring these non-linear processes into adequate theoretical representations is a challenge not only of the methodological, but also of the conceptual kind. Contemporary culture shows a remarkable lack of imagination in addressing this challenge; it favors instead the predictably plaintive refrains about the end of ideologies, run concurrently with the apology of the 'new'. Nostalgia and hyper-consumerism are two faces of the same coin of neo-liberal restoration. In opposition to this I want to argue that we need cartographies of subjectivity, which adequately reflect the processes of flows, fragmentation, mutual interdependence, and mutations that mark our era. In ethics, as in social and political theory, we need to learn to think differently about ourselves and our systems of values, starting with the accounts of our embodied and embedded subjectivity.

I propose a non-unitary vision of the subject as not coinciding with rational consciousness, but rather as a dynamic, time-bound, embodied and embedded subject in process: a nomadic or rhizomic subjectivity. As I have argued (Braidotti 2002; 2006) the nomadic subject is not a prescriptive but rather a cartographic figuration; it evokes a conceptual form of self-reflexivity, which is specifically addressed to the subjects that occupy the center – one of the many centers that dot the web of the scattered hegemonic powers of advanced post-modernity. Speaking from my own location, paraphrasing Deleuze (1978) I would define the hegemonic vision of the subject as: 'male, white, heterosexual, educated, able-bodied, speaking a standard language, living in an urban center and owning property'. To reflect this specific location in the age of global flows and transformations requires that we shift the priority from concepts to processes. The fact that nomadic thought is an activity that reflects the spaces in-between does not, however, make it a view from nowhere in either spatial or temporal terms. Quite on the contrary, my nomadic subject is strictly connected to the issue of locations and to the forms of self-reflexivity and accountability that they entail. Locations are not only topological sites, but also temporal zones: embodied memories that can be activated against the grain of the dominant and traditional visions of the subject. The interaction between these two aspects of locations constitutes the site of co-production of the subject as a spatial and temporal process of interaction and exchanges.

2 Flows and Processes of Becoming

A poststructuralist position therefore assumes: a break from modernist visions of the unitary subject; a break from the teleological view of history; a break from Eurocentric modes of reading modernity and development. The emancipatory project of modernity entailed a view of the 'knowing subject' (Lloyd 1985) which excluded a number of structural others: the sexualized other, or women; the racialized ethnic or native others and the natural environment as a whole – animals and plants. They constitute the three inter-connected facets of structural otherness that posit difference as pejoration and as such they play an important specular role in defining the norm, the normal and the normative view of the dominant subject. Their exclusion form the enlightened circle of reason has been instrumental to the institution of masculine, euro-centric self-assertion (Braidotti 2006).

To say that the structural others of modern subject re-emerge in post-modernity amounts to making them into a paradoxical and polyvalent site. They are simultaneously the symptom of the crisis of the subject – and for the conservatives allegedly even its 'cause' – but they also express positive alternatives. It is a historical fact that the great emancipatory movements of post-modernity are driven and fuelled by the resurgent 'others': the women's rights movement; the anti-racism and decolonization movements; the anti-nuclear, pacifist and pro-environment movements are the voices of the structural others. Their emergence therefore inevitably marks the crisis of the former 'center' or dominant subject position. In the language of philosophical nomadology which I analyzed elsewhere (Braidotti 2002), they express both the crisis of the majority and the patterns of becoming-minoritarian of both the majority and of its margins. The rejection of dualism, specifically the mind/body culture/nature dichotomy is replaced in fact by Deleuze (1962; 1968), by a Spinozist political ontology of monism which posits a notion of Being as univocal and immanent. As a consequence, the relation of the majority to its margins is unhinged from any dialectical oppositional logic and becomes a matter of fluctuations and mutual specifications in processes of flow and exchange. The challenge consists in being able to tell the difference between qualitatively different flows or lines of motion. The criteria by which such differences can be coded and established are a matter of forces and values and hence of ethics. I shall return this in the next section.

This analysis of our historical condition also raises serious methodological issues in trying to deal with the illogical, non-linear and often quite simply irrational structures of advanced, post-industrial systems and their networks of power. It is a challenge that seriously tests the resources and the methodological stamina of social critics, like the poststructuralists, who are committed to provide adequate cartographies of contemporary culture. Meeting such a challenge requires some creative efforts that go beyond the traditional call of methodological duties: it also involves the creative quest for more adequate representations for the kinds of subjects we are becoming.

One of the forms taken by this analysis is bringing to the fore a political economy of affects in advanced post-modernity. Massumi (1992) has argued, for instance, that a state of perennial fear of an imminent catastrophe or a fatal accident is central to the affective economy of contemporary post-industrial, terror-crazed societies. The social imaginary surrounding the imminent threat has shifted from the nuclear to the ecological disaster, with special emphasis on the fear of genetic mutation or immunity breakdown. This state of constant anticipation of a bio-accident that is due to happen and whose unfolding is only a question of time, introduces high levels of anxiety in our societies not only about the present, but also about the possibility of actually sustaining a future. I want to argue that it is important to acknowledge this insight without precipitating into the manic-depressive mode favored by contemporary culture: the mixture of paranoia and frenzy, which express the modus vivendi of capitalism as schizophrenia (Deleuze and Guattari 1972; 1980). An ethical stance has to combine a lucid analysis with a commitment to action, as I will outline in the next section.

The first partial conclusion however is that, to confront the challenges of our historicity requires creativity, as well as intellectual and moral courage. It forces us to take seriously the conditions of our historicity, thus resisting the traditional move that disconnects philosophy from its immediate context. This move entails the assumption of responsibility or accountability so that one can engage actively with the social and cultural conditions that define one's location. Only such a process of full immersion into one's here and now, actual present location can offer the spaces of elaboration of possible modes of resistance to the schizoid logic of our time. The next step, however, consists in elaborating powerful alternatives to the dominant schizoid political economy of affects and thus resists both the neo-determinism of biogenetic capitalism and the techno-utopianism of the converted. This is the project of an ethics of sustainability.

The humanistic notion of the subject, as well as the logocentric vision of consciousness, which hinges on the sovereignty of the 'I', have been displaced. It can no longer be safely assumed that consciousness coincides with subjectivity, nor that either of them is in charge of the course of historical events. Both liberal individualism and classical humanism are disrupted at their very foundations by the social and symbolic transformations induced by our historical condition. Far from being merely a 'crisis' of values, I think this situation confronts us with a formidable set of new opportunities. Renewed conceptual creativity and a leap of the social imaginary are needed in order to meet the challenge. I want to argue that classical humanism, with its rationalistic and anthropocentric assumptions is of hindrance, rather than of assistance, in this process. I propose a post-humanistic brand of non-anthropocentric vitalism, inspired by philosophical nomadism, as one possible response to this challenge. My quarrel with humanism, in such a context, has to do with the limitations of its applicability and hence relevance in the present historical context.

3 Steps Towards a Nomadic Ethics

The moment the issue of ethics is posited in these terms, within a monistic view, the question on non-human, pre-human and post-human forces must be raised. The flows, exchanges and relations a subject encounters in his/her patterns of becoming encompass not only the social domain, but also the whole of the natural environment. As Lloyd put it in her commentary on Spinoza: we are all part of nature (Lloyd 1994). This opens up an eco-philosophical dimension, which inaugurates alternative ecologies of belonging. More importantly, it adds another layer to the post-human condition mentioned above, namely it marks a shift away from anthropo-centrism, towards a new emphasis on the inextricable entanglement of material, bio-cultural and symbolic forces that co-produce the subject. This post-human twist has implications for the discussion of ethics in that it forces a re-consideration of bio-centered egalitarianism (Ansell-Pearson 1997b), of 'the politics of life itself' (Rose 2001) and of planetary political and ethical agency (Guattari 1992).

Contemporary, embodied and embedded subjects are both techno and eco-logical units. Like all other living organisms, they are marked by the interdependence with their environment through a structure of mutual flows and data transfer that is best configured by the notion of symbiotic relations, viral contaminations (Ansell-Pearson 1997b), or intensive inter-connectedness. This nomadic eco-philosophy of belonging is complex and multi-layered, but also very materialist and concretely situated.

This environmentally bound subject is also a collective entity, moving beyond the parameters of classical humanism and anthropocentrism. The human organism is an in-between that is plugged into and connected to a variety of possible sources and forces. As such it is useful to define it as a machine, which does not mean an appliance or anything with a specifically utilitarian aim, but rather something that is simultaneously more abstract and more materially embedded. The minimalist definition of a body-machine is an embodied affective and intelligent entity that captures, processes and transforms energies and forces. Being environmentally bound and territorially based, an embodied entity feeds upon, incorporates and transforms its (natural, social, human, or technological) environment constantly. Being embodied in this high-tech ecological manner means being immersed in fields of constant flows and transformations. Not all of them are positive, of course, although in such a dynamic system this cannot be known or judged a priori.

Last but not least, the specific temporality of the subject needs to be re-thought. The subject is an evolutionary engine, endowed with her or his own embodied temporality, both in the sense of the specific timing of the genetic code and the more genealogical time of individualized memories. If the embodied subject of bio-power is a complex molecular organism, a bio-chemical factory of steady and jumping genes, an evolutionary entity endowed with its own navigational tools and an in-built temporality, then we need a form of ethical values and political agency that reflects this high degree of complexity.

To defend this position, I start from the concept of a sustainable self that aims at endurance. Endurance has a temporal dimension: it has to do with lasting in time – hence duration and self-perpetuation (traces of Bergson, here). But it also has a spatial side to do with the space of the body as an enfleshed field of actualisation of passions or forces. It evolves affectivity and joy (traces of Spinoza), as in the capacity for being affected by these forces, to the point of pain or extreme pleasure – which comes to the same. It means putting up with hardship and physical pain.

Apart from providing the key to aetiology of forces, endurance is also an ethical principle of affirmation of the positivity of the intensive subject. Endurance is the joyful affirmation as potentia. The subject is a spatio-temporal compound that frames the boundaries of processes of becoming. This works by transforming negative into positive passions through the power of an understanding that is no longer indexed upon a phallogocentric[1] set of standards, but is rather unhinged and therefore affective. This sort of turning of the tide of negativity is the transformative process of achieving freedom of understanding, through the awareness of our limits, of our bondage. This results in the freedom to affirm one's essence as joy, through encounters and mingling with other bodies, entities, beings and forces. Ethics means faithfulness to this *potentia*, which is my definition of the desire to become.

Affectivity is intrinsically understood as positive. It is the force that aims at fulfilling the subject's capacity for inter-action and freedom. Affectivity is Spinoza's *conatus* or the notion of *potentia* as the affirmative aspect of power. It is joyful and pleasure-prone and it is immanent in that it coincides with the terms and modes of its expression. This means concretely that ethical behaviour confirms, facilitates and enhances the subject's *potentia*, as the capacity to express her/his freedom. The positivity of this desire to express one's innermost and constitutive freedom is conducive to ethical behaviour. However, it only leads to ethical behaviour if the subject is capable of making the positivity of desire to last and endure, thus allowing it to sustain its own *potentia*. Unethical behaviour achieves quite the opposite: it denies, hinders and diminishes that *potentia*. Thus, unethical behaviour is unable to sustain becoming.

This introduces a temporal dimension into the discussion that leads to the very conditions of possibility of the future, to futurity as such. For an ethics of sustainability, the expression of positive affects is that which makes the subject last or endure: it is like a source of long-term energy at the affective core of subjectivity.

Deleuze's 'nomadology' (1972; 1980) as a philosophy of immanence rests on the idea of sustainability as a principle of containment and tolerable development of a subject's resources, understood environmentally, affectively and cognitively. A subject thus constituted inhabits a time that is the active tense of continuous 'becoming'. Deleuze defines the latter with reference to Bergson's concept of 'duration', thus proposing the notion of the subject as an entity that lasts, that is to say that endures sustainable changes and transformation and enacts them around him/herself in a community or collectivity. Deleuze disengages the notion of 'endurance' from the metaphysical tradition that associates it to the idea of essence, and hence also of permanence and links it instead to a form of transcendental empiricism or of anti-essentialist vitalism. In this perspective, even the Earth/Gaia is posited as a partner in a community that it still to come, to be constructed by subjects who will interact with the Earth differently. This is in some ways close to 'deep ecology', but radically anti-essentialist in its understanding of the structure and location of the human within it.

3.1 What, Then, is this Sustainable Subject?

It is a slice of living, sensible matter. A self-sustaining system activated by a fundamental drive to life: a *potentia* (rather than *potestas*) – neither by the will of God nor the secret encryption of the genetic code. Yet, this subject is psychologically embedded in the corporeal materiality of the self. The enfleshed intensive or nomadic subject is rather an in-between: a folding-in of external influences and a simultaneous unfolding-outwards of affects. A mobile entity – in space and time – an enfleshed kind of memory, this subject is in process, but is also capable of lasting through sets of discontinuous variations, while remaining extraordinarily faithful to itself.

This 'faithfulness to oneself' is not to be understood in the mode of the psychological or sentimental attachment to a personal 'identity' that often is little more than a social security number and a set of photo albums. Nor is it the mark of authenticity of a self ('me, myself and I') that is a clearinghouse for narcissism and paranoia, the great pillars on which Western identity predicates itself. It is rather the faithfulness of mutual sets of inter-dependence and inter-connections. The sustainable subject is thus made up of sets of relations and encounters. Those multiple relationships encompass all levels of one's multi-layered subjectivity, binding the cognitive to the emotional, the intellectual to the affective, and connecting them all to a socially embedded ethics of sustainability. Thus, the faithfulness that is at stake in nomadic ethics coincides with the awareness of one's condition of interaction with others; one's capacity to affect and to be affected. Translated into a temporal scale, this is the faithfulness of duration, the expression of one's continuing attachment to certain dynamic spatio-temporal coordinates.

In a philosophy of temporally inscribed radical immanence, subjects differ. But they differ along materially embedded coordinates: they come in different mileage, temperatures and beats. One can and does change gears and more across these coordinates, but cannot claim all of them, all of the time. The latitudinal and longitudinal forces that structure the subject have limits of sustainability. By latitudinal forces Deleuze means the affects a subject is capable of, following the degrees of intensity or potency: how intensely they run. By longitude is meant the span of extension: how far they can go. Sustainability is about how much of it a subject can take. Ethics can be understood as geometry of how much bodies are capable of.

3.2 What, Then, is this Threshold and How Does it Get Fixed?

A radically immanent intensive body is an assemblage of forces, or flows, intensities and passions that solidify – in space – and consolidate – in time – within the singular configuration commonly known as an 'individual' self. This intensive and dynamic entity does not coincide with the enumeration of inner rationalist laws, nor is it merely the unfolding of genetic data and information encrypted in the material structure of the embodied self. It is rather a portion of forces that is stable enough to sustain and to undergo constant, though, non-destructive, fluxes of transformation.

On all scores, it is the body's degrees and levels of affectivity that determine the modes of differentiation. Joyful or positive passions and the transcendence of reactive affects are the desirable mode. The emphasis on 'existence' implies a commitment to duration and conversely a rejection of self-destruction. Positivity is inbuilt into this programme through the idea of thresholds of sustainability. Thus, an ethically empowering option increases one's *potentia* and creates joyful energy in the process. The conditions which can encourage such a quest are not only historical; they all concern processes of self-transformation or self-fashioning in the direction of affirming positivity. Because all subjects share in this common nature, there is a common ground on which to negotiate the interests and the eventual conflicts.

So how does one know if one has reached the threshold of sustainability? By trial and error and by experiment. This is where the non-individualistic vision of the subject as embodied and hence affective and inter-relational is of major consequence. Your body will tell you if and when you have reached a threshold or a limit. The warning can take the form of opposing resistance by falling ill, feeling nauseous or by somatic manifestations, like fear, anxiety or a sense of insecurity. Whereas the semiotic-linguistic frame of psychoanalysis reduces these to symptoms awaiting interpretation, I see them as corporeal warning-signals or boundary-markers that express a clear message: 'too much!'. One of the reasons why Deleuze and Guattari are so interested in studying self-destructive or pathological modes of behaviours, such as schizophrenia, masochism, anorexia, various forms of addiction and the black hole of murderous violence, is precisely in order to explore their function as markers of thresholds. This assumes a qualitative distinction between on the one hand the desire that propels the subject's expression of her/his *conatus* – which in a neo-Spinozist perspective is implicitly positive in that it expresses the essential best of the subject – and on the other hand the constraints imposed by society. The specific, contextually determined conditions are the forms in which the desire is actualised or actually expressed. To find out about thresholds, you must experiment, necessarily, relationally or in encounters with others. We need new cognitive and sensorial mappings of the thresholds of sustainability for bodies-in-processes-of-transformation.

Deleuze's reading of Spinoza supports this. Another word for Spinoza's *conatus* is self-preservation, not in the liberal individualistic sense of the term, but rather as the actualisation of one's essence, that is to say of one's ontological drive to become. This is not an automatic, nor an intrinsically harmonious process, in so far as it involves interconnection with other forces and consequently also conflicts and clashes. Negotiations have to occur as steppingstones to sustainable flows of becoming. The bodily self's interaction with her/his environment can either increase or decrease that body's *conatus* or *potentia*. The mind as a sensor that prompts understanding can assist by helping to discern and choose those forces that increase its power of acting and its activity in both physical and mental terms. A higher form of self-knowledge by understanding the nature of one's affectivity is the key to a Spinozist ethics of empowerment. It includes a more adequate understanding of the inter-connections between the self and a multitude of other forces, and it thus undermines the liberal individual understanding of the subject. It also implies, however, the body's ability to comprehend and to physically sustain a greater number of complex inter-connections, and to deal with complexity without being overburdened. Thus, only an appreciation of complexity and of increasing degrees of complexity can guarantee the freedom of the mind in the awareness of its true, affective and dynamic nature. In this respect, sustainability is about decentring anthropocentrism and the ultimate implication is a displacement of the human in the new, complex compound of highly generative post-humanities.

4 Re-grounding Universalism

A non-unitary vision of the subject endorses a radical ethics of transformation, thus running against the grain of contemporary neo-liberal conservatism. This amounts essentially to a rejection of individualism, which however asserts an equally strong distance from relativism or nihilistic defeatism. A sustainable ethics for a non-unitary subject proposes an enlarged sense of inter-connection between self and others, including the non-human or 'earth' others, by removing the obstacle of self-centred individualism. This is not the same as absolute loss of values – as we shall see in the next section. It rather implies a new way of combining self-interests with the well being of an enlarged sense of community, which includes one's territorial or environmental inter-connections. This is an ethical bond of an altogether different sort from the self-interests of an individual subject, as defined along the canonical lines of classical humanism. It is a nomadic eco-philosophy of multiple belongings.

This position does not reject universalism, but rather expands it, to make it more inclusive. Contemporary science and biotechnologies affect the very fibre and structure of the living, creating a negative unity among humans. The Human genome project for instance unifies all the human species in the urgency to organize an opposition against commercially-owned and profit-minded technologies. Franklin, Lury and Stacey refer to this situation as 'panhumanity' (2000: 26), that is to say a global sense of inter-connection between the human and the non-human environment, as well as among the different sub-species within each category, which creates a web of intricate inter-dependences.

Most of this mutual dependence is of the negative kind: 'as a global population at shared risk of global environmental destruction and united by collective global images' Franklin, Lury and Stacey (2000: 26). There are also positive elements, however, to this form of post-modern human inter-connection. Franklin et al. argue that this re-universalization of one of the effects of the global economy and it is part of the recontextualization of the market economy currently under way. They also describe it in deleuzian terms, as the 'unlimited finitude', or a 'visualization without horizon' and see it as a potentially positive source of resistance.

The paradox of this new pan-humanity is not only the sense of shared and associated risks, but also the pride in technological achievements and in the wealth that comes with them. Nicholas Rose (2001) has written eloquently about the new forms of 'bio-sociality- and bio-citizenship' that are emerging from the shared recognition of the bio-political nature of contemporary subjectivity. We need to define the parameters of this new eco-philosophy of belonging in terms of share ethical sensibility: a new zoe-ethics is in the making. In a more positive note, there is no doubt that 'we are in this together'. Any nomadic philosophy of sustainability worthy of its name will have to start from this assumption and re-iterate it as a fundamental value. The point, however, is to define the 'we' part and the 'this' content, that is to say the community in its relation to singular subjects and the norms and values for a political eco-philosophy of sustainability.

Far from being a symptom of relativism, I see them as asserting the radical immanence of the subject. They constitute the starting point for a web of intersecting forms of situated accountability, that is to say an ethics. The whole point is to elaborate sets of criteria for a new ethical system to be brought into being that steers a course between humanistic nostalgia and neo-liberal euphoria. An ethics of sustainable forces that takes life (as *bios* and as *zoe*) as the point of reference not for the sake of restoration of unitary norms, or the celebration of the master-narrative of global profit, but for the sake of sustainability.

Note

[1] In spite of the many bad jokes made around it, this term is actually quite useful to describe a form of power that combines the concepts of phallus and logos, which expresses the belief that such power is structurally connected to masculinity and hence bound up in male identity.

References

Ansell-Pearson, Keith. *Viroid Life: Perspectives on Nietzsche and the Transhuman Condition*. London and New York: Routledge, 1997b.

Bauman, Zygmunt. *Postmodern Ethics*. Oxford: Blackwell, 1993.

Bauman, Zygmunt. *Globalization: The Human Consequences*. Cambridge: Polity, 1998.

Braidotti, Rosi. *Metamorphoses: Towards a Materialist Theory of Becoming*. Cambridge: Polity, 2002.

Braidotti, Rosi. *Transpositions: On Nomadic Ethics*. Cambridge: Polity, 2006.

Critchley, Simon. *The Ethics of Deconstruction: Derrida and Levinas*. Edinburgh: Edinburgh University Press, 1992.

Deleuze, Gilles. *Nietzsche et la philosophie*. Paris: Presses Universitaires de France, 1962. (English translation: *Nietzsche and Philosophy*. Hugh Tomlinson and Barbara Habberjam (trans.). New York: Columbia University Press, 1983.)

Deleuze, Gilles. *Spinoza et le problème de l'expression*. Paris: Minuit, 1968. (English translation: *Expressionism in Philosophy: Spinoza*. Martin Joughin (trans.). New York: Zone Books, 1990.)

Deleuze, Gilles. *Un nouvel archiviste*. Paris: Fata Morgana, 1972a. (English translation: "A New Archivist." In *Theoretical Strategies*. Peter Botsman (ed.). Sydney: Local Consumption, 1982.)

Deleuze, Gilles. "Les intellectuels et le pouvoir. Entretien Michel Foucault – Gilles Deleuze." *L'arc*, 49, 3–10, 1972b. (English translation: "Intellectuals and Power." D. Boudiano (trans.). In *Language, Counter-memory, Practice*. D. Bouchard (ed.). Ithaca: Cornell University Press, 205–17, 1973.)

Deleuze, Gilles. "Philosophie et minorité." *Critique*, 369, 154–5, 1978.

Deleuze, Gilles and Guattari, Fielix. *L' anti-Oedipe. Capitalisme et schizophrénie I*. Paris: Minuit, 1972. (English translation: *Anti-Oedipus. Capitalism and Schizophrenia*. Robert Hurley, Mark Seem and Helen R. Lane (trans.). New York: Viking Press/Richard Seaver, 1977.)

Deleuze, Gilles and Guattari, Felix. *Mille plateaux. Capitalisme et schizophrénie II*. Paris: Minuit, 1980. (English translation: *A Thousand Plateaus: Capitalism and Schizophrenia*. Brian Massumi (trans.). Minneapolis: University of Minnesota Press, 1987.)

Deleuze, Gilles and Guattari, Felix. *What is Philosophy?* New York: Columbia University Press, 1992. (Translation of *Qu'est-ce que la philosophie?* Paris: Minuit, 1991.)
Deleuze, Gilles. "L'immanence: une vie. . .." *Philosophie*, 47, 3–7, 1995.
Derrida, Jacques. *The Work of Mourning*. Chicago: University of Chicago Press, 2001.
Foucault, Michel. *Les mots et les choses*. Paris: Gallimard 1975. (English translation: *The Order of Things*. Alan Sheridan (trans.). New York: Pantheon Books, 1980.)
Foucault, Michel. *Histoire de la sexualité I: La volonté de savoir*. Paris: Gallimard, 1976. (English translation: *The History of Sexuality, vol. I*. Robert Hurley (trans.). New York: Pantheon, 1978).
Foucault, Michel. *L'ordre du discours*. Paris: Gallimard, 1977.
Foucault, Michel. *Histoire de la sexualité II: L'usage des plaisirs*. Paris: Gallimard, 1984a. (English translation: *History of Sexuality, vol. II: The Use of Pleasure*. Robert Hurley (trans.). New York: Pantheon Books, 1985.)
Foucault, Michel. *Histoire de la sexualité III: Le souci de soi*. Paris: Gallimard, 1984b. (English translation: *History of Sexuality, vol. III: The Care of the Self*. Robert Hurley (trans.). New York: Pantheon Books, 1986.)
Franklin, Sarah, Celia Lury, Jackie Stacey. *Global Nature, Global Culture*. London: Sage, 2000.
Guattari, Felix. *Chaosmose*. Paris: Galilée, 1992. (English translation: *Chaosmosis: An Ethico-Aesthetic Paradigm*. Paul Bains and Julian Pefanis (trans.). Sydney: Power, 1995.)
Haraway, Donna. *Modest_Witness@Second_Millennium.FemaleMan©_Meets_ OncoMouse*TM. London and New York: Routledge, 1997.
Irigaray, Luce. *L'éthique de la différence sexuelle*. Paris: Minuit, 1984. (English translation: *An Ethics of Sexual Difference*. Carolyn Burke and Gillian Gill (trans.). Ithaca: Cornell University Press, 1993.)
Levinas, Emmanuel. *Totality and Infinity*. Pittsburgh: Duquesne University Press, 1969.
Lloyd, Genevieve. *The Man of Reason*. London: Methuen, 1985.
Lloyd, Genevieve. *Part of Nature: Self-Knowledge in Spinoza's Ethics*. London and New York: Routledge, 1994.
Massumi, Brian. "Anywhere you want to be: an introduction to fear." In *Deleuze and the Transcendental Unconscious*. Joan Broadhurst (ed.). Coventry: Warwick University Press, 1992.
Rose, Nicholas. "The Politics of Life Itself." *Theory, Culture & Society*, 18, 6, 1–30, 2001.
Sokal, Alan and Bricmont, Jean. *Intellectual Impostures*. London: Profile Books, 1998.

Genetics, a Practical Anthropology

Christoph Rehmann-Sutter

Genetics is a branch of the life sciences. In university organograms it does not appear under the social sciences and humanities, like anthropology does. But actually genetics, in a nontrivial sense, also deals with what humans can or think they should do in their lives. It shapes their practices, and produces new dilemmas and moral responsibilities. And in addition to all the facts and figures about the composition and sequence of the genome, about risks and susceptibilities, it also guides our understanding of those dilemmas and responsibilities by providing content-rich, interpretative patterns explaining what mutations, genes or genomes signify in our practical life. This, however, goes beyond pure science. In this essay, I want to discuss some selected aspects of these interpretative patterns from the point of view of practical philosophy.

Most salient in the development of genetics in the last few decades has been a transition in our understanding of the role of the genome in the organism, from the concept of the genome as 'genetic program' to a 'systems' view. The 'anthropological' side of genetics has gone through a development that affects the significance of genetic information in the actor perspective. I start with a brief discussion of the extra-scientific uses of the qualifier 'genetic' (Section 1). Next, I explain some of the main biological counter-evidences to the genetic program view (Section 2) and sketch an alternative, systemic view of genetic information (Section 3). A gene (a mutation, a genome etc.) is no longer only a part of our cells but, as a result of the possibilities offered by testing, has suddenly become a component of our relationships and practical life. A patient's narrative ('Cora's story') will provide insights into the social complexities that are connected to the biological complexities of the genes in the body (Section 4). In the final Section (5), I suggest that we should look at the 'corporeal' and the 'social' complexities of genetic information together. How we *see* the body 'doing' things with genes and what people *practically do* with genes are closely interrelated.

C. Rehmann-Sutter
e-mail: Christoph.Rehmann-Sutter@unibas.ch

M. Düwell et al. (eds.), *The Contingent Nature of Life*,

1 It's Genetic

In an essay of 1998 'On Breads, Bibles and Blueprints', sociologist Barbara Katz Rothman offers a functional analysis of the statement 'It's genetic' in its colloquial, extra-scientific use. It is 'often offered as an excuse', she writes, 'a kind of throwing up of one's hands, helpless before a larger force'. 'If somebody can't lose weight or gain weight, can't help eating too much, or tends to get addicted (to whatever), and says "it's genetic", she or he is refusing personal responsibility for the action concerned'. People can't help what they do because the genes cause, even force, them to do it.

In the current biological understanding of genetics, she continues, 'genes' are stretches of DNA that code the production of specific proteins. But the term 'gene' was introduced into the language of science long before the discovery of the role of DNA as the carrier of inheritable genetic information: '"Gene" was the name given to the force that transmits qualities from parent to child, whether among people or among plants' (Rothman 2001). How can this be one and the same idea? On the one hand, the biological meaning seems too narrow and concrete to meet the practical need to explain the 'genesis' of any traits: it explains only the biosynthesis of a macromolecule. On the other hand, it seems too ambitious and hypothetical to meet the demands of the molecular system-functional approach that asks for a concrete and identifiable cause (as opposed to a hypothetical one), which can be made responsible in a sweeping retrospective. Anyway, even if science could bring these levels into coherence, genetic knowledge and genetic tests remain a far cry from the colloquial demands of genetic explanations.

But there is a commonality between the colloquial 'gene' and the gene in its pre-Watson/Crick use: both refer to an unknown hypothetical entity that may fulfil the explanatory (and exculpating) demand. Wilhelm Johannsen, in 1909, had no idea of what could be the molecular representatives of those 'factors' that explain the inheritance of traits. This explanatory role has remained in the post-Watson/Crick era, and even after the Human Genome Project.[1] For example, a recent study on twins concluded that the feeling of loneliness must have 'genetic contributions' of 48%, because monozygotic twins describe this feeling more frequently in similar terms than dizygotic twins (Boomsma and Dorret 2005). We have no idea which genes are involved and what they do; the study speaks of the 'genetic architecture of loneliness' just in the sense of inheritability.

On both levels of genetic explanation, the nano and the social, the reference to 'genetic' causes is a discursive figuration that introduces very special meanings to factors underlying a phenomenon, and explains its generation. At the nano/molecular level it is the sequence of DNA, the genetic information responsible for the generation of a protein molecule under certain circumstances; at the social/colloquial level it is something that is passed down the generations.

There is a temptation here, a temptation that sometimes led the genetic discourse of the twentieth century astray. When such underlying causes can be used for explanatory purposes, some people are tempted to project too much on them, the whole basket full of unfulfilled needs and desires for explanations, so to speak. And in the realm of life, being, becoming, nature, evolution etc. there are plenty of these. The genome, this verbal embrace of the totality of the genes that are available for an organism during its lifetime, turned into a projection screen for the possible fulfilment of many desires for explanations. The scientific idea of the genome has taken over many functions that were attributed to the soul in earlier times; thereby, importantly, the genome has mutated back to a metaphysical entity that serves ontological functions, to explain things like 'the being' of living beings, the 'principle' or even the 'essence' of life etc. (cf. Keller 2000; Hubbard and Wald 1997; Nelkin and Lindee 1995).

Not surprisingly, in this metaphysical turn of the double helix, ideologies came in, which are rooted in fields other than pure science.[2] Barbara Katz Rothman sees similarities to the ideology of patriarchy: 'that we are of our fathers, that our fathers make us from their seed, and that we unfold from our fathers' loins while curled in the safety of our nurturing mothers. The patriarchal assumption places our essence in a seed' (Rothman 2001: 135). Of course, genetics includes the female component in its modern concept of the seed, which is the genome made up of a diploid set of chromosomes. But the structure of the idea has remained the same: the 'essence' organizes from within, not through relationships, not on the level of the organism or its context; contexts are conditions or additions and the genome contains all the information that is essential.

There is one powerful word that catches these metaphysical desires most effectively: the term 'genetic program'. If the genes are built together like an instruction manual for the cells that tells them what to do in order to become an organism (a *Drosophila* fly, a human child, a female or a male adult), the genome is nothing less than the principle that organizes matter during development, the organizer that makes the living beings which populate the biosphere.

But there is also an irony in the history of genetics (Keller 2000). In the last decades of the twentieth century, when genomics celebrated its triumphs in the form of the Human Genome Project, molecular genetics provided increasing evidence that is drastically at odds with the program hypothesis of the genome. Thinking, in current molecular biology, has become contextual and makes much more use of systemic ideas than of reductionist gene centricism. I summarize some of the main reasons for this in the following section.

2 Scientific Counterevidence for the Program View of the Genome

Plants and animals are eukaryotes, their cells have a clearly distinguishable nucleus containing the chromosomes, which is separated by a nuclear membrane with pores, through which communication between the contents of the nucleus and the surrounding cytoplasm is possible. The synthesis of proteins in the cell is guided by sequences of DNA, but only indirectly. In the nucleus, relatively short fractions of the chromosomes are copied, which are needed by the developing organism at a distinct moment and place. Those copies consist of RNA, a single-stranded polynucleotide molecule, which functions as a messenger between the interior of the nucleus and the cytoplasm. These molecules are therefore called 'messenger RNA' (mRNA). If we look carefully at the words, we see that in this name, a particular picture of the relation between DNA and the organism is anticipated. DNA can be seen as the provider of the information needed for the function of the cell machinery, after all for development, and mRNA as a sort of outgoing mail, sent to the cytoplasm by the chromosomes, through the nuclear pores. The title 'messenger' is still plausible in a much narrower sense: mRNA contains the raw sequences that can be translated into the sequences of amino acids in proteins. But the proteins for which mRNA provides the codes are not organisms, and therefore, there is an important and philosophically extremely significant difference between the information content of a protein sequence (of amino acids) and the 'blueprint' or 'architecture' of the organism. Lenny Moss (2003) draws attention to this by distinguishing between two uses of the term 'gene': Gene-P for 'explaining' phenotypic traits, and Gene-D for explaining the immediate molecular interactions after a DNA sequence is activated. Eva Neumann-Held has introduced a similar distinction between the 'gene' as a difference-maker on the level of phenotype and the 'gene' as a process of interactions on the molecular level (Neumann-Held 1999).

In the description of protein synthesis, another key term from molecular biology appears: the 'genetic code'. It is crucial to keep in mind what this term does and does not signify. The genetic code can be represented as a table correlating base triplets of mRNA, i.e. all 64 possible combinations of three out of four RNA 'bases' (adenine (A), uracil (U), cytosin (C) and guanin (G))[3] with the 20 different amino acids, as well as the start and stop signs (see standard textbooks, e.g. Alberts et al. 1983: 98–111). The code therefore connects DNA and amino acid sequences and does not connect organic information (like hair colour, body size, the form of organs or the probability of diseases) with the nucleotide sequence of RNA (or DNA). Features of the cell and the body are far down the scale of living processes. However, the original hope of molecular biologists in the 1950s and 1960s was finally to be able to explain the organism and its development (ontogeny) through an extended application of this coding relationship. The metaphor of the genetic program, generated and brought into circulation in the late 1950s by eminent biologists such as Jacques Monod, François Jacob and Ernst Mayr (Kay 2000: Chapter 5), expresses this hope. The intuition was: We need to know and learn to understand the information content of the genome in order to be able to predict large portions of the structure and functioning of the whole organism.

It was always known that environmental factors also play a role. For instance, the direction of the light directs the growth and shape of a plant. The term 'genetic program' does not imply a view in which the environment plays no role, because a program can also be thought of as containing the prescription or a norm, how the organism should *react* to selected environmental factors in a complex environment, and which factors these should be (Van der Weele 1999: 91).[4]

In my view, however, the real scientific problem with the genetic program perspective came with the vast evidence for the multi-functionality of genes. One and the same gene can have different functions and play different roles in the concert of metabolic processes, according to the time, and the exact place of the cell within the developing multi-cellular organism. Below, I give a few examples of such instances of multi-functionality.[5]

Alternative Splicing: mRNA molecules are not used in their native state. Before use, parts of their sequences are cut out. The pieces that are cut out are called 'introns', the remaining fragments that are composed to a 'mature' mRNA molecule are called 'exons'. The process of cutting and pasting is called 'splicing'. The surprising fact, however, is that the pattern of splicing seems not to be predetermined by the genome. Depending on the situation during development, it is possible that several different, alternatively spliced mRNA molecules can be made out of one gene; and these different mRNA molecules lead to different proteins. The old rule 'one gene – one protein – one function', has been replaced by 'one gene – many different proteins – even more different functions'.

Overlapping genes: the same stretch of DNA is 'used' by different genes.

Alternative reading frames: triplets of three DNA-'letters' that each code for one amino acid are 'read' in one stretch with a shifted beat so to speak, e.g. not as . . . AAT TTG CCT . . . but as . . . A ATT TGC CT . . . .

Trans splicing: some exons are brought in from other open reading frames.

Anti-sense transcript: some of the exons are 'read' in a reverse direction from the other two single strands of DNA.

mRNA editing: after transcribing DNA into mRNA, the cell changes the sequence of the mRNA before protein synthesis, thereby generating a different protein.

Selective methylation: for the regulation of differential gene activities, particular DNA stretches are changed covalently by adding methyl groups.

Multiple, 'place'-specific function of genes and proteins: one and the same gene or protein has different functions at different times in the developmental process and/or at different places in the micro-architecture of the organism, even within one and the same cell.

Evolution genes: there are enzymes that serve the 'purpose' of biological evolution only, not the life of the individual going from one generation to the next. Some of them are enzymatic variation generators, others act as modulators of the frequency of genetic variation (Arber 2005).

The phenomenon of evolution genes is special, because it is not an example of the 'one-locus-multiple-product dilemma', as Thomas Fogle (2000: 8) has titled the others. It is rather a use of DNA by the organism that encompasses genes that are not functional for the individual life, which may even be dysfunctional for the individual when helping to create less well-suited individual variants. But they are there nonetheless, because they help to ensure the adaptability of the population over many generations. The genome, in this picture, can no longer be described comprehensibly as a 'list of instructions for making that structure' of the organism, which can be transmitted to the next generation (Smith and Szathmáry 1999: 2), but it is rather an information-storing component of the cell that is utilizable in different ways, which has acquired functions even beyond stabilizing the life of the individual.

The principle of multi-functionality also holds for gene products. There are examples in developmental genetics where a gene product (mRNA, protein) is used at one side of the egg cell of *Drosophila* to help determine the anterio-posterior axis, by inducing surrounding follicle cells to adopt special features as posterior follicle cells, and is subsequently drawn to another part of the egg cell to help determine the dorso-ventual axis, by inducing another area of surrounding follicle cells to adopt the features of *dorsal* follicle cells.[6]

If genes are multi-functional (via a wide variety of mechanisms that can be described in detail), the assumption that DNA sequences, the genes, or the genome function as a list of instructions and act as a program for development and life, is not plausible anymore. This implausibility is also manifest in the term 'instruction'. The content of an 'instruction' would have to exist independently of the process that is instructed. But this does not seem to be the case, at least in many instances. The examples of the phenomenon of multi-functionality demonstrate rather that the functional 'meaning' of DNA is dependent on contexts, in particular on cytoplasmatic interactions and processes. The activity of DNA and the function of the genes derive their content (or 'meanings') only within the context of a dynamic cellular or multi-cellular system in a certain kind of environment. For the operative functioning of living systems, i.e. their development and their life performance, the coded information of DNA is necessary but not sufficient. As Fogle (2000: 19) explains: 'The mutual dependency of DNA and protoplasmic interactions bedevils a simplistic labelling scheme for expressed segments of hereditary information.' I agree with this statement. The assumption of 'instructions' and 'genetic programs' encoded in DNA sequences has proved to be an oversimplified description. It is thwarted by the empirical evidence of molecular biology, despite many cases in which a strong correlation still exists between mutation (genotype) and phenotype.

3 What is the Alternative?

I now want to sketch out an alternative view of the genome-organism relationship that is consistent with these newer findings, and which at the same time respects the key role that is undoubtedly played and fulfilled by DNA in the living process. The alternative, if it should be philosophically rewarding, in my view, is not just to de-emphasize the genes and take refuge in the, sometimes striking, influence of environmental and social factors in the achievement of features or capacities in individuals. The stance that can lead to a new and consistent picture of the role of molecules in living processes is, as Susan Oyama has argued convincingly (Oyama 1985), not on the other side of the nature-nurture divide, neither it is an enlightened interactionism emphasizing the interplay between genetic and environmental factors; it is *beyond* that divide altogether. But how can such an alternative division of work between genes and the rest of the organism be conceived without using a nature-nurture dichotomy?

The key term is genetic information. I follow closely Oyama's argument in this regard. She stresses the point that the information content of DNA, i.e. its sequence, does not mean anything before it becomes involved in the processes of regulation or transcription. 'Yet information "in the genes" or "in the environment" is not biologically relevant until it participates in phenotypic processes. Once this happens, it becomes meaningful in the organism only as it is constituted by its developmental system. The result is not more information but significant information' (ibid.: 13). The implication of this approach is that significant information, i.e. the information which *in-forms* developmental steps, does *not pre-exist* the processes that give rise to it. Information, in the sense of developmentally relevant information, is *produced* in the course of the interactive processes of the cell, the organism and the environment. Or, as Oyama puts it: information has an ontogeny.

This view fully respects the impressive significance of DNA, currently being investigated in detail by molecular biology, particularly genomics. But the interpretation of the experimental results at the level of the 'basic picture' of the role of the genome within living processes is strikingly different. To replace the 'program' as the guiding metaphor by the 'system' is much more than a slight change of words or a shift in images. It is rather a totally different view of life, of inheritance, development and of the *identity* of living beings. This difference appears in the following parallel statements by John Maynard Smith and Eörs Szathmáry, who express the baselines of the program view, and Susan Oyama who expresses an argument in favour of a developmental systems view:

> The basic picture, then, is that the development of complex organisms depends on the existence of genetic information, which can be copied by template reproduction [. . .]. What is transmitted from generation to generation is not the adult structure, but a list of instructions for making that structure.
>
> (Smith and Szathmáry 1999: 2)

> What is transmitted is *macromolecular form*, which, though it is necessary for the development of phenotypic form, neither contains it nor constitutes plans for it, and *developmentally relevant aspects of the world.*
>
> (Oyama 1985: 22)

To Oyama's statement I would add that the micro-architecture of the cell (egg cell, nucleus) and the character and form of its surrounding cells (in the case of multi-cellular organisms) also belong to 'the world' of the genome. They are also transmitted from generation to generation.

A systems theory of the genome can be consistently conceived when starting from the following four basic statements:

1. The successive steps of development follow each other in a 'historical' logic: each step is determined by the structure and dynamics of the previous one, not by a program or blueprint that has existed from the beginning. Each step establishes a causal network that informs and determines the next.[7]

2. No factor or element, such as DNA, is privileged a priori as an ontologically superior cause, or as the essential part that is fundamental in such a way that it requires all the other factors or elements as means to its end.[8]
3. Genetic information, i.e. the information that is significant for performing a developmental step, is generated by interactive processes between DNA and other factors within the organism, as well as selected environmental factors. Genetic information therefore (unlike DNA) has an ephemeral existence, i.e. it only exists in the actual performance of the step that it informs.[9]
4. The dynamic structural aspects of the organism, i.e. the architectural form (which is a constant movement) 'seen' from a point within the developmental system, appear as one type of cause in explanations of developmental steps. Causal processes, where genes are involved, are therefore placed within dynamic microstructures, which have elements that are not only biochemical and biophysical, but are also genuinely morphological (molecules that are 'here', not 'there'; flow patterns in space, etc.).[10]

This view of the organism as a developmental system, sketched here only very crudely by its basic philosophical assumptions, uses metaphysics much more sparingly than the view of the organism as a phenotypic expression of a genetic program. But it is still a content-rich philosophical theory of life.[11] I would concede that it cannot work completely without metaphysical assumptions, for example because a system is seen as an entity that establishes itself via a differentiation between self and non-self. These terms (like systems) are not deductions from experiments, but interpretative concepts that are used to *understand* the empirical evidence. These assumptions, however, are minimal and less sweeping than the assumption of the 'essential' nature of DNA sequences in the context of programme genomics.

The basic ideas behind the alternative approach can perhaps be better understood when we compare DNA with elements of everyday life. Barbara Katz Rothman, in the essay that I have quoted at the outset, refers to the recipe for making bread. She tells how her challah is different each time she makes one ('smooth as silk', 'heavier this time'), starting from the same recipe and using the same ingredients, 'I end up with something different each time' (Rothman 2001: 24). So, even if DNA were something like a recipe (which I do not assume in a systemic view), this would not support the essentialist assumption, because the essence of the challah is not in the recipe but in its actual phenotypic presence as a bread with a character. The significant information for the challah is not what stands there in the recipe book but how it is interpreted and performed by the baker of the bread. But then, Rothman uses another comparison: 'DNA as a musical score, notes on a page, but capable of so many nuanced interpretations' (ibid.: 25). Here too, there is an anti-deterministic emphasis, leaving room for the interpretation that I favour. I would say: yes, if genes were something like a text with prescriptions and instructions, then they are indeed more like a musical score, because in this comparison, as with the recipe, it becomes clear that 'life' (the essential bit) is not the prescription but what somebody makes out of it: 'Is it possible that this static thing, these notes on the page, this string of ATCGs, *is* life? Or is life the process itself in which these – and other – notes are played?' (ibid.) Yes, it is. Life is the process; the organism is the process, the *performance*, not the notes used for it. Significant information, in Oyama's sense, is the interpretation that is both used and *created* in the performance. The information contained in the score is, of course, still important, but not in the sense that it contains the music. The score is rather one element in a complex series of interaction, where many other things play a role, like the instrument used, the knowledge, the style and the emotions of the performer. Glenn Gould's interpretation of Bach's Goldberg Variations, for instance, is certainly very different from the Goldberg Variations in performances of Bach's time, when no modern Steinway or Bösendorfer was available. The music is what is *done* by the performer and how it is heard by the audience. Significant information, we could say, was not there on the score before it was played, even if Gould still played Bach's piece, not his own. This paradox of being there already, and not being there already, holds true for genetics too. But in many other ways, of course, the comparison of DNA with the musical score fails. In the case of DNA there are more rearrangements, like alternative splicing and mRNA editing, which influence the composition of the score itself, before it is read by the performer.

In genetics, we have a close analogue to reading: genetic testing. In genetic testing, pieces of genetic information are disclosed, understood in a certain way and introduced into a world of human self-understanding, medical consequences and social interactions. Let us look more closely at one example.

4 Cora's Story

In a recent research project on the social aspects of genetic tests, we used a qualitative interviewing technique to find out how people who had to decide whether or not to take a genetic test perceived the implications of this decision. We were particularly interested in the temporal implications, i.e. what 'time' signified to the participants in the context of that decision. The main results are published in Scully et al. (2007) and Scully (2006). One participant, we call her Cora, told us what she anticipated when she was offered a test:

> I wanted to have that done, so that I'd know if I had this gene mutation, so I could be tested more often.[12]

Cora was 38 years old at the time of the interview. She came from a family with hereditary colon cancer, and had breast cancer herself before the interview, as well as an operation and chemotherapy. The test for colon cancer predisposition was offered to her, in order to get information to help her make decisions about doing more tests for preventive reasons. The interview was performed seven years after the test, i.e. the story was told in retrospect.

However, her anticipation of what the results from the test would signify for her (the basis for decision making about more frequent cancer check-ups), proved to be a crude simplification. She struggled with the (positive) results of the genetic test and said:

> I had difficulties coming to terms with it.... I wanted a kind of new identity. I didn't want to be who I was... lifestyle and everything. I struggled really hard for 3, 4 years....
>
> My own body was alien to me.
>
> In these 3, 4 years I was afraid and didn't know what to do and worried about everything and couldn't relax. Suddenly I said to myself: Stop! I'll try to enjoy my life now, as long as I am healthy.... Cancer is always here. But now I can handle it. Live with it....

The unanticipated complexity involved not only her own body, self and well-being, but also her family relationships. Two relationships seemed to be most directly affected: the relationship with her father, from whom she had inherited the mutation, and the relationship with her husband, from whom she felt alienated because of the impossibility of sharing her experiences with him:

> For some time I had difficulties with the fact that I'm a member of this family, that I inherited this gene from my father. But I loved my father more than anything and he died when I was 15. So I was pulled to and fro. I had inherited something, but a bad thing, that could kill me.
>
> ... in the beginning I had trouble falling asleep. I saw my own funeral, read my own obituary... and I couldn't tell my husband.

Cora mentioned a series of what we called integration tasks in this interview, all of which concerned connecting the test results with her life world (ranging from coming to terms with a new perception of her own body and her social self to psychotherapy). Reading through the interview, I listed the following tasks, each illustrating one aspect of the social complexity of genetic information:

- Embodiment-identity
- Social identity
- Family relationships
- Reclaiming agency in the present, putting the future back in its proper place
- Learning something from 'it'
- Psychotherapy

The information provided to Cora by the test was an answer to a distinct question: 'Do I have the mutation?' – the test did say: 'Yes, you do.' But it was not 'just' a truth telling: this information about the actual composition of her genes developed into a challenge for her self-perception, her ideas about the future, her life with her family, her identity. Genetic information is particular because it adds many nuances. Re-integrating the social self was a task that started with a crisis and took her several years. It brought her into a situation of distress, a *Grenzsituation* as Rouven Porz puts it: it confronted her with 'the absurd' in Camus's sense (see Porz 2004).

Cora's story is unique, because what happened to her will not happen to others in exactly the same way. But within the particulars there are also more or less general features that may characterize other life histories with genetic tests. (i) The expected significance of the test result (or the reasons for taking a test) can be different from the real implications of the test, which become manifest only *after* the test. (ii) The actual meaning of genetic information, in other words the actual content of this information, depends not only on DNA sequences that can be tested, but also on the social relationships, and the internal (psychological) dynamics of the person who has to come to terms with the results of the test.

Genetic information for the users, in a word, also depends on social complexities. These hypotheses (i and ii) can be developed out of Cora's story, and are also consistent with other qualitative interviews we made.[13]

5 'Doing the Genes'

From the perspective of a potential user of genetic tests, genetic information is relevant at different levels (see Table 1). Most directly, the user is confronted with the result of a genetic test (level 3), either expected or unexpected. Underlying the results of a test is level (2), DNA sequences in the chromosomes, the direct object of a test, what a test 'shows'. But knowing the sequence or mutation is not the ultimate aim of the test.

Most medical genetic tests look for the likelihood of a certain disease or for a diagnosis. This information, however, is beyond the rough sequence information, and involves knowledge about the interactive processes between DNA and many other factors and processes in the cells of the body that generate developmentally significant genetic information. This is the level (1) where the 'significance' of a mutation or a sequence for the body is investigated. But there are at least two different levels of interpretation on the other side. At level (4), hermeneutic frameworks are used to understand the implications of genetic information for the perception of the body. On this level, genome theories, metaphors and expectations play a key role. For instance, it makes a difference whether one sees a mutation that indicates a genetic risk for developing a cancer as a section of the genetic program, i.e. as an instruction to the body to make cancerous tissue, or, in the framework of a systems approach, as a factor that could become involved in a process leading to cancerous tissue. The latter understanding indicates a higher level of likelihood, but it does not

Table 1 Levels of genetic information in testing and the inter-disciplinarity of the field of genetics: at levels 1 and 2 genetic information is the object of empirical, scientific studies and experiments; at levels 4 and 5 the participant's hermeneutic skills, together with cultural and social science methodology are key.

What the body does		What people do		
1	2	3	4	5
Interactive processes between DNA and other factors generating developmentally significant 'genetic information' in the cells of the body	DNA sequences (a mutation, a 'gene')	Results of a genetic test	Hermeneutic framework for understanding the biological significance of genetic information in the body	Social complexities and processes of social interactions generating 'genetic information' from the perspectives of the actors

imply that the genes in the cells of the body are like little springs wound up to release and make tumours (Rehmann-Sutter 2000). There is a further level (5), at which genetic information is reinterpreted as a meaningful and consequence-laden message in social contexts that take part in processes of social interaction. Other actors are involved in a joint practice of making genetic information a socially meaningful and practically pertinent entity.

The levels of understanding (4) and social integration (5) both depend on the scientific descriptions of the body processes. But the level of social integration also depends on the hermeneutic framework used for interpreting the data. Therefore, the table also represents a map of interdisciplinary cooperation between empirical scientific (experimental) methods, as developed in biology, and hermeneutic interpretative methods, as developed in the humanities, the cultural and social sciences, and also in ethics.

Ethics is involved because the processes on nearly all levels are not just natural events but also practices. Level (1) is the level of what I call 'organic practices'. These practices are activities of the body as a living organism. They are not actions that could be chosen by a moral subject, but they can still be considered to be more than just biochemical or biophysical events, just because they are living processes. Understanding them as practices in the Aristotelian sense (Rehmann-Sutter 2006) is, I believe, possible. As such, their functional goal is not only productive (*poietic*, as Aristotle said), to bring about certain effects (e.g. to synthesize a protein) but also, or rather primarily, their goal is the performance itself, because it is part of the living process, of 'being in the world' as a living thing. Developmental processes do not simply lead to the next steps or the next intermediate states, but are – if we adopt such an organic practice view – themselves intrinsically significant. They are parts (steps) of a being's continuous *presence* in the world as the subject of a life.[14]

On level (3), ethics is involved because there are decisions to be taken about which tests to perform and when. Genetic counsellors and their patients may find themselves in sometimes demanding ethical dilemmas of many different kinds. These are explored in a growing literature about the ethics of pre-implantation, prenatal or pre-symptomatic genetic testing.[15]

However, a comprehensive understanding of the ethical implications of a certain decision about testing is only possible if the participants (doctors, counsellors, patients, partners) on level (3) take levels (4) and (5) into account. It can happen (as in Cora's case) that the personal and social *reality* of genetic information is different from its anticipation, which was what motivated or justified the test. Anticipating this as a possibility cannot be irrelevant for a free and informed decision about taking or not taking a certain test. In an individual case, however, it may be difficult or even impossible to anticipate what will happen concretely with the test results on level (5). But systematic sociological and empirical ethical studies can at least provide some knowledge about what *can* happen in typical cases, and how ethical dilemmas can be modulated by this for the different persons involved.

There is another common feature among the various levels of genetic information: genetic information is not just *there*, lying around in order to be picked up. It is produced by activities in a variety of ways. In other words, we *do* genes in a variety of ways.[16] On level (1) it is the body as a developmental system that is in a sense the author of significant genetic information (a systemic reading of genomics is presumed here; it would not be so in a genetic program account). On level (3) there is always a choice about which tests to perform, how, at which time in life, and under what conditions (e.g. before pregnancy as pre-implantation genetic diagnosis, or during pregnancy at different stages; a single gene test, or a standard genetic check-up using different tests simultaneously, etc.). On level (4) we actively interpret the meaning of genetic information. There are different interpretative frameworks and metaphors that could be utilized in order to create that meaning. And on level (5), a texture of social interaction is involved in a process of co-production of genetic knowledge, and in shaping its practical implications for the partners in relationships.

6 Conclusion: Contingency of Life

It should not escape our attention that such an emphasis on the constitutive practical aspects of genetic information connecting the various levels under an overarching principle of 'doing genes' also has a consequence for the contingency problem. Contingency, the condition of being subject to chance or the happening of something by chance (Webster's Dictionary), can have two different meanings: (1) epistemological: information that is accidental, not deducible, unpredictable; (2) ontological: a process or event that occurs by chance, or without intent. In both senses, being contingent means that things fall together by some unpredictable interaction that will not necessarily take place.

In the epistemological sense, a systems approach to genetic information assumes that development is indeed a contingent process. DNA information alone cannot predict the structures of the organism that develops with it.[17] They are not 'programmed'. But ontologically, development is *not* contingent. The organism constitutes its own highly structured environment for each cell. Each cell, even each molecule within a cell, is a *place* and *in* a place within a complex micro-architecture. The conditions of this place determine what impact DNA sequences will have and which bits of them will be selected. The picture is deterministic *from step to step*, but it is not deterministic in a transitive sense (from the beginning to the end). A systems approach to development does therefore not imply the contingency of developmental steps. Genetic information is generated during development, but this does not mean that development is subject to chance. There are, of course, other factors that may or may not be present, which are indeed contingent for successful development: enabling outer conditions that allow us to live or influence our health.

Genetic information may be a bad surrogate for the meaning of life, but the consequence of the deconstruction of the gene myth is not that the future of human bodies will simply be at the mercy of contingent human desires and interests – like the desire for immortality, for full functionality, for the ideal and able body, for the enhancement of intelligence etc. – and nothing else. A new understanding of the genome also re-opens a new space for the ethical. What is a good balance between what is chosen and what is not chosen? What is the value of being different from each other by non-chosen contingencies? Why is it better to be a being with finite capabilities and limited powers? Such questions will continue to be on our minds.[18]

Notes

[1] Contemporary biology has not yet explained life in all its forms, this would be too high a claim, as Lenny Moss (2006: 526) states correctly. Otherwise it would not be reasonable to proceed with funding molecular biology on the level of fundamental research. The genes sometimes play a role as 'blanket' explanations for all we do not yet understand in molecular terms. 'Somehow' the genes are responsible for it.

[2] One recent example of a metaphysical reading of the double helix is environmental philosopher Holmes Rolston III. In a series of papers and books he makes a surprising mystical teleological entity from the molecule: information that provides the ends, the *causa finalis* of life. See Rolston (2006), critically Rehmann-Sutter (2004).

[3] DNA contains the same bases A, C, G, except U, which is thymidin T in DNA.

[4] I disagree with Van der Weele (1999) that such an ecological perspective, which would be more directly interested in environmental causes in development, and would make extensive use of the conceptual tool of the reaction norm, would be an alternative 'integrative framework', that is 'an alternative to a genetic program perspective' (ibid.: 121). In my view, the program perspective itself is capable of integrating environmental causes: by using the conceptual tool of the reaction norm. The alternative must differ more substantially.

[5] Cf. Griffiths/Stotz 2006; Stotz et al. 2006; Shapiro 2002; Fogle 2000; Neumann-Held and Rehmann-Sutter 2006.

[6] This example is described in detail, with references in Rehmann-Sutter (2002).

[7] The term 'historical logic' has been suggested by Gunther S. Stent (1981).

[8] This thought has been worked out for the concept of the gene by Eva M. Neumann-Held (1999), resulting in an account of the gene as the *process* that leads to one particular polypeptide.

[9] This idea, which can be regarded as a key to the others, was introduced by Susan Oyama (1985).

[10] This point was stressed by biological structuralism. See Webster and Goodwin (1982, 1996) .

[11] Paul Griffiths and Karola Stotz (refs. already given) don't seem to engage in philosophical theory building but restrain themselves to a deconstructive work. Sulmasy (2006: 537) deplores this; however, his definition of the genome ('A genome is the heritable information, specified in nucleic acid sequences, upon which an organism draws in generating its phenotypic response [through the production of peptides or the regulation of their production] in specific cellular, organismal, developmental, phylogenetic, and ecological contexts.') can neither really add to a rich philosophical account, but – unspectacular as it is – it is at least on the level of contemporary developmental genetics and genomics.

[12] The interview was originally in German. English translation: Jackie Leach Scully.

[13] They are, however, not tested in a study with a statistically representative sample.

[14] In Rehmann-Sutter (2006) I argue that the adoption of such an organic practice view is an ethical choice: a choice about the mode of understanding and perceiving developmental systems.

[15] See, for example, the entries in the *Encyclopedia of Bioethics* (Post 2004).

[16] I borrow the expression 'doing the genes' from Annemarie Mol's concept of 'doing the body' (Mol and Law 2004).

[17] So far I agree with the contingency claim of developmental systems theory (DST). See Oyama, Griffiths, Gray 2001: 3.

[18] I am grateful to Jackie Leach Scully and Rouven Porz for inspiring discussions while working on the project 'Time as a contextual element in ethical decision making in the field of genetic diagnosis' (Swiss National Science Foundation grants 11-64956.01 and 101311-103606), to Eva Neumann-Held for collaboration in a project on the philosophical issues of 'genomes', to Georg Gusewski for documentation work, and to Rowena Smith for English language revision.

References

Alberts, Bruce, et al. Molecular Biology of the Cell. New York/London: Garland, 1983.

Arber, Werner. "Dual Nature of the Genome: Genes for the Individual Life and Genes for the Evolutionary Progress of the Population." *IUBMB Life*, 57, 263–266, 2005.

Boomsma, Dorret I. et al. "Genetic and Environmental Contributions to Loneliness in Adults: The Netherlands Twin Register Study." *Behavior Genetics*, 35, 745–752, 2005.

Fogle, Thomas. "The Dissolution of Protein Coding Genes in Molecular Biology." In *The Concept of the Gene in Development and Evolution: Historical and Epistemological Perspectives*. Peter Beurton et al. (eds). Cambridge: Cambridge University Press, 3–25, 2000.

Griffiths, Paul E. and Stotz, Karola. "Genes in the Postgenomic Era." *Theoretical Medicine and Bioethics*, 27, 499–521, 2006.

Hubbard, Ruth and Wald, Elijah. *Exploding the Gene Myth*. Boston: Beacon, 1997.

Johannsen, Wilhelm. *Elemente der exakten Erblichkeitslehre. Mit Grundlagen der biologischen Variationsstatistik*. Jena: G. Fischer, 1909.

Kay, Lily E. *Who Wrote the Book of Life? A History of the Genetic Code*. Stanford: Stanford University Press, 2000.

Keller, Evelyn F. *The Century of the Gene*. Boston: Harvard University Press, 2000.

Mol, Annemarie and Law, John. "Embodied Action, Enacted Bodies: The Example of Hypoglycaemia." *Body & Society*, 10(2–3), 43–62, 2004.

Moss, Lenny. *What Genes Can't Do*. Cambridge: Cambridge University Press, 2003.

Moss, Lenny. "The Question of Questions: What Is a Gene? Comments on Rolston and Griffiths & Stotz." *Theoretical Medicine and Bioethics*, 27, 523–534, 2006.

Nelkin, Dorothy and Lindee, M. Susan. *The DNA Mystique: The Gene as a Cultural Icon*. New York: Freeman, 1995.

Neumann-Held, Eva M. "The Gene is Dead – Long Live the Gene! Conceptualizing Genes the Constructionist Way." In *Sociobiology and Bioeconomics*. Peter Koslowski (ed). Berlin: Springer, 105–137, 1999.

Neumann-Held, Eva M. and Rehmann-Sutter, Christoph (eds). *Genes in Development. Re-Reading the Molecular Paradigm*. Durham: Duke University Press, 2006.

Oyama, Susan. *The Ontogeny of Information: Developmental Systems and Evolution*. Cambridge: Cambridge University Press, 1985. (Rev. and exp. ed.: Durham: Duke University Press, 2000).

Oyama, Susan; Griffiths, Paul E. and Gray, Russell D. (eds). "Introduction: What Is Developmental Systems Theory?" In *Cycles of Contingency: Developmental Systems and Evolution*. Cambridge: MIT Press, 1–11, 2001.

Porz, Rouven. "Das Absurde erleben. Grenzsituationen, Sinnfragen und Albert Camus' Absurdität im Bereich der Gendiagnostik." *Folia Bioethica* 30. Basel: Schweizerische Gesellschaft für Biomedizinische Ethik, 2004.

Post, Stephen G. *Encyclopedia of Bioethics*. 3rd ed. New York: Macmillan, 2004.

Rehmann-Sutter, Christoph. "Die Interpretation genetischer Daten: Vorwort zu einer genetischen Hermeneutik." In *Die Zukunft des Wissens. XVIII. Deutscher Kongress für Philosophie*. Mittelstrass, Jürgen (ed.). Berlin: Akademie Verlag, 478–498, 2000. (Repr. in Rehmann-Sutter 2005, Chapter 5).

Rehmann-Sutter, Christoph. "Genetics, Embodiment and Identity." In *On Human Nature: Anthropological, Biological, and Philosophical Foundations*. Armin Grunwald et al. (eds). Berlin: Springer, 25–50, 2002.
Rehmann-Sutter, Christoph. Book Review: Holmes Rolston III: Genes, Genesis and God. *Ethical Theory and Moral Practice*, 7, 95–98, 2004.
Rehmann-Sutter, Christoph. *Zwischen den Molekülen. Beiträge zur Philosophie der Genetik.* Tübingen: Francke, 2005.
Rehmann-Sutter, Christoph. "Poiesis and Praxis: Two Modes of Understanding Development." In Neumann-Held and Rehmann-Sutter, (2006), 313–334.
Rolston III, Holmes. "What Is a Gene? From Molecules to Metaphysics." *Theoretical Medicine and Bioethics*, 27, 471–497, 2006.
Rothman, Barbara Katz. *The Book of Life: A Personal and Ethical Guide to Race, Normality, and the Implications of the Human Genome Project*. Boston: Beacon, 2001.
Scully, Jackie Leach. "Time, Tests, and Moral Space." In *Zeithorizonte des Ethischen. Zur Bedeutung der Temporalität in der Fundamental- und Bioethik*. Georg Pfleiderer and Christoph Rehmann-Sutter (eds). Stuttgart: Kohlhammer, 151–164, 2006.
Scully, Jackie Leach; Porz, Rouven and Rehmann-Sutter, Christoph. "'You don't make genetic decisions from one day to the next' – Using Time to Preserve Moral Space." *Bioethics*, 21, 208–217, 2007.
Shapiro, James A. "Genome Organization and Reorganization in Evolution: Formatting for Computation and Function." *Annals of the New York Academy of Sciences*, 981, 111–134, 2002.
Smith, John Maynard and Szathmáry, Eörs. *The Origins of Life: From the Birth of Life to the Origin of Language*. Oxford: Oxford University Press, 1999.
Stent, Gunther S. "Strength and Weakness of the Genetic Approach to the Development of the Nervous System." *Annual Review of Neuroscience*, 4, 163–194, 1981.
Stotz, Karola; Bostanci, Adam and Griffiths, Paul E. "Tracking the Shift to 'Postgenomics'." *Community Genetics*, 9, 3, 190–196.
Sulmasy, Daniel P. "The Logos of the Genome: Genomes as Parts of Organisms." *Theoretical Medicine and Bioethics*, 27, 535–540, 2006.
Van der Weele, Cor. Images of Development. Environmental Causes in Ontogeny. Albany: SUNY Press, 1999.
Webster, Gerry and Goodwin, Brian. "The Origin of Species: A Structuralist Approach." *Journal of Social and Biological Structures*, 5, 15–47, 1982. (Repr. in Neumann-Held and Rehmann-Sutter 2006, 99–134).
Webster, Gerry and Goodwin, Brian. *Form and Transformation. Generative and Relational Principles in Biology*. Cambridge: Cambridge University Press, 1996.

Science, Religion, and Contingency

Dietmar Mieth

1 Religious Motivation and Ethical Reflection

Religion is often linked to progress in the biosciences in specious ways. For example, it has been asserted that in East Asia religion has less reservations about related ethical issues. Or reference has been made to isolated statements by Islamic scholars, who have no objections to destructive embryo experimentation or to research cloning. Or conditions in Israel conducive to scientific progress have been named, in which religion and biotechnology seem to have a propitious relationship. Conversely, in the United States a broad alliance has evolved among Christian fundamentalists of various persuasion for the right-to-life and against 'playing God'. And in South Korea and in Italy the Catholic Church continues to protest the extension of assisted reproductive technologies.

On closer consideration the position of the religions on bioethical questions is not so easy to determine. In many areas of life a society oriented to economics and knowledge functions without any apparent association with religion. Differences and parallels between the religions are often difficult to specify due to innerdenominational or innerconfessional divergences.

This is less prevalent if religions, like Roman Catholicism, have a uniform teaching authority (*magisterium*) (see Hilpert and Mieth 2006). Uniformity in questions of morality, i.e. in concrete convictions and modes of practical conduct, is to be anticipated in cases where 'ethics', i.e. the reflection on the rightness of moral rules as a secular discourse, does not replace or augment the moral authority of religion. Religion can exercise influence on morality through denominational ties and traditions. In the case of ethical reflection in a secular state and a pluralistic society, if religion attempts to do this exclusively on the basis of authority and without the use of rational arguments appealing to general insight, then it isolates itself on islands in the currents of progress (see Lem 1978).

Throughout its history the Christian religion has provoked, sustained, and at times suppressed ethical reflection. One of the driving forces of Christianity is the

D. Mieth
e-mail: FamilieMieth@t-online.de

M. Düwell et al. (eds.), *The Contingent Nature of Life*,

idea of creation. The conception of the contingency of the human being, i.e. our finitude and fallibility, derives from this idea. But this insight follows equally from daily human experience. Evidently, there are insights in which general human experience and a religious motive can concur. This also holds true for the experience of and the respect for the diversity of life.

Counterpoint to the motivating force of divine creation is the responsibility for the sovereign composition and shaping of the world. Yet the idea of creation does not specify how far the human being can and is allowed to go in the process. Here the responsibility implicit in the entrustment of creation is in tension with the experience of contingency and finitude. Moreover, in the Abrahamic religions the conception of the human being as an image of God is derived from the creation of the human being. Since this is valid for humans, regardless of our qualities, capabilities and constitution, the idea of the human being as a likeness of God is also invoked as a theological motive for human dignity and human rights. In Christianity this motive is reinforced by the belief in the 'incarnation of God'. In that God assumed human nature in Jesus Christ, i.e. not only became an individual human being but also manifested the dignity of the human as the 'flesh' of God, the human being has, according to the teachings of the Church Fathers, at the same time become capable of 'deification' (cf. Haug and Mieth 1992). This encompasses, as Immanuel Kant expressed it, the respect for humankind in every human being,[1] inasmuch as Christ 'revealed' human nature as the place of the inner-directedness to God in every human being. This also encompasses the idea of the spontaneous overcoming of self-interest through the awareness of the other, the unconditional acceptance of the dependent, and the responsibility for the well-being of the weak.

All of these motivations also initiate ethical thinking. Conversely, ethical reflection – through the self-assertion of reason – also places the motives of religion in question. Often it is less a matter of the motivations than their fixation in norms, norms with temporally determined justifications that are first noticeable in times of radical upheaval. The relationship between religion and ethics is productive and tension-ridden at the same time (see Düwell 2007; Mieth 2005: 282–293). This becomes clearer in times experiencing the anti-humanistic and even terrorist tendencies of religions: religion requires ethical reflection based on reason for the purification of its motivations; conversely, the potential that religion holds for the discovery of ethically relevant motivations is of lasting significance. It is not different in bioethics and will not be in the future.

2 Seven Prerequisites for Assessing the Relationship Between Science and Religion

1. Scientific progress functions to a certain degree like a religion. This is particularly valid for the public promotion of scientific progress and the inherent normative power of the fictive. One example is the concept of therapy initially founded purely on uncertain options in so-called 'therapeutic' cloning. The dilemma here

is that science without anticipatory fantasy gradually loses its impetus, while this fantasy also constructs its economic basis. This basis has extraordinarily contingent foundations. Moreover, this contingency becomes all the more visible the more science is dependent on exorbitantly expensive laboratories, a dependency apparently vulnerable to the acquisition of technology and economic power. How extensively contingency is present in science is also evident in the fact that every advance in scientific knowledge, if explained with scientific integrity and precision, simultaneously entails an advance in the broadening of the awareness of ignorance. Responsible popular-scientific accounts should communicate this contingency, rather than offer sensational stories of tidings of salvation one day and replace them with the apocalyptic fears the next, etc.

2. The concept of religion which we assume here is not entirely apt as a generic term indicating forms of belief, which always differentiate themselves from the general concept of 'religion' in reference to their *proprium* as well. This is why there has always been a tradition of 'critique of religion' in Christianity. On the other hand, in my opinion it is not incorrect to use religion as a generic term, but we must be aware of the fact that the different religions cannot and do not comprehend themselves simply as its variants or variables. Otherwise theology would deteriorate into religious studies.
3. Morality as a system of modes of conduct, preferred moral attitudes and convictions is to be found in every concrete religious community or denomination. In Judaism religion is a compendium of practical conduct per se. There are also various Christian traditions, possibly even with denominational nuances. It is, of course, imperative to distinguish this concept of existing modes of conduct, which if jointly practised are possibly sanctioned, from the standards of moral reflection. A morality that actually exists and is practised is unquestionably a significant indication of reflection; conversely, moral reflection – in its communicable, generally accessible, democratic, and discursive structure – influences actual modes of conduct. The social discourse is an expression of these tensions, which are to be endured and utilized but not to be resolved.

   Consequently, ethical reflection is often called for where morality can no longer be taken for granted. This is true for philosophical as well as for theological reflection, including a historical and critical confirmation of moral developments in the context of religion.
4. Christian theology has a unique relationship to reason and, accordingly, to rational justifications in ethics. Reason can simultaneously offer a critique of reason as well as self-enlightenment, and in this way deal constructively with destructive powers which, according to Christian tradition, also encompass reason. To assert Christian convictions as authorities beyond the scope of reason causes religion to return to a pre-Enlightenment state. Equally problematic is the direct utilization of the authority of religion without self-enlightenment. As may be expected, in the Christian tradition applied ethics is, on the basis of approach, i.e. from the question of the conception of the human being, incompatible with certain philosophical-ethical approaches. (This becomes clear in the discussion on utilitarianism and pragmatism in Christian bioethics.) The tradition of 'natural

law', which is above all particular to the Catholic Church, has indisputably undergone transformations: on the one hand there has been a dramatic shift in epistemology, according to which 'nature' is not something discernible which reason passively perceives but is only to be comprehended through the categories of reason. On the other hand, pre-modern natural law incorporates a teleological conception that rests on the Aristotelian teaching on actuality and potentiality, correspondingly finalized all things and living beings, and theologically assigned the 'finus ultimus', God, to them as the ultimate perspective of perfection. In contemporary ethical argumentation 'nature' (see Fraling 1990) tends to appear as a conceptual indication of problems, to the extent that, according to Jonas and Habermas, we encounter new questions of responsibility in the context of the displacements of nature by human practice and in the attempt to surpass the contingency of nature (see Jonas 1984: 245–255; Habermas 2001a).

5. The appeal to Kant's ethical anthropocentricity gained acceptance above all, but not exclusively, in Christian ethics. We could even speak of a 'Christian-Kantian' continuum. Kant is occasionally attacked for this 'cryptic theology', a criticism which I do not find convincing.
6. A critically reciprocal relationship exists between theological reasons for action (motives, motivations, experiences, elements of the search for meaning, lived convictions, denominational ties), and rational proofs. Particularly Paul Ricoeur has demonstrated this, although it is also found in the thought of Alfons Auer as the 'reciprocity' between an 'ethos of salvation' (an ethos applied to the religious search for meaning) and a 'world ethos' (an ethos applied to individual worldwide action), to be distinguished from a comprehensive 'global ethic' as 'world ethos' in the sense of Hans Küng's concept of a cross-cultural ethos of understanding that is more necessary than ever today.
7. Christian belief cannot be reduced to ethics. On the one hand, this means that even action motivated by faith must remain open to criticism through an autonomously conceived ethics; on the other hand, it also means that ethics can profit precisely from the fact that it is not the final point of reference for the human being in search of meaning.

3 Life Between Worldly Wisdom and Life Science

The question of the interpretation of life is invariably connected with the deepest form of reflection, philosophy or the love of wisdom, and with religion, the deepest commitment to the meaning and experience of life. In ancient Greece the concept 'bios' originally had to do with the way a human being conducted his or her life. The life of plants, animals, and human beings was considered 'animated', i.e. inherently moving, in contrast to the machine, which first has to be set in motion. The Bible portrays the unique life of every human being as infused with the life-giving breath of God. Seen together with the plants and animals in a graduated way, the human being is a co-creature, even created on the same day as the animals. Obviously, the 'spirit-soul' of the human being is not simply an incremental progression nor a third level of the sensitive life of the animals and the vegetative life of plants but a new order. The feelings of the human being are different than those of the animal, the interplay of nerves and muscles in the human being are not merely the expression of a vegetative scheme. Human and pre-human life are not simply homologous in the sense of a common development, yet not without commonality either. They are analogous, i.e. their properties are not given to them in the same way, despite their similarity.

Why do we continue to speak of this past conception today, despite the theory of evolution and the methodological materialism of the natural sciences? Because today there are still correspondences with the ancient teachings on life. Today, ideas like the 'art of living' have become important again. In conjunction with the human way of life and the meaning of life, we still comprehend the ancient languages of worldly wisdom or we comprehend them again and again. But do we comprehend and mean the same thing today when we speak of life in the sense of the 'life sciences' (in German: *Lebenswissenschaften*; in French: *sciences de la vie*)? This is a concept equivalent to 'biosciences'. Why was the prefix 'bio-' replaced with the word 'life' in recent years? Two reasons have been given: first, it is not only a matter of biology. The life sciences extend from biophysics to medicine. Second, the public acceptance which one had hoped to secure for the innovative sciences seems to be more easily attainable through the word 'life' than through the technological comprehension of the laboratory sciences associated with the word 'bio'. In fact, bio-logy, with the introduction of the expression by Laplace and others around 1800, had been transformed from a science of collection and observation into a science of intervention and experimentation. This change had not yet reached the public school system in the 1950s, when experiments were only carried out in physics and chemistry classes, but today the biosciences have triumphantly entered the schools via 'bio-mobiles', by means of which pupils are lured into a new world.

In methodology the life sciences are monistic, i.e. they represent one uniform materialistic connection between, as it is put, 'nonliving' and 'living' matter. 'Life', it is said, 'came into being', and the conditions under which it came into being are known, although not yet replicated in their causality, i.e. according to cause and effect, in experimentation. We can do much with life, but we have not (yet) created it.[2] Evolution has been placed in a theoretical context without the presence of 'creation'. For scientists who adhere to the belief in creation held by the 'religions of the book', God's plan is a kind of 'invisible hand' (Max Planck, Albert Einstein), an awesome set of blueprints for the laws of nature.

But is God an ideal physicist? Or is divinely bestowed 'life' perhaps something other than the material laws of cause and effect?

'Are we sad, because we cry?' asks neuroscientist Niels Birnbaumer. Here crying is understood as a neurological schematism of stimulus and response. The philosopher might prefer to claim that we cry because we are sad.

We can formulate the question differently: Is the locus of philosophical and religious wisdom, which sees life as a mystery and not as a sequential course of the most refined mechanisms, the 'natural order' or is it 'freedom'? Whoever assumes freedom, a primeval experience of the human being with himself or herself, has the tendency to consider the experience of God, which takes place within the human being, as more important than the objectifiable side of nature, of which he or she is a part, even as a research subject, apparently without exception.

Must we pay taxes to wisdom or to science? (At universities this is not a metaphorical question.) The disciplines of philosophy and theology did not participate in the evaluation of the life sciences in the German state of Baden-Württemberg in 2001. Does 'life' belong to the methodological-materialistic monoculture of the biosciences? (This is not to imply that scientists are materialists.) Or does 'life' only stand for health, in the sense that medicine leads us out of the materialistic straits, which it utilizes at the same time, for the welfare of the human being? Or do we reduce 'life' to 'living matter,' so that the entire realm of 'nature', including the human being as a physical being, belongs to the classification of things?

In any event these are not questions of science but questions of wisdom. Wisdom builds the bridge between life, knowledge, and truth. In Christianity 'life' and 'truth' are directly contiguous (cf. John 14: 6).

Is this the reason why we speak two different languages, the language of wisdom and the language of science? Do we really live in two distinct cultures as is often claimed?

It is, of course, easier to raise these questions than to answer them. Strikingly, Albert Einstein and Albert Schweitzer had different answers: respect for the laws of nature (Einstein) or 'reverence for life' (Schweitzer). Does life acquire another, greater vitality through the laboratory or does it lose its vitality through experimental objectification? Are they mutually exclusive alternatives, without the hope of ever bringing them together?

It would seem prudent to contemplate these questions before we find ourselves confronted with the moral questions – both ecological and biomedical – which arise when dealing with life. Otherwise it is conceivable that there may be reasons other than moral reasons why communication is no longer possible when we argue over the course society should take into a world of innovation.

The question of the beginning of human life and the beginning of the life of the individual human being must be differentiated initially. The first is a question of the general theory of evolution and of palaeontology; the second is a question of the development from embryo to neonate. Neither question, raised without ulterior motives, is necessarily a moral question. When we inquire about the origins of humankind or the development in the womb (or in the test tube) we must not immediately formulate this question as a question of 'moral status'. This question first arises out of practical responsibility and is part of the framework of ethical inquiry. Here it is perhaps relevant to re-examine such questions by reflecting on how we have come to ask ourselves such questions in the first place. Have we crossed a Rubicon when we are forced to define ourselves, although, from a religious perspective, we are not capable of doing this at all? Will we someday dangle like marionettes from the end of the questions which we are incapable of answering but must answer?

Karl Rahner considered the self-manipulation of the human being as so unquestionable that Peter Sloterdijk refers to him in the context of his proposals for a 'human park'.[3] However, Rahner was equally opposed to a total instrumentalization of the human being. Today, a generation later, we are preoccupied with the question: how can we sustain a balance between these two assumptions?

4 The Conflict Between Science and Religion

Science and religion – this is a history of conflict and alliance. The conflict is often better known than the alliance. It is characterized by names like Galileo Galilei, Giordano Bruno, Charles Darwin, and Rudolf Virchow (his famous words: 'I dissected the entire human brain and did not find the soul'). Virchow coined the expression '*Kulturkampf*' in the nineteenth century because he viewed the confrontation with religion as a fight for culture. Today this conflict is still celebrated in fundamentalist Bible study groups as well as in the biosciences, where the percentage of agnostics is said to be extremely high.

Earlier the pressure of Church power burdened science; since the eighteenth century at the latest, the opposite is the case. The battle of the Church against scientific worldviews, the Inquisition, always proceeded under the premise: 'There are truths which may not belong to revelation but are so closely connected to it that belief would be difficult without them.' Even today the Catholic Church, as a consequence of the anti-modernist oath (under Pius X at the beginning of the twentieth century), demands an oath of fidelity from those holding ecclesiastical offices which binds them not only to the profession of faith but also to those truths that ostensibly constitute a historical or factual prerequisite of belief. The conflict between authority and scientific or scholarly freedom has by no means been resolved. The exercise of power has, of course, become much more subtle, and the scope of Church jurisdiction is limited to the guild of theologians. Currently the Church 'mandate' (the official permission to teach Catholic theology) may still be denied when, for example, findings in the humanities and social sciences call for a reappraisal of the ethics of relationships in the Church. The freedom of research in the arts and sciences, a human right, is an achievement of secular – in France and Italy they would say 'laical' – democracy.

We frequently overlook the fact that this conflict is sometimes carried out without restraint on the part of science as well. The attempt to banish religious disciplines from the academic world, the discriminatory characterization of theological disputants in questions of scientific ethics as 'servile' or 'irrational', the dogmatization of scientific paradigms, although science without the possibility of falsification is not a science – all of these are experiences that can be easily confirmed by examples. The cardinals of the Church and the cardinals of science (for example, the Nobel laureates) delight in speaking *ex cathedra*! While the former would like to impose their worldview on science, the latter attempt with their scientific – or ostensibly scientific – worldview to domesticate the social environment. Imagining a fictive conversation between Pope Benedict XVI and James Watson, the co-proponent of the DNA double helix, can certainly make this clear.

5 Beyond the Conflict: The Remaining Tension

The conflict continues to smoulder like the coals under the ashes of heretic Giordano Bruno's pyre. Although the possibilities for mediation have constantly increased, although the dialogue is constantly sought by both sides, and although the barriers and obstacles are being eliminated on both sides, a tension does remain between social convictions and scientific explanations. We can characterize this tension either as destructive or as constructive.

In many French discussions I have experienced the coexistence of testimony (*témoignage*) and argument, and philosopher Paul Ricoeur has attempted to show that convictions derived from tradition and everyday experience can just as readily be a source of morality as argumentative constructs for the justification of norms (cf. Ricoeur 1996; Mandry 2002).

6 The Irrevocable Alliance Between Religion and Science

Given our focus on the conflict we would like to defuse and on the tension that we would like to maintain with good reason, we should not overlook the existence of an alliance between religion and science in the name of truth. The beliefs of the great 'religions of the word', particularly of Christianity, freed the world of antiquity from the omnipresence of the divine. This was the *first* step toward secularization. The *second* step was the invention of the discipline of theology in the Middle Ages, ensuing from the Christian philosophy of antiquity. The justification of belief before the forum of reason was so successful that today every theologian immediately notices when a scholar or scientist abandons reason – and when reason abandons the theologian. Theology as a field of scholarly inquiry, as a 'science', is the most powerful forum for the critique and the justification of belief that I am aware of. Hardly anyone can make things more difficult for theologians than they do for themselves. The *third* step was the gradual development and establishment of Humanism and the Enlightenment at the beginning of modernity. It essentially consists, to put it in a simplified way, in a secularization of hope: 'We want to create heaven here on earth' (*Wir wollen hier auf Erden schon das Himmelreich errichten*), as Heinrich Heine expressed it. In actuality the re-contextualization of the liberating and redeeming hope in the hereafter, which again dominated daily human life in unending religious rituals and practices and Church feast days, had led earthly existence to become a place of lethargy, inertia, narcotization, and the disciplining of political will. Today this is still recognizable in those societies in which religion asserts its worldview against science. Although the Occidental revolution in science began at Islamic universities, it was an Islamic reaction which has blocked the way to modernity since the fourteenth century.

Catholic theologians are well aware of the fact that Popes of the nineteenth century, above all Pius IX, condemned the freedom of religion, the freedom of conscience, and the freedom of scientific and scholarly inquiry, and placed certain areas of scientific and scholarly inquiry on the Index, together with books forbidden for believers. (As recently as 1960, when I began to study Catholic theology, I needed special permission to read so-called dangerous books.)

The radical transferral of hope to life on earth exceeds the alliance between science and the Christian religion in which a liberal theology, from Scholasticism to Humanism and the Enlightenment, contributed to the elimination of prejudices. Albertus Magnus, Nicolas von Cusa, Nicholas Steno (bishop *and* anatomist), Gotthold Ephraim Lessing, Friedrich Wilhelm Joseph von Schelling and untold others are witnesses to this alliance between '*fides et ratio*' (the title of one of the last encyclicals of Pope John Paul II).

7 Science as Religion? A Quasi-religious Disempowerment of Society Through Science?

The alliance, and this is my third venture in approaching the topic of the transformation of the relationship between science and religion – following the discussions of the conflict and the alliance in the name of reason – has of course given rise to questionable tendencies on the part of science as well. Instead of maintaining a productive tension with religious convictions, science has also established itself as an authoritative, quasi-religious disempowerment of society. This begins with the power over language in society, particularly with the power to define scientifically (or apparently scientifically) induced problems. From science and, in its wake, technology and the economy, most people today expect survival and the viability of a better life in society. At the same time these expectations are internalized in the scientific mentality. The temptation is great for science to exploit public relations to establish itself as a quasi-secular equivalent of the 'Congregation for the Doctrine of the Faith' for the scientific community. This begins with the linguistic authority invested in the contemporary counterpart of Latin, i.e. globalized English. Through language the social context of a utilitarian pragmatism influencing science dominates even the first words with which a scientific innovation is 'proclaimed' as gospel. When I hear a member of a commission using this language, indicating that 'risks' and 'benefits' must be weighed against each other, I do find this correct, pragmatically speaking; however, it conceals the fact that extra-scientific, socially relevant criteria – to be examined scientifically as well – are also necessary to determine what is to be seen as a risk for whom and what is not.

The pragmatism of an ethics of science is either introverted – then it is only a matter of the exactness of the scientific method and the soundness of the scientific data – or offensive: then areas of the sciences assume the primary responsibility for society and attempt to evade the social responsibility of science. It is undoubtedly imprecise to speak of science in the sense of a unified subject, since experts' opinions are often contradictory. However, this is only noticeable to the average person when a public controversy has already erupted.

Announcements of findings or breakthroughs in the natural sciences lacking monitoring discourses in the humanities and the social sciences or discourses on norms are often dangerous simplifications, which may possibly resurface as promises of salvation in the language of politics and the language of economic promotion. Objective misrepresentations can arise very quickly, for example, when research involving cloning in the test tube without clear therapeutic prospects is called 'therapeutic', or when the use of embryonic stem cells is expressly predicted for Alzheimer's disease, misrepresentations which politicians and journalists parrot (for example, in the clarification of a concept by the Deutsche Presse Agentur, the leading German press agency in 2004), and which are then firmly anchored as a new 'hereafter' in the minds of those religiously trusting science. I call this the 'normative power of the fictive.' How difficult it is to correct such predefinitions and the ensuing normative power of the fictive is well-known. Expressions like 'reproductive' cloning have even found their way into the Charter of Fundamental Rights of the European Union (2000), where the restriction of the prohibition of cloning to the propagation of children gives the impression that experimental attempts to clone embryos have nothing to do with reproduction, although they presuppose reproductive medicine. In this way a hidden subtext is created, which again limits the ban on cloning. The political repercussions extend as far as the German position on a procedural measure to be voted on by the Legal (Sixth) Committee of the General Assembly of the United Nations in October 2003. There the motion, which proposed a two-year postponement of the planned negotiations on a worldwide ban on cloning, was supported by the German delegation; in the process the original resolution of the German Parliament to support a 'comprehensive ban on cloning' (*umfassendes Klonverbot*) was interpreted, against its intentions (namely, to ban all forms of cloning), to mean a ban on cloning ideally having the consent of all member countries. Language is a highly adaptable and supple medium when it is a matter of avoiding unambiguous positions.

Today science must accept responsibility for its power over and through language. The times are long past when science as the mighty executor of social wishes could withdraw to a supposed ivory tower, where, in the isolation and freedom of the laboratory, methodological precision and a climate of openness and honesty among colleagues were cultivated. Science must learn to accept its role as a social force just as the media and the economy must. What this means specifically, however, is: to evaluate and monitor information as well as public impact and public reaction; to avoid blending personal, subjective views of the world with scientific options, and to refrain from promising what could or may be impossible to realize.

8 A New Alliance Instead of Hegemony?[4]

We live in a world in which the powers of belief that have been Church-established and Church-controlled to date either float freely or have sunk under the surface, admittedly, without having lost their subcutaneous effectiveness. To put it succinctly:

science imitates religion. The critic of science has become a critic of religion, the historian of science an enlightener on ambivalent 'enthusiasms', which often accompanied paradigms that are now obsolete.

We could imagine, for example, human beings who suffer due to various, mostly health-related restrictions in their lives. They often receive assistance as the result of enormous scientific efforts. However, far more often they only receive a perspective for the future, a future unattainable for them, whether it is the promise of direct democracy through Internet access or new environment-friendly breakthroughs or, for example, predictive or regenerative medicine. I live a better life in the awareness of the beginning of a future, in which I do not yet live. Or as the advertisement for a financial magazine proclaims: 'Today tomorrow is better than today.' The normative power of the fictive cannot be expressed more perfectly. What distinguishes the post-Christian disciple of science, whose 'today' is what is not yet present, from the heaven-oriented observer of religious practices? Hans Jonas has warned against overburdening the future. His 'principle of responsibility' proposed that in decision-making we should not subscribe to the hermeneutics of hope but to the hermeneutics of fear.

Much about the future is just as otherworldly as heaven. Are, for example, the participants in a Maria procession in Cologne more realistic than the politicians who believe the visionary promises of embryonic stem cell researchers? German *Bundestag* member Christa Nickels raised this question in conjunction with the parliamentary debate there on stem cells in 2002.

The alliance, which modern society – with the blessings of a tolerant religion – has entered into with science, technology, and the economy, should be a contract and not a hegemony. This is true of both directions. Yet the danger of hegemony today, despite extant weak relics from the times of the Inquisition, which are only a vexation for contemporary theologians, no longer issues from the social context to which religion belongs and in which it expresses its witness. The hegemony of science in the explication and the standardization of the world is clearly not to be contained by fleeing the scientific world, although there, like everywhere else, alternatives to the powers that be are greatly advantageous. The alternative to science however is always a scientific alternative, the alternative to technology is always a technological alternative, the alternative to economy is always an economic alternative. Science cannot be replaced. But it must accept new dimensions of responsibility, and scientists must be trained to do so. For whoever has the privilege of explaining the world to us has power over our world.

Can we demand of scientists who participate in social, ethical and legal discourses that they have an equal responsibility to acquire professional approaches to the relevant areas of knowledge and thought just as, conversely, scholars in the humanities, social sciences, ethics and law have the responsibility to inform themselves through scientific opinions and expertise? If we answer this question in the affirmative, we have tremendous political challenges ahead of us in the areas of research and education. Particularly in conjunction with research in the European Union we will discover that there are no relevant conceptual frameworks allowing for alternative, complementary, and ethical research on many questions. In the

European Parliament the areas of research and industry even overlap in the official Committee on Industry, Research, and Energy (ITRE). Ethics is possible, yet it is clearly limited by such structural givens both in direction and in procedures.

9 An Open Discourse

This is why we urgently need an open discourse on worldviews and explanations of the world. The social discourse is the only system of monitoring and controlling supremacy known to an open society. Frequently it is a weak instrument of monitoring and control, yet it is clearly preferable to hegemonic instruments. The discourse is not unlike the market: equal access must be granted and fair conditions created.

Certain earlier points of contention between science and religion have disappeared today, for example, when theologians accept Darwin and scientists develop the hypothesis that we all may be descendants of a human Eve, possibly a black Eve in Africa.

Our world can only be construed multilaterally. Religion is sustained by the memory of finiteness, of failure, and the realistically human visions surviving these failures. It works towards a culture of recognition and acceptance, of 'compassion' and of the 'interruption' of continuities in powers hostile to humankind. Here religion needs the alliance with the scientific world.

Conversely, science can use the religious experience in scientific self-enlightenment. Max Planck insisted that one needs fantasy and belief in science – but what one needs should also be examined with a sceptical apparatus, the same apparatus which enabled modern theology, under the guidance of science and scholarship, to become a modern discipline.

In closing I will consider two aspects of the relationship between science and religion related to the concept of the boundaries of knowledge and to the question of the meaning of life, and thus digress from the question of the relationship between science and religion in the context of bioethics and bio-politics that has dominated up to now.

10 Contingency in Science and Religion

Scientific inquiry in the natural sciences proceeds from a method that establishes its own conceptual framework. In order for a problem to be solved it must first be isolated. Then an experimental protocol can be assigned to it. The scientific findings, positive or negative, are valid within the conceptual framework that was established in advance. If this specific framework is eliminated or ignored, accuracy is lost. This is the reason why every scientific insight which does not comprehend itself as a revolutionary discovery, has limitations from the beginning: these determine its contingency and by definition guarantee some day it will be obsolete due to new constellations of advances in knowledge that establish the conceptual framework differently. This boundary can be redefined by re-examining the insight in additional areas. A conclusion of a more general type is more likely to be a falsification than the verification of a hypothesis. Moreover, every advance in knowledge, as I mentioned earlier, is accompanied by an advance in the knowledge of ignorance. The answer to a question always raises new questions. This is *progressus in infinitum*, rather than an infinite regression as is criticized in religion, for example, by Hans Albert. But is the difference in the experience of contingency really so great?

Every research paradigm has its limits, every theory has its limits, every application in a non-delimited area of reality, where everything is related to everything else, must expect the unexpected. Scholars in the humanities and experts in the area of hermeneutics assume the non-isolability of an insight and thus cannot extract a problem from its contexts. They are acutely aware of the boundaries of language in the representation of reality, perspectivism as a way to the whole, the observer-oriented form of insight, and, finally, that transcendence of boundaries traditionally called meta-physics. It is conceivable that the dialogue between science and religion can be better carried out on the concept of limits, of boundaries rather than on the concept of truth. For religion, too, possesses the whole only in infinite details, which point from themselves to the whole.

Blaise Pascal asserted the structural analogy of the forms of knowledge in various dimensions of knowledge (*esprit de géométrie, esprit de finesse, esprit du coeur*) (Pascal 1978). Progress takes place through 'aberration' and through the corresponding transformation in the cybernetics of the system, which is not be mistaken for reality but instead reflects observer-oriented functionalism (cf. Rombach 1987).

A new facet of the question of the significance of religion and religions has more recently been broached by Jürgen Habermas, among others: is it possible without lived experiences of meaning, without their bonds in social or religious contexts, and without their transposition in lived moral convictions in society, for there to be sufficient 'material' social and lived morality, which can then be introduced in the formal discourses on establishing norms in a democratic society?[5] While they do not directly determine what is to have validity, without recourse to these lived experiences of orientation and forms of orientational knowledge the discourses are hollow.

We should not overlook the fact that the Christian religion has entered an alliance with science, without which Christian theology in the contemporary sense would be nonexistent. On the other hand, society cannot exist without sources of religion. Of course, these do not have an exclusive claim to the experience of meaning and to knowledge for orientation. Given the progressing individualization of the religious and the biographical dissolution or conscious modification of religious bonds, society would then either surrender to authoritarian relicts or to an atomistic-individual pluralism. In discourses on norms the latter would allow only the lowest possible common denominator or favour those who are capable of introducing their interests in a rational discourse with professional and rhetorical sophistication.

In my opinion, this would have a highly detrimental effect on vulnerable groups at the beginning and at the end of life as well as on ways of life suffering from serious limitations. A society which does not have recourse to solidarities which are more than miscellaneous and passing solidarities based on interests will no longer be able to develop solidarity itself, and will instead only have recourse to contracts involving interests. This 'cold' society is what awaits us if we choose to leave quasi-religious impulses to the paradigm of science and to ignore the authentic religious impulses together with their powers of forging community and generating lived orientation.

Notes

[1] cf. Kant 1977: 600–601 Bruch 1998: 32–33.
[2] The conflict over the patenting of the so-called onco-mouse seemed to assume this, since the mouse was approved – objections overrided – as a 'product patent' in the US and Europe, although not in Canada.
[3] cf. Rahner 1966: 45–69. For the laudatory appraisal of Karl Rahner by Peter Sloterdijk, see Sloterdijk 2000: 97–116. cf. Mieth 2004.
[4] cf. Roy, Wynne and Old 1991.
[5] cf. Habermas 2001b. However, this is often misused as an argument for authority.

References

Bruch, Richard. *Person und Menschenwürde*. Münster: LIT, 1998.
Düwell, Marcus: "Theologie und Ethik, Anmerkungen zu einer problematischen Beziehung." In *Autonomie durch Verantwortung. Impulse für die Ethik in den Wissenschaften*. Jochen Berendes (ed.). Paderborn: Mentis, 79–98, 2007.
Fraling, Bernhard (ed.). *Natur im ethischen Argument*. Freiburgim Breisgau: Herder, 1990.
Habermas, Jürgen. *Die Zukunft der menschlichen Natur*. Frankfurt a.M.: Suhrkamp, 2001a.
Habermas, Jürgen. *Glauben und Wissen*. Frankfurt a.M.: Suhrkamp, 2001b.
Haug, Walter and Mieth, Dietmar (eds.). *Religiöse Erfahrung, Historische Modelle in christlicher Tradition*. München: Fink, 1992.
Hilpert, Konrad and Mieth, Dietmar. *Kriterien biomedizinischer Ethik. Theologische Beiträge zum gesellschaftlichen Diskurs*. Quaestiones Disputatae 217. Freiburg/Basel/Vienna: Herder, 2006.
Jonas, Hans. *Das Prinzip Verantwortung*. Frankfurt a.M.: Suhrkamp, 1984.
Kant, Immanuel: *Die Metaphysik der Sitten*. Werkausgabe, Vol. VIII. Frankfurt a.M.: Suhrkamp, 1977.
Lem, Stanislaw. *Sterntagebücher*. (English transl.: *The Star Diaries*). Frankfurt a.M.: Suhrkamp, 1978.
Mandry, Christof. *Ethische Identität und christlicher Glaube*. Mainz: Grünewald, 2002.
Mieth, Dietmar. "Der operable Mensch. Karl Rahners Beitrag zur Selbstmanipulation des Menschen (1966) im Disput." In *Stimmen der Zeit*, 222, 807–817, 2004.
Mieth, Dietmar. "Ethik." In *Handbuch Theologischer Grundbegriffe*. Peter Eicher (ed.). München: Kösel, 282–293, 2005.
Pascal, Blaise. *Penseés*. E. Wasmuth (ed.). Heidelberg: Schneider, 1978.
Rahner, Karl. "Experiment Mensch." In *Die Frage nach dem Menschen. Festschrift für Max Müller*. Heinrich Rombach (ed.). München: Alber, 45–69, 1966.
Ricoeur, Paul. *Das Selbst als ein Anderer*. München: Fink, 1996.
Rombach, Heinrich. *Strukturanthropologie*. Freiburg/München: Alber, 1987.
Roy, David J. et al. *Bioscience and Society*, Schering Foundation Workshop. Vol. 2. Chichester etc.: Wiley, 1991.
Sloterdijk, Peter. "Der operable Mensch." In: *Der imperfekte Mensch*. Hygiene Museum Dresden, 97–116, 2000.

Further Literature

Auer, Alfons. *Autonome Moral und christlicher Glaube*. 2nd ed. Dusseldorf: Patmos, 1984.
Auer, Alfons. *Zur Theologie der Ethik*. SthE 66. Freiburg Breisgau: Herder, 1995.
Düwell, Marcus; Hübenthal, Christoph and Werner, Micha (eds.). *Handbuch Ethik*. 2nd ed. Stuttgart: Metzler, 2006.
Gaziaux, Eric. *Morale de la foi et Morale autonome*. Leuven: Leuven University Press, 1995.
Haker, Hille. *Ethik der genetischen Frühdiagnostik*. Paderborn: Mentis, 2002.
Mieth, Dietmar. *Was wollen wir können? Ethik im Zeitalter der Biotechnik*. Freiburg/Basel/Vienna: Herder, 2002.

Part II
Ethical Theories and the Limits of Life Sciences

Bioethics and the Normative Concept of Human Selfhood

Ludger Honnefelder

The course of public debate as well as common issues presently discussed among ethicists make it clear that developments in the *life sciences* and their application in modern medicine are confronting the human kind with questions that definitely surpass the usual – and in themselves already frighteningly complex – problems concerning the regulation of innovative technologies. These questions pertain to the core of our self-understanding as human beings and concern the foundations of morals and ethics on which we base our lives.

True, this does not refer to all discoveries in the field. Many developments in the *life sciences* are completely novel and require a high degree of regulation. Still, as experience has shown, even though the process is not a simple one, or one capable of resolving all questions raised, in many areas of the life sciences and their medical applications it suffices to refer to widely accepted moral norms at hand in order to reach a consensus regarding the extent of necessary regulation.

However, some questions remain which can be resolved neither by the application of accepted norms, nor by the generation of new norms on the basis of more abstract accepted principles; and it is these questions which presently attract attention within the public at large as well as within the scholarly debate. What makes these questions so important is neither the novelty of the application context, nor the complexity of the consequences involved; it is their intimate link to the foundations of our morals and to our normative self-image – that self-image which basically guides us in our quest to develop and to instigate new norms. This has a direct impact on the normative self-understanding of human beings, which is based on our status as responsible agents and as bearers of elementary rights and obligations.

If we try to identify the specific fields in which we are faced with such "fundamental" questions, then surely innovations concerning possible intervention in the fields of genetics and reproductive medicine are to be included, such as the cloning of human beings, germline manipulation or embryo selection. The key issue to be addressed is the following: Is the implementation of these novel possibilities on human beings consistent with the basic norms of accepted universalist morals or ethics and the criteria of responsible action manifested therein, or does it question

L. Honnefelder
e-mail: honnefelder@drze.de

M. Düwell et al. (eds.), *The Contingent Nature of Life*,

the basic norms and thereby the very endeavour of ethics, which is founded on the reciprocal recognition of autonomous subjects? How can we determine the status of the questions themselves and to what extent can we reach a consensus concerning the answers?

1 *From Chance to Choice*: The Approach of "Liberal Eugenics"

Progress in molecular biology and medicine has undoubtedly led to means of intervention which have not only considerably expanded – or at least promise to expand – the scope and magnitude of medical diagnosis and therapy, but which have also enabled us to manipulate the genetic endowment of humans in a way which would have a direct impact on human identity and human nature – a possibility which has thus far been completely out of reach. Even proponents of *germinal choice technology* such as *G. Stock* consider the possibility of intervening in or determining the genome of future individuals as "the greatest challenge", since it is inextricably related to "what it means to be a human being" and inevitably "changes our image of ourselves" (Stock 2002: 110, 196, 155). What, however, is the precise change under scrutiny and which challenge does it raise?

R. Dworkin, the Anglo-American human rights theorist, argues that it is the possibility of displacing the boundary between *inherited* and *produced* nature that has caused the cloning of human beings and the intervening in or the determining of the genome of future individuals to meet with such widespread intuitive resistance. For, according to Dworkin, the boundary between *chance* and *choice* is nothing less than the "spine of our ethics and our morality" (Dworkin 1999, 2000: 444) and is as such fundamental to our distinction between "what nature has created" and "what we do in that world" (Dworkin 2000: 443). It allows us to distinguish between what simply happens to us and what we are responsible for, and thus "structures our values as a whole" (ibid.: 444). The fact that we are who we are by coincidence and not by choice does not mean that we have to somehow justify the "genetic lottery" in which all participate. On the contrary, we must show solidarity with it (cf. ibid.: 445). If this boundary becomes displaceable, then the nature thus far prescribed to us will no longer be "the absolute paradigm" (ibid.: 444) for something which is important to us for the simple reason that it lies beyond our scope of power and responsibility. Human kind is afraid of losing its footing, and our fear of thus being deprived of security and stability expresses itself in our concern that with such a significant genetic modification, as is made possible by contemporary molecular medicine, we shall make a start in "playing God" (ibid.).

According to Dworkin, what causes intuitive resistance is therefore not only the genetic modification, but primarily its impact on our value system. For it affects values which, being intrinsic to the objects and events in question, are to be regarded as *detached values*. These are to be distinguished from *derivative values*, which result either from a recourse to interests, from cost-benefit analysis or from social compromise (cf. ibid.: 427f.).

If, as in the case of displacing the boundary between *chance* and *choice*, the validity of intrinsic values is affected, we should not be surprised when "deep moral uncertainty" results and the fear of a "moral free-fall" spreads (ibid.: 445f.).

No matter how gravely Dworkin is inclined to describe the situation, he is not prone to draw the consequence that we should completely refrain from making use of these novel possibilities; for this would be "cowardice in the face of the unknown" (ibid.). The proper reaction can only be one of further developing our morals with regard to the new challenge.

With this result, which Dworkin himself only sketches, he concurs with the much more exhaustive diagnosis and substantiated approach put forward by a group of prominent American bioethicists, A. Buchanan et al. (2000), in their book *From Chance to Choice*. This group also considers the genetic modification of humans by means of cloning, germline intervention and embryo selection on the basis of genetic testing to raise a radically new type of challenge. What renders this challenge so fundamental is the potential not only to develop forms of therapy for treating hitherto incurable diseases, but also to "shape some of the most important biological characteristics of the human beings we choose to bring into existence" (ibid.: XV). After reaching the necessary level of technical development, there are several reasons which could lead us to make use of such a potential: the implementation of individual rights, especially the right to reproductive freedom, the desire of future parents to have as perfect a child as possible, but also motives resulting from public health care or job market interests – not to speak of the concept of "genetic communitarianism", propagated by some social groups with recourse to the freedom of religion.

What gives weight to the moral challenge lying in such scenarios is, according to the mentioned group of authors, not only the concern that our capacity for moral judgement and implementation might well not suffice to effectively draw the necessary boundaries in time, especially considering the temptations raised by the potential of these innovations. Because, for this purpose we would have to be certain about the values that allow us to distinguish in this field between what we *can* do, and what we *should* do. However, it is precisely this certainty which is missing. We must therefore ask: "What are the most basic moral principles that would guide public policy and individual choice concerning the use of genetic interventions in a just and humane society in which the powers of genetic intervention are more developed than they are today?" (ibid.: 4f.).

As the question reveals, the diagnosed moral challenge goes far beyond questions concerning mere application. It concerns the very foundations of morality; however, this is no reason for a general ban, but rather – so the argument for permitting therapy – a reason and incentive for a thorough review and further development of our fundamental moral principles. According to the authors, this is also not contradicted by a historical analysis. For, according to the authors, the "shadow of eugenics", with which the second chapter of the study extensively deals (cf. ibid.: 27–60), does not end up labelling eugenics – which is precisely what the discussed genetic modifications are all about – as simply illegitimate. But what could the basic moral principles be which would permit a feasible distinction between legitimate and illegitimate eugenics and could thus effectively tackle the outlined challenge?

In attempting to answer this basic question, the authors do not depart from the field of deontological ethics, as expressed in the language of fundamental rights based on elementary demands. In an appendix (cf. ibid.: 371–382) dedicated solely to methodological questions, the authors emphasize the necessity to foster a broad and balanced discourse to mediate between our diverging basic moral intuitions, and specify the ideas considered fundamental to the "liberal moral-political theory" (ibid.: 373) which should be adhered to: moral individualism, the fundamental equality of all people, the ability to criticise and revise individual concepts of the good with respect to justice in basic institutions and the necessity to distinguish between the public and the private spheres (cf. ibid.: 379).

On the backdrop of a system of morals based on these ideas, the authors regard the new options made possible by genetics as an extension of individual rights and freedoms and consider the very act of endorsing and protecting these individual rights, such as the right to reproductive freedom, as drawing the necessary line to rule out objectionable practices, including state-controlled eugenics (cf. ibid. especially the final Chapter: 304–345). The right to reproductive freedom with respect to making use of the possibilities that genetic technology and reproductive medicine may offer, is naturally constrained by the formal requirements of justice, equal opportunity, the principle of nonmaleficence, and the recognition of the freedoms of others. What follows from this is the obligation to observe the welfare of yet nonexistent human beings affected by our actions, including avoiding suffering and promoting care. This implies that in some cases the deployment of medical innovations would be refused, in others it would be permitted, which would also not completely rule out cases of enhancement. As for individual freedoms and justice, the state has the obligation to protect the right to reproductive freedom, to enable equal access to the opportunities in question, and to accompany the impact for handicapped people through a "morality of inclusion". Prohibition is only acceptable on the grounds of avoiding foreseeable damage or upholding equal opportunity, inevitably leading to a limited "genetic stewardship" by the state with respect to "the genetic well-being of future generations" (ibid.: 336f.). "Moral firebreaks" such as the distinctions between positive and negative eugenics or therapy and enhancement do not offer an adequate solution to the challenge raised by these new technologies. Rather the endorsement of reproductive freedom should be pursued, which in exceptional cases may overrule the interests of third parties including those of the offspring, as long as the principle of nonmaleficence as well as justice is upheld.

2 The Debate Concerning the Ethical Self-Image of the Human Race

A diagnosis of the situation differing in several important aspects from the study of the American group of authors, and thus drawing different conclusions with respect to permissible therapeutic measures, is presented by J. Habermas in his latest bioethical publication (2001). In accordance with Dworkin and the cited American study (to which Habermas refers in his notes), he agrees that the possibility of genetic modification of human beings would radically displace the boundary between nature as we find it and nature as we ourselves create it, between *chance* and *choice*, a boundary which is constitutive of the human condition. Such a displacement would entail putting the system of norms on which our morals are based into question. Contrary to the authors mentioned, Habermas argues that the basic norms would not only be challenged, but rather directly affected. For if it is inherent to the human condition that the determination of one's genetic individuality be immune to manipulation by third parties in a way which surpasses all common possibilities of intervention (such as the choice of a partner), then the deliberate selection or modification of an individual's genome by means of genetic technology would change the nature of the entire species. For such a modification would "unilaterally and irreversibly intervene in the formation of a future person's identity" and breach "the boundary-sustaining, deontological sanctuary. . .which guarantees one's personal inviolability, individuality and the unrenounceability one's own subjectivity" (cf. Habermas 2002: 287). It would affect the personal identity of the person in question who, due to such foreign intervention, would no longer experience himself or herself as the sole author of his or her own biography. It would also affect the *moral community*, since it would raise members confronted with the "scenario of a dislocated future" (ibid.) who would have a different relation to their own inception than all others (cf. Habermas 2001: 72ff.). In contrast to all postnatal socialization efforts by others, these prenatal foreign interventions infringe on one's biography in a way to which the affected person can no longer relate (cf. ibid.: 93ff.). If the intervention is irreversible and all attempts at revision ruled out, then the reciprocity and symmetry constitutive to moral equality is destabilized. What is then affected by the prospect of genetically modifying future human beings is the close association between personal inviolability and "the sanctity of a person's natural physical development and embodiment" (ibid.: 41). The decision to dissociate the two would constitute, according to Habermas, not the displacement of just one of the many boundaries of human influence that have thus far been drawn by nature; it would constitute nothing less than a "self-declaration concerning the ethical self-image of our species. . .which in turn determines whether we may further regard ourselves as authors of our own biographies and recognize each other as autonomous agents and persons" (ibid.: 49).

Against the background of such a diagnosis, it should come as no surprise that the conclusions Habermas draws regarding permissible therapeutic measures differ from those of the American authors cited earlier. If modifications are not possible without the stipulated consequences and without our status as autonomous agents and the equality in choosing a life-plan as cornerstone of our moral heritage being affected, then a deployment of such measures would not be possible without abolishing our moral foundation. The idea of "liberal eugenics", as put forward by Dworkin and the others, is self-contradictory from Habermas' point of view (cf. ibid.: 86ff.). According to him, the only justifiable application is in non-instrumental cases which are inherent to the "logic of healing" (ibid.: 79), such as permitting an intervention in cases without any possibility of obtaining a subject's consent if and only if it serves the purpose of treating or avoiding serious disease and if one would rationally expect the subject to consent to such treatment. Of course, such an attempt to safeguard reciprocity and symmetry presupposes that the unborn human is to be regarded as a second person (cf. ibid.: 66ff.). In addition, a purely therapeutic intervention for which we would be justified to expect the subject's consent, made possible, however, – as in the case of preimplantation diagnostics – only through the elimination of other human lives, nevertheless remains committed to the principle of protecting those deemed worthy of protection; the latter group, according to Habermas, includes prenatal human life, thus sheltering it from instrumental exploitation (cf. ibid.: 56–69).

Asking how and why the principles of just coexistence, as expressed in the concept of the inviolability of human dignity, may be extended to encompass the entire species as well as unborn human life, according to Habermas, one can only refer to the "ethical self-image of the human race" in the ethos of the species ("Gattungsethik") inherent to our moral convictions. It is in this self-image that "the abstract rational morals of human-rights subjects themselves. . .find their footing" (ibid.: 74; cf. ibid.: 96) and which calls – on this side of the public debate concerning the moral status of the embryo – for an anticipation of the subject-status of unborn human beings and thereby for an extension of the right to protection from free and equal subjects to prenatal life as well.

3 Key Anthropological Questions

Despite their differences, all authors discussed so far agree that the options now open to us have a direct impact on the foundations of the human condition. The reason for this is that intervening in human nature influences the identity of a person in a way that directly affects his autonomy and social equality. The moral relevance of an intervention of that kind lies not in the fact that it modifies human nature as such, but rather in the resulting displacement of the boundary between *naturally developed* and *created*.

If it is inherent to the human condition that the boundary between *naturally developed* and *produced* is highly relevant to the identity of a human being and his self-understanding, and at the same time, this boundary is not fixed but can instead be significantly displaced, then we must ask ourselves what distinguishes the new prospect of modifying the genome of future human beings from other displacements of the boundary between nature and culture, or between *chance* and *choice*, that have accompanied the history of mankind as a cultural being thus far? Is this transformation not inherent to man's quality of transcending his own nature (Rahner 1967: 286–321), perhaps even with the consequence that "playing God" will sooner or later be our destiny (cf. Peters 1997)? The authors mentioned so far supply us only with an indirect answer to the question as to what boundary, from a moral point of view, the displacement of the boundary between *naturally developed* and *produced* should itself respect. According to Habermas, the American group of authors regards the nature affected by genetic modification as a kind of "inner environment" to which the subject in question can himself relate (cf. Habermas 2001: 89).[1] The plea in favour of "liberal eugenics" would in fact lose its plausibility without such a premise. For the general relation between a subject and a person, this would not be convincing if we presuppose an inextricable union between the mental and the organic system, unless of course we restrict genetic modifications to areas that are not crucial to personal identity, but rather belong to a kind of "inner environment". In this case, the normative upshot of "liberal eugenics" would already be inherent in its anthropological premises and would dismiss all genetic modifications to which a subject could no longer relate. This leaves the question unanswered whether and to what extent there are genetic modifications that have such a restrictive character and whether the interpretation of human nature as an "inner environment" is at all plausible.

In contrast to the position discussed, Habermas regards the displacement of the boundary between *naturally developed* and *produced* by means of genetic modification as being morally contentious, because it involves manipulation by third parties who intervene in the self-understanding constitutive to a subject's identity in such a way that – with the exception of the therapeutic case – the subject can no longer retrospectively relate to. For this reason it must be regarded as an unjustifiable infringement of personal autonomy and the principle of equality. This leads to the conclusion that the heteronomy of nature is to be respected since it maintains autonomy and equality more strongly than the intervention by third parties, except when such an intervention conforms with the "logic of healing" in which the patient is regarded as a second person whose consent we would be justified in expecting to obtain. Although the reference to autonomy and equality relieves Habermas from recurring to strong anthropological premises, it also forces him to make speculations regarding the extent to which genetic modifications would indeed result in a deficiency in autonomy and equality (cf. Siep 2002: 116ff.).

4 The Moral Challenge

It may be helpful to summarize the points stated thus far.

1. What constitutes the *moral* challenge according to the authors mentioned is the fact that the novel possibilities of intervention raise questions concerning premises which did not have to be addressed in traditional moral discourse, and that meeting the challenge requires additional premises which most likely cannot be formulated on the basis of traditional morals; these premises concern the general ethical framework as well as specific contexts of moral behaviour.
2. The American group of authors asks whether our system of morals can be based on the reciprocal recognition of autonomous and equal subjects if genetic modifications affect the *nature* constitutive of a subject's identity, in addition to the fact that these modifications are inflicted on not yet existing subjects. If so, how can consensus be reached regarding regulation, if even displacing state-controlled

eugenics to the individual decision of parents does not solve the general problems associated with eugenic application of genetic technology? If, as opposed to the "public health model", the "personal service model" is not able to solve the problems, then one must ask whether there is a *third approach* which might allow the use of genetic modification without infringing on equality, autonomy or personal inviolability (cf. Buchanan et al. 2000: 11ff.).

3. The answer developed by the American group of authors suggests that this is possible if one applies the constraints of nonmaleficence and equal opportunity, as guaranteed to born humans, to future humans as well (cf. ibid.: 242–257). But the question remains as to how such a principle can be introduced to an individual case without having unacceptable consequences in other areas. And how could we justify not being able to grant future and unborn human beings a status rendering them worthy of protection?
4. If one includes forms of genetic *enhancement* as an area in which one may legitimately apply the principles of nonmaleficence and equal opportunity, then one faces the problems of identifying what is to be considered as "enhancement", of determining for which forms of enhancement one can presuppose obtaining the patient's consent (cf. ibid.: 219ff.), and of deciding how to avoid a social "colonization" of natural inequalities and their consequences (cf. ibid.: 82ff.). If one wishes to achieve this through restrictions on the basis of the principles of nonmaleficence and adequate care and curtail reproductive freedom by means of state measures, one at least would have to justify the state action by appealing to criteria such as disease relation.
5. Concerning the extension of entitlements to justice granted to autonomous and equal subjects to not yet existing humans, the authors face the problem that, according to accepted morals, not yet existing humans are not regarded as legal subjects and it has not yet been possible to reach a social consensus regarding the moral status of unborn human beings on the basis of these morals.
6. Habermas sees clearly that the moral intuition to the effect that cloning, germline intervention and embryo selection are seen as a violation of the right to self-determination and the principle of equality can only be adequately established if one presupposes an "intrinsic value of human life before birth" (Habermas 2001: 61). However, he considers an interpretation of human development in favour of an unconditional moral status even of unborn humans as "reasonably controversial" (ibid.: 60f.). If "human dignity" is, strictly speaking, contingent upon the symmetry of relationships, then its "inviolability" could only hold for legal subjects (cf. ibid.: 62).
7. This, however, does not rule out for Habermas that the members of a legal community may mutually make a moral commitment to grant unborn human beings, though not "inviolability", but "undisposability" (ibid.: 59), and "as a reference point for our obligations [to grant them] legal protection" (ibid.: 66). Beyond mere appreciation for what it is, pre-personal life, though not yet "addressable in its *prescribed* role as a second person, [has] an integral value for the entirety of an *ethically* constituted form of life", so that it – and this is the suggestion – should

be granted protection on the basis of the "dignity of human life", though not due to "human dignity, which is legally guaranteed to all persons" (ibid.: 67).

5 Critical Evaluation and an Alternative: The Recourse to Human Dignity and Human Rights

If we proceed on the basis of the arguments presented thus far, we are faced with grave problems and doubts regarding both the diagnosis and the therapeutic measures associated with the positions discussed.

Let us begin with the diagnosis.

As the discussion thus far has clarified, the innovative possibilities of genetic intervention into the nature of the human subject must not only bring a universalist system of morals, which serves as a foundation for both positions, to the limits of its capacity, but must also question important premises underlying it.

a. Due to the fact that moral systems are constituted in relationships of mutual recognition, morality becomes dependent upon the existence of legal subjects and remains restricted to such. The intuition that unborn or future human beings are to be included can only be indirectly accounted for, either by an *extension* of the concepts of equality and justice beyond the previously defined strict members of a possible moral community, or by *introducing additional criteria for being worthy of protection*. If, as is necessary in the first case, one extends the (strong) concept of a legal subject, as posited in the moral systems in question, to include unborn and future human beings, this, as was shown, inadvertently leads to counterintuitive consequences, such as the unrestricted subsumption under the criterion of equality (which encompasses all subjects). Other possibilities of bestowing moral claims to unborn or future human beings, such as by a principle of anticipation with recourse to a species-related ethos, may clearly only be established at the price of expanding the concept. However, such an expansion would be inconsistent with the fundamental contractualist character of such a moral system. For the anticipation can only be reasonably applied to an unborn human if an identity of the unborn human with the born legal subject is stipulated, which presupposes the extension which shall be introduced by such an anticipation in the first place. Moreover, it is not very convincing if the call for expanding the number of addressees of a universalistically conceived moral system can only be accounted for by a species-related ethos which *by definition* is not justifiable in a universalist way.
b. Furthermore, the presupposed concept of an autonomous, responsible subject implies a relationship of the subject to its own nature as that of a system to its (inner) environment. This, however, amounts to a latent form of Cartesianism (Meilaender 1996), which is itself cast into serious doubt precisely due to the possibility of intervening into a subject's nature by the novel means of genetic technology. Modifications of nature by others which directly affect the identity of

a subject and its relation to itself and which furthermore touch the fundamental equality between subjects in a severe manner, change the subject in a way to which it can no longer relate. The vulnerability of the subject which becomes evident by the options of genetic manipulation presupposes a concept of the unity of a subject and its nature, or more precisely, an interweaving of identity and non-identity, of subject and nature, which cannot be properly accounted for within the concept of an autonomous, responsible subject without surrendering its constitutive function to moral systems.

c. The precarious character of the relationship between subject and nature becomes evident (as discussed by the American group of authors) when we try to apply justice claims under the conditions of possible modifications of a subject's nature. If justice involves (if not in all cases, then in general) correcting natural inequalities, what then could serve as a measure for justice if nature, including all previous "natural inequalities", becomes the object of human production? A modification affecting not only this or that property, but rather the entire ensemble, would constitute not only "colonization" of natural inequalities, but rather a kind of total expropriation. In the extreme (and perhaps fictional) case of producing nature in its entirety, it becomes evident that the boundary between *naturally developed* and *produced* can be displaced even further, and that this cannot continue arbitrarily without fundamentally transforming the relations involved. A measure for the limits to the displacement of the boundary cannot be derived from the concept of an autonomous subject. The recourse to the "logic of healing", i.e. to the criterion of disease and health, is intuitively appealing. However, it presupposes as point of comparison a concept of nature which in its entirety is intact; in addition, such a recourse is plagued with the recurring problem of proper delineation.

d. If the boundary deemed necessary is to be drawn in form of legally binding regulations in lieu of the unborn or future subjects affected by the displacement, we must ask what criteria such a stand-in protection should adhere to. Since in the case of unborn and future subjects the recourse to their stipulated intentions actually relies on what we ourselves deem to be the good or obligatory, we are in need of something like an objective order of goods and claims. Habermas relies on an "objective legal order". However, such an order inevitably emerges from the given relationship of a subject to its nature. How could the recourse function if this relationship itself is rendered an object of modification? Moreover, we are faced with key questions regarding an order of rights and goods, such as the priority of a life without genetically caused handicaps over physical life itself – a question which a contractualistically founded universal moral system can only address on the basis of the self-determination of the affected person.

So far for the diagnosis. Let us now proceed with the therapy.

If we wish – as is the case in the positions discussed – to allow for the intuition that the basic requirements of universalist morals be extended to future and unborn human beings, then in founding such a moral system we will not be able to avoid introducing premises that go beyond what is acceptable within a contractualist framework. Since it has become clear that the intuition calls for such premises and that these are – at least to some extent – implicitly advocated in the positions discussed, it would fully correspond with the method of establishing a reflective equilibrium between our fundamental moral intuitions – as propagated by the American group of authors – and our theoretical ethical conceptions if we were to introduce an appropriate revision in the founding of a universalist moral system.

a. The basic intuition calling for an extension of the group of moral addressees to future and unborn humans is, in my opinion, nothing but the basic intuition on which the idea of human rights is based and which secures this idea's binding character. It is the belief that all living beings that we refer to by means of the sortal predicate "human being" have an intrinsic or unconditional value which bars them from being evaluated in comparison to other goods. This means that all morality is based on a fundamental practical judgement stating that a human, as a living being equipped with the natural capacities of reason and free will, is an intrinsic or unconditional good, and that humans have this value simply by being humans – i.e. regardless of any other property except for the property of being human, expressed by their being referred to by means of the sortal predicate "human" (Sulmasy 2002).
b. This fundamental practical judgement can itself be accounted for in more detail, to which purpose there are several approaches. A rather apparent explanation is that we presuppose such a value judgement in all contexts of action and communication, and that its denial would lead to the dissolution of the social framework regarded by all participants as binding. In this sense we are justified in claiming that a contractualist founding of a moral system presupposes such a fundamental practical judgement – rather than being able to found or replace it.
c. If, in accordance with the intention underlying the idea of human rights, the fundamental practical judgement which assigns human beings their intrinsic value is not to be arbitrarily restricted, it has to refer to human beings *as human beings*; i.e. it has to refer to the same object to which the sortal predicate "human being" refers. However, the sortal predicate "human being" refers to a certain type of living being during the complete time of its existence. Modern enlightenment took this into account, and e.g. in the *General land law for Prussian states (Allgemeines Landrecht in den preussischen Staaten)* of 1794 all humans were placed under the protection of law from their birth to their deaths. Kant also states that we are obliged to maintain the concept of substance in practical philosophy and to regard human beings as a living beings in the sense of persisting entities and correspondingly as goods worthy of protection (Kant 1781: 365).
d. If intrinsic value must be assigned to human beings as human beings, and if the human being is to be understood as a specific unity of subject and nature, then the natural dispositions, which must be regarded as necessary conditions for subjectivity, are to be protected as well. This is so in the case of the idea of human rights when under the label of "human rights" we protect – from intervention by

the state or by others – certain natural frame-conditions such as bodily integrity and the right to life. This could be considered as a departing point for designating the boundary beyond which intervention by means of genetic technology and reproductive medicine would be regarded as illegitimate, insofar as they wuld affect future or unborn human beings which are not able to give their consent.

Acknowledgments This paper is based on the German article "Bioethik und Menschenbild" (Honnefelder 2002). The author especially thanks Rimas Cuplinskas for preparing the English translation.

Note

[1] The idea that there might be some sort of Cartesianism inherent in this conception of owning one's body or having an "inner environment" is developed in (Meilaender 1996).

References

Buchanan, Allen et al. *From Chance to Choice: Genetics and Justice*. Cambridge (Mass.): Cambridge University Press, 2000.
Dworkin, Ronald. "Playing God." *Prospect*, 41, 37–41, 1999.
Dworkin, Ronald. *Sovereign Virtue: The Theory and Practice of Equality*. Cambridge (Mass.): Harvard University Press, 2000.
Habermas, Jürgen. "Auf dem Weg zu einer liberalen Eugenik? Der Streit um das ethische Selbstverständnis der Gattung." *Die Zukunft der menschlichen Natur. Auf dem Weg zu einer liberalen Eugenik?* Frankfurt: Suhrkamp, 34–125, 2001.
Habermas, Jürgen. "Replik auf Einwände." *Deutsche Zeitschrift für Philosophie*, 50, 283–298, 2002.
Honnefelder, Ludger. "Bioethik und Menschenbild." *Jahrbuch für Wissenschaft und Ethik*, 7, 33–52, 2002.
Kant, Immanuel. *Kritik der reinen Vernunft*. 1st ed. (A). Riga: Johann Friedrich Hartknoch, 1781.
Meilaender, Gilbert. *Body, Soul, and Bioethics*. Notre Dame: The University of Notre Dame Press, 1996.
Peters, Ted. *Playing God? Genetic Determinism and Human Freedom*. New York: Routledge, 1997.
Rahner, Karl. *Schriften zur Theologie VIII*. Einsiedeln/Zürich: Benziger, 1967.
Siep, Ludwig. "Moral und Gattungsethik (zu Jürgen Habermas, Die Zukunft der menschlichen Natur)." *Deutsche Zeitschrift für Philosophie*, 1, 111–120, 2002.
Stock, Gregory. *Redesigning Humans: Our Inevitable Genetic Future*. Boston/New York: Houghton Mifflin, 2002.
Sulmasy, Daniel P. "Death, Dignity, and the Theory of Value." *Ethical Perspectives*, 9, 2–3, 103–118, 2002.

Human Cognitive Vulnerability and the Moral Status of the Human Embryo and Foetus

Deryck Beyleveld

1 Traditional Grounds of Divergent Views of Moral Status

Moral theories generally maintain that beings become a matter of moral concern and are owed moral respect because they possess some qualifying property. However, there is no consensus about what this property is. For example, the following are just some of the properties that have been proclaimed:

(a) being alive;
(b) being sentient (having the capacity to experience pain/pleasure);
(c) being a member of the human species, biologically defined;
(d) being self-conscious (or having personhood);
(e) being a rational agent (in the sense of having the capacity to act for reasons) (or as Kant described it, "a rational being with a will". See e.g. Kant 1785/1948: 89–91).
(f) having the potential to develop one or other of these properties;
(g) having the potential to develop rational agency, or the past possession of rational agency, as well as the possession of rational agency itself *within* the context of a teleology. One example is the view that all human beings possess moral status as members of the human species, characterised, centrally, by possession of rational agency, because this must be viewed in the context of human beings existing only to fulfil God's purpose.
(h) being a vulnerable rational agent (which is my own position; vulnerability being necessary, because beings that cannot be harmed can hardly require the concern of others). (While this is probably taken for granted in most other views, I have argued that it is worth taking explicit note of it, because attention to it has important implications for moral theory, some of which will be touched on in this paper. See Beyleveld and Brownsword 2001: 114–117).

D. Beyleveld
e-mail: deryck.beyleveld@durham.ac.uk

M. Düwell et al. (eds.), *The Contingent Nature of Life*,

2 Metaphysical Nature of These Views

The general justification for locating moral status in the possession of a particular property is not something that I will focus on in this paper. Clearly this is tied to the justification for whatever wider moral theory is involved. Instead, one of the things I shall focus on is an issue that arises when we try to apply a moral theory in which the property that is held to ground moral status is a mental state or other subjective property, which sentience, self-consciousness and rational agency are, either entirely, or at least in part. The issue in question arises because, that I am self-conscious, in pain, or acting for a reason is something that only I can know directly. When I conclude that others are self-conscious, etc., this is an inference from their behaviour or from the existence of certain biological structures that are correlated with their capacities to behave in characteristic ways. Of course, there are theories that will identify, e.g., experiencing pain with specific behaviour when this is coupled with the possession of certain biological structures, and others that will regard pain as epiphenomenal upon certain neurological or other biological processes. However, while there is undoubtedly a correlation between these things in me, this correlation can be explained by numerous different theories about the relationship between mind and body that are metaphysical in the sense that no experiment or empirical evidence can ever prove or disprove them. Consequently, the possibility must be admitted that I am the only self-conscious, etc., being that exists, and that others that behave as though they are self-conscious, etc., might merely be automata with no minds, or even complete figments of my imagination. Hence, though the theory I espouse might hold that all sentient beings have moral status, or that all rational agents have moral status, I might try to evade having to grant moral status to any being other than myself on the grounds that there are no other sentient beings (or rational agents) other than myself.

3 Recognition of Human Cognitive Vulnerability Requires a Turn from Ontology

Vulnerability, as I have intimated above, is an essential feature of morality. Only vulnerable beings need the protection of rules. But vulnerability, specifically cognitive vulnerability, is also important for meta-ethics, especially moral epistemology if this concedes that rational justification for moral positions and judgements is possible. However, it must be noted here that rational justification has both a positive and a negative aspect. Rationality, I suggest, demands (negatively) not only that beliefs should be held with a sense of conviction that is in proportion to their justification, so that beliefs that have no rational justification should be avoided wherever possible, but also (positively) that the fact that certain beliefs cannot be known or justified rationally should be accepted as a positive premise in theoretical thinking. This is a very large claim and it has very wide application. Here I shall elaborate on it in relation to two more specific claims that I will apply to the issue of the moral status of the human embryo and foetus. These claims are:

(a) At specific points, primacy must be given to the results of procedural rationality, in which the concern is not with the truth or rationality of the beliefs to be subject to rational critique, but with the rationality of the procedures by which decisions are to be made as to which beliefs to operate with.
(b) Within the context of moral theories that hold that moral prescriptions are categorically binding and that those who have moral status have this status on the basis of their possession of capacities that are, or include, essentially subjective states, the human embryo and foetus is to be accorded a moral status in proportion to the probability that it might have the relevant capacities in question even though it does not display these capacities in full. This, however, is not asserted because the embryo or foetus has the potential to develop the relevant capacities, or because it has them in part, but because the categorically binding nature of morality demands that precaution be exercised in not denying moral protection to those who might have moral status.

4 The Primacy of Procedure

Morality, at least as human beings are capable of comprehending it, is not merely for vulnerable beings, it is also propounded/comprehended by vulnerable beings (which is what human beings are), whose vulnerability extends to the cognitive and rational capacities through/by which they propound/comprehend morality. One of the consequences of this is that commitment to a morality that is guided by reason is that it must, as both a rational and moral demand, be something that, in large part, admits of error, uncertainty and doubt about its correct application and about its very formulation and justification, even when, in principle, these are knowable or determinate. And, when various of its propositions (metaphysical propositions) are not, in principle, securable by reason, then they may not be treated as propositions to which assent must be given or from which conclusions are to be inferred that others who do not share commitment to them must give assent. It follows from this, I contend, that metaphysical propositions must be avoided in morality to the extent that this is possible, and propositions of this nature should be replaced by a premise of uncertainty. The consequences of this, I have argued elsewhere, are a radical agnosticism (to be distinguished from atheism) in ethics (ibid.: 134–141). But, to the extent that such propositions are not avoidable, how are conflicts that arise from different beliefs about them to be handled? Similarly, how are conflicts to be handled about the rational justification of what can, in principle, be justified rationally? These questions point to issues about the limits of rational justification in ethics. I have argued elsewhere that such conflicts can be handled in part by commitment to a procedural morality, which is a morality that commits its adherents to the results of adjudication processes that satisfy specific procedural requirements, despite the fact that the results of this process might be prescriptions with which the persons involved do not agree. I have also argued, however, that such a process cannot be completely open-ended. At a political level, we need to note that persons generally attach more importance to some of their values than to others, and in these terms it is rational for them to give up less important values for the sake of important values. Correlatively they ought to be prepared to accept the results of a procedural process if and only if it is for the sake of and does not threaten their most important commitments. Alongside this, a procedural ethic can only yield rationally compelling results if there are some values that all rational persons must accept, whatever else they might disagree about.[1] At this level, there are certain possibilities, one of which is taken by discourse ethicists (when they embrace, or their contentions are subjected, to a transcendentalist interpretation), who contend that the idea of an intersubjective discourse presupposes specific values. In effect, disagreement about values supposes a commitment to certain values. And, another possibility lies in developing Gewirthian thinking, not as propounding a set of values that all agents must accept on pain of contradicting that they are agents, but as a set of values that all who claim rights to conduct activities of any kind must accept.[2] Applied to the dispute about the moral status of the embryo and foetus, the principal implication is that those positions that rest their stance on specific metaphysical theses (such as the existence of God, a teleological scheme of nature, materialism, epiphenomenalism, etc.) cannot seek to impose their views on others except through a procedural ethic, which in essence requires disagreement to be allowed wherever possible and where not possible, rules and law to be propounded as the result of an open, good faith, accountable process that attends to the rational limits of reasoning in these matters. Alongside this it demands a deep and honest reflection on what values we really regard as important, for ultimately dispute about the most important values justifies war. Here, I will say only that if the embryo and foetus really is to be accorded the same status as that of, e.g., an adult human being, then research on embryos is a crime equivalent to that of those committed in the holocaust and a similar response is not only permitted by demanded. If those who propound this view are really sincere about it then, at the very least, they cannot live in political union or fellowship of any kind with those who do not. If they are prepared to do so then it can only be because their belief is not sincere, they do not really believe it, and they actually attach more value to the values of adult human beings, and, perhaps, implicitly, they recognise the limits of human reason in a way that applies the need for a procedural approach.

5 Precaution and the Proportional Moral Status of the Embryo and Foetus

I have written about this in numerous other places (see e.g. Beyleveld and Pattinson 2000; Beyleveld and Brownsword 2001: 119–134) and I do not think it necessary to detail my reasoning here, especially since it is contained in the paper that I wrote for the last Euresco conference on the theme of the current conference.[3] Suffice it to say that being as sceptical about things as we ought to be when we realise the limits of our reason, we must accept that we do not know with certainty that there are any rational agents, subjects of a life, persons, or even sentient beings other than ourselves. However, if we operate under the premise of a categorical imperative that demands that we respect equally one or other categories of such beings in addition to ourselves, then we are driven to accept that we must so treat all those who *might* be such beings *when it is possible* for us to do so (simply because if we err in according status to a being that actually happens to lack it we do not violate our categorical imperative, but we do violate our categorical imperative if we err in denying status to a being that actually happens to have it). Hence, we must treat all beings that behave as though they have the pertinent property (which will vary according to the particular theory we espouse) as having the property. Further, because those who do not behave as though they have the property might still have the property, we are released from the duty to treat them as having the property only to the extent and the degree that we cannot treat them as having the property. From this it follows that all beings who do not behave as though they were agents, or sentient beings, or whatever, must be accorded a moral status proportional to the degree to which it is possible to treat them as agents or sentient or whatever. I have previously argued this specifically in relation to Gewirthian theory (see e.g. Beyleveld and Pattinson 2000; Beyleveld and Brownsword 2001: 119–134), but I see no reason why it does not have equal applicability to any theory that accords moral status on the basis of possession of a property that only I can, with certainty, know that I possess, provided that it involves the idea that morality is categorically binding.

Finally, to tie in these reflections with those about the primacy of procedure, rationality does not come up against limits merely in justification of an ethical theory. It also has limits in the application of an ethical theory. Thus, supposing acceptance of the proportional status of the embryo and foetus, questions arise in applying any theory to quantification of the proportional status and hence to the way in which conflicts over duties to those with full status and proportional status are to be handled when different values are at stake. Here, too, there may be (indeed, there are likely to be) indeterminacies that can only be handled procedurally, and this is worth mentioning because internecine disputes can be every bit as bitter (often more so) than those between "faiths".

6 Some Concluding Remarks

The broad message behind these reflections is that because ethics is not only for the vulnerable, but is propounded by the vulnerable, humility is a primary value in any ethics driven by reason. Rationalistic ethical views that propound rational necessities or other "absolutes" are sometimes accused of being authoritarian, even arrogant, the suggestion being that the avoidance of these things requires moral relativism. This, however, is not the case at all. Indeed, humility only makes sense within a context in which error is objectively possible, for when all standards are in the final analysis relative no error is, in the final analysis, possible and one need not defer to anything other than one's private commitments. Correlatively, if ethics requires recognition of the value of others it requires acceptance of the existence of a common reason that transcends contingent commitments or the lack of them.

Notes

[1] Solter et al. 2003: 232–237, and most fully, Beyleveld and Brownsword (2007: Ch. 10).
[2] This is suggested by Gewirth himself (1978: 63). For full development of the idea, see Beyleveld 1996 or Beyleveld and Brownsword 2001: 79–86.
[3] See Beyleveld 2006. (This paper also contains some earlier, less developed reflections, on proceduralism within Gewirthian ethics).

References

Beyleveld, Deryck and Brownsword, Roger. *Consent in the Law*. Oxford: Hart, 2007.
Beyleveld, Deryck. "Legal Theory and Dialectically Contingent Justifications for the Principle of Generic Consistency." *Ratio Juris*, 9, 1, 15–41, 1996.
Beyleveld, Deryck and Pattinson, Shaun. "Precautionary Reason as a Link to Moral Action." In *Medical Ethics*. Michael Boylan (ed.). Prentice-Hall: Upper Saddle River, 39–53, 2000.
Beyleveld, Deryck and Brownsword, Roger. *Human Dignity in Bioethics and Biolaw*. Oxford: Oxford University Press, 2001.
Beyleveld, Deryck. "Rationality in Bioethics: Reasonable Adjudication in a Life and Death Case of the Separation of Conjoined Twins." In *Bioethics in Cultural Contexts: Reflections on Methods and Finitude*. Christoph Rehmann-Sutter, Marcus Düwell and Dietmar Mieth (eds.). Dordrecht: Springer, 145–162, 2006.
Gewirth, Alan. *Reason and Morality*. Chicago: University of Chicago Press, 1978.
Kant, Immanuel. *Groundwork of the Metaphysics of Morals*, 1785. Translated with an introduction by Herbert James Paton as *The Moral Law*. London: Hutchinson, 1948.
Solter, Davor et al. *Embryo Research in Pluralistic Europe*. Berlin: Springer-Verlag, 2003.

Needs and the Metaphysics of Rights

Bernard Baertschi

1 To Each According to His Needs

As everybody knows, Marxists would have written on the flags of the classless society: "To each according to his needs". It is the marxist formula of justice, but it reaches far beyond Marxism. Bernard Williams has proposed the same formula in the domain of health care and, at least in Europe, we can say with confidence that a majority of people consider it to be an appropriate formula for the distribution of basic goods like health care. Famously, Robert Nozick does not concur: for him, although needs determine who is susceptible to use the goods that satisfy these needs, this does not give the needy the *right* to have the goods they need; otherwise tomorrow we shave for free (cf. Nozick 1974: 233–234).

I don't want to enter this dispute,[1] but only to emphasize a critique often directed at Nozick's position: his argument, it is said, is not valid, because it confuses needs and preferences; though preferences (*mere* preferences as it is often added) do not generate rights, needs (*real* needs) do. As those expressions show, there is a link between needs and rights. Of course, not everybody acknowledges this, because there is no consensus on the nature of rights, and it is quite possible to disconnect rights and needs, for example if you conceive rights as entitlements conferred upon citizens by a political authority. I do not hold this last position, because I think that morality is not a political affair but is foremost linked to human flourishing; and to flourish, it is necessary to have real needs satisfied. This is why I want to investigate here the relationship between goods and rights and more precisely to elucidate this rather vague expression "generate" when we say that needs *generate* rights. As it will appear, it must not be confused with the question of justification: "Needs *generate* rights" does not mean "Needs *justify* rights": the first is an ontological or metaphysical question, the second an epistemological one; but of course, they are not independent, and I will contend that, here as elsewhere, metaphysical questions are foundational.

B. Baertschi
e-mail: bernard.baertschi@lettres.unige.ch

M. Düwell et al. (eds.), *The Contingent Nature of Life*,

My analysis will focus on positive rights and for the sake of the argument, I wish to generalize Williams' thesis to all positive rights: they all depend on needs. It is probably not true, but by the end, I hope to convince you that my analysis remains correct for any item you would put in the place of needs.[2]

The model I shall use to elucidate the generation of rights is neither new nor original, it is the model of supervenience whose sketch has been devised by G.E. Moore and Richard Hare long ago. But it will be adapted and modified on certain crucial points.

2 The Supervenience of Rights

For Hare, moral features supervene on nonmoral features, that is on ontological features, as he says in this well-known passage: "Suppose that we say 'St. Francis was a good man.' It is logically impossible to say this and to maintain at the same time that there might have been another man placed exactly in the same circumstances as St. Francis, and who behaved in exactly the same way, but who differed from St. Francis in this respect only, that he was not a good man" (Hare 1952: 145). More generally, valuational properties supervene on ontological properties: A knife is good if its blade's edge is sharp. As we can see from these two examples, and as Jaegwon Kim has very clearly shown, supervenience is a relation of asymmetrical dependency: if SUP is the supervenient property and BAS is the ontological basis, it is not possible that something changes in SUP without a corresponding change in BAS, but the reverse is not true.[3] So, two identical moral actions – two generous actions – can have a different ontological basis, but from such an ontological basis it is not possible to generate an immoral action – a mean one: two different actions cannot have the same ontological basis.

So far, so good; but I think it has not escaped your attention that I have somehow changed my topic: I should investigate the relations between needs and rights, and so far I have addressed ontological items and values. But if "to have a need" is an ontological or natural property, "to have a right" is not an evaluative one, as is "to have goodness". So, from the thesis that values supervene on the natural, it does not follow that rights supervene on the same basis. From "St. Francis has such and such properties that make him a good man" it does not follow that he has rights. True, but that is not the problem: I don't want to know if and how St. Francis' rights depend on his goodness – it would be a very peculiar and non-standard ethical theory – but to explore the relation between needs and rights. The difficulty of this project is located in one feature of rights, namely that it is not an evaluative concept, but a deontic one, and as we know the grammar of both is not the same (cf. Castañeda 1975: 185–190, 335–336).

This difference is perhaps not so important, and we would be better off frankly asking the following question: Are rights supervenient on needs? It seems possible to proceed in this way, because "to have a need" and "to have a right" are both properties of persons. So we can indeed ask: What is the relation between these two properties? Let us postulate that it is supervenience; what will this mean? BAS is need, SUP is right; if rights supervene on needs, it implies that two identical rights can have other needs as their bases, but that two identical needs cannot result in two different rights (or the first in a right, the second in a no-right). Is this possible? It depends on how you construe needs and rights.

Two persons have the same need regarding health care; have they both the same right to it? Not necessarily: only one of them may have this right, if he alone is in possession of health insurance, or – but this is controversial – if he alone is not responsible for his condition. So supervenience is possible only if (i) rights are not conceived as institutional privileges only and (ii) needs are conjoined with other pertinent features.

Do rights so understood supervene on needs or on *basic* needs? Basic needs, of course. But how to discriminate between needs that are basic and needs that are not? The obvious answer is the importance of *goods* that persons need. Health (or health care) is an important good, to be two meters high is not (let us accept Norman Daniels' criterion: needs are basic when their satisfaction is "necessary to achieve or maintain species-typical normal functioning". Daniels 1985: 26). But the notion of "important good" is normative (let us say that the normative covers the deontic and the evaluative); it is even twofold evaluative:

(A) X is a *good* = X has value

(B) X is an *important* good = X has a great value

We are now confronted with another question: rights supervene on needs, but do they supervene on the factual properties of important needs or on their evaluative properties (to be tied to goods)?

3 Supervenience and the Role of Values

Here, we must be careful not to misunderstand our question. Goods that are the objects of needs *are not* moral goods. They are ontological goods, that is goods in virtue of what they *are*. Even if you are an antirealist with regard to axiological properties, if you think that we project values upon things, you must concede that the goodness or value of something depends at least partially on what it is: if bread is good for us, it is because of its natural properties (and of our natural properties). So the problem I address now is not the meta-ethical problem of the priority of evaluative properties on deontic ones, or vice versa – often misnamed the problem of the priority of the right on the good – I am not concerned here with the question of the *moral* good. Therefore I will not investigate the relationships between deontic predicates and moral-evaluative predicates – a problem that divides deontologists and teleologists – but only between deontic predicates – to have a right – and ontological-evaluative predicates – to have nonmoral value.[4] The reason why is very simple: "right" is a deontic concept and "need" is not a moral one.

Now we can return to our question: do rights supervene on the factual properties of important needs or on their evaluative properties? What is a need? In their search for transcultural and universal needs, Len Doyal and Ian Gough remark that when such a need is not satisfied, humans *as such* risk serious harm: "If such needs exist, they must be shown to constitute goals which all humans have to achieve if they are to avoid serious harm" (Doyal and Gough 1991: 45).

Needs are linked with "species-typical normal functioning"; if they are not satisfied, this functioning is put in jeopardy. We can describe this in biomedical terms, even perhaps partially in chemical terms, but of course it acquires human significance when we do it with expressions like "serious harm", that is with evaluative terms. So it is not plausible to tie rights with a biomedical description of man: they supervene on the evaluative component of needs.

The evaluative component – i.e. value – is therefore situated somewhere between biological reality and rights. If you believe, following the Scripture's lessons, that man does not live by bread alone, you can add a spiritual-metaphysical reality to the biological one; with respect to this, my argument is quite tolerant. I do not wish to investigate here the relation between the two components of needs; it is probable that supervenience too can be invoked, but it is not my topic. So values form the basis (BAS) of rights. But simple value is not enough: every need does not generate a right, only basic needs do. That is, needs that are satisfied by goods which have important value. This is not surprising: rights are generally correlative of duties – duties of assistance or of non-interference – and one of the most accepted definition of a right is: "A right or claim, then, is the [moral] position created through the imposing of a duty on someone else" (Kramer 1998: 9). But I can impose a duty on someone only if my interest is important, and basic needs are good candidates for such interests. If the interest is not so important, I can help the person to satisfy it, it is morally *good* if I help, but it is not my duty, it is not obligatory. The ontological values are always at bottom, but only some of them can generate and ground rights.

4 Supervenience and Justification

Only certain needs can *ground* rights: supervenience or justification? Both, in my opinion. Kim notes rightly: "We believe in the supervenience of epistemic properties on naturalistic ones, and more generally, in the supervenience of all valuational and normative properties on naturalistic conditions" (Kim 1993: 235). The reason is simple: normative disciplines cannot hang in the air, they must be grounded in reality, the reality of our needs, of our desires or of our goals. This is an ontological dependency; but with it comes an epistemological one, as Kim adds: "If a belief is justified, that must be so because it has certain factual, nonepistemic properties".[5] As we know since Tarski – and beyond since Aristotle – "It is raining" is true if and only if it rains. Of course, the conditions of justification are more complex than the conditions of truth – a justified belief is not only a true belief –, but the pattern is the same. It is therefore not surprising that *all* ethical doctrines mention nonmoral facts at the foundations of their conceptions when they are summoned to justify their principles. They ground their justifications in human nature (eudaimonism), divine volition (theological voluntarism), human volition (contractualism), facts of reason (kantism) or psychological facts like preferences (utilitarianism).

Supervenience is in fact ubiquitous where there is a relation of asymmetrical dependence. Mental events supervene on biological ones, biological ones on physical ones; justified beliefs supervene on facts, moral values on features of actions, behaviours and persons, rights on evaluative features of needs, and those values on biological features of needs. Moreover, it is easy to see that supervenience is often a transitive relation (think of the relations between the mental, the biological and the physical). Thus, we are confronted to a web of relations between different levels of reality, but always one-way.

Of course, this ubiquitous nature of supervenience does not give it a strong role in explanation, and indeed its role is much smaller than that of causality. But for our problem, that is to understand the relation between needs and rights, it nevertheless allows us to advance the following theses:

1. Rights are not groundless, for they depend on needs.[6]
2. If rights depend on needs, it is because needs have a valuational feature.
3. If someone objects to a right – a moral right, not a legal one – the only good answer will be to mention a real basic need: supervenience grounds justification. Of course, it is possible for us to disagree as to what such a need is, but here, the answer is no longer ethical, it is biological, anthropological or metaphysical (it does not mean that the answer will be any easier to give or that the controversy will end any sooner).

So far so good, a skeptic about rights will think, yawning: all this makes sense only if rights exists; but why postulate their existence? We are born with two arms and two legs, but not with rights! Are they not "nonsense upon stilts", as Bentham so aptly said? And what morally interesting entity could supervene on stilts? In the remaining section, I shall make some remarks on this subject, which is so important when we examine the *metaphysics* of rights.

5 The Normative Unreality of Rights

The presence of rights in the furniture of the world has been doubted for long. Hare denies their reality in those terms: "They are not part of the fabric of the world, and do not exist *in rerum natura*, if those terms are used in the strict sense" (Hare 1985: 48) – but he is an utilitarian and a antirealist (his denial extends to all moral items). On the other side of the ethical fence, Loren Lomasky acknowledges that rights could be dispensable in principle: the ethical work can be done without them; they have only rhetorical force, but it is an important one for our morality as rhetoric is the art of putting something, here certain values, in a prominent place: "The very vigor and insistence of rights advocates may lead us to conjecture that the language of right has an importance which would not survive a shift of idiom" (Lomasky 1987:

10). Rights focus our attention on features of our moral life we deem important in our liberal conception of human being, but they don't *create* these features.

Are these philosophers right? If we listen to the common moral language and to official discourses, doubts will arise regarding their position: rights (and especially human rights) are ubiquitous and invoked at every moment; every kid proclaims, as soon as he can speak: "I have all my rights!" Which rights? He couldn't say, but for sure, he has them!

Then, who is right? Once more, I think that a little reflection on supervenience will show us the right way.

In the mind-body problem, the criterion of reality is causal power: mentality is real only if it has causal power, as physicality has. For physicalists, identity theorists or functionalists, only particles have real causal power, so only particles exist, even if mentality supervenes on its physical basis. And if Kim is right, if supervenience is a kind of reductionist relation, notwithstanding the opinion of many philosophers whose hope was to escape reductionism via supervenience, then physicalism is true (because it is better that two possible alternatives: epiphenomenism and eliminativism) (cf. Kim 1998: 119–120). Let us accept this position for the sake of the argument – and because it could well be true. Does it imply that morality is reduced to the status of unreality like mentality? Yes it does. However, this does not mean that it is reduced to unreality *in a relevant sense*: like mentality, its unreality only implies that it is reduced qua *physical* entity (and remember: reduction is not elimination or erasure). Like mentality, it does not belong to the furniture of the world. But physical reality is not the only reality. In the normative domain, we have another reality and therefore another criterion of reality: it is not causal power but practical justifying power – that is giving *reasons* to do.[7] An entity has normative reality if it is irreducible in the domain of practical justification. Are rights irreducible? No, and it is very easy to see why: rights supervene on the axiological feature of human needs; all the justifying work can therefore be done by this feature; therefore values may replace rights altogether at the level of normative reality – sometimes, they are even identified and, interestingly, we put rights in the place of values: when we talk of human rights, we sometimes consider them not as imposed duties on others, but as ideals, objects of (rational) desire.[8] But where rights strictly considered are irreducible, it is in the domain of rhetoric: rights have rhetorical reality, that is a very crucial reality in the realm of human relations and social life, a reality that plays a great role in the actual processes of justification (if you are not convinced, ask a barrister).

But has not Hare given an argument to the contrary? Deontic concepts have a prescriptive force absent in axiological-ontological concepts: "Moral words have [...] a commendatory or condemnatory or in general prescriptive force which ordinary descriptive words lack" (Hare 1981: 71), and axiological-ontological concepts are descriptive ones (cf. Jackson 1998: Ch. 6). So you can't reduce the prescriptive to the descriptive, you can't eliminate prescriptive features from the normative reality. It is nevertheless easy to answer this objection, at least as far as rights are concerned: rights protect *important* needs and needs are important because of the *great value* of the goods they aim at. So if prescription is expressed *only* in deontic concepts, it is obviously not grounded in them, but in evaluative ones: it indicates the importance of the value, whence their rhetorical force. And if prescription is reducible to value importance, as values supervene on natural properties, Hare's objection evaporates.

Let me finish in summarizing my metaphysics of rights in a few sentences.

If rights supervene on needs and if supervenience is a relation of reduction, then:

1. Unlike needs, rights have no physical reality
2. Unlike values, rights have no normative reality
3. Rights have rhetorical reality, and this is why they are so important in our social and moral life, but more important in the first than in the second, because rhetoric is a social phenomenon.

Notes

[1] I have dealt with this subject in Baertschi 2003.

[2] I think that it could be extended to negative rights or liberties too. The general reason in favour of this is simple: we *need* liberties to flourish. But I will not dwell on liberties here.

[3] Cf. Kim 1998: Ch. 1. Supervenience can be defined as neither symmetric nor asymmetric, but so defined it is not a relation of interest to us. Cf. also Kim 1993: 67.

[4] For the first question, see Baertschi 2004, in which I argue in favour of the priority of moral values on deontic norms.

[5] Cf. also Kim 1993: 166 "Valuations must terminate in non-valuational grounds".

[6] That is on certain important interests.

[7] In the normative domain we have motivational power too, but we will let it outside the picture lest it becomes too complicated.

[8] We have various theories to explain the justifying force of values, some are rationalist, some empiricist. In my opinion, this force comes from their emotive power, tied to our desires and emotions: values are what is to be desired because of what its substratum is.

References

Baertschi, Bernard. "La place du normatif en morale." *Philosophiques*, 1, 69–86, 2001.

Baertschi, Bernard. "Exclusion et allocation des ressources médicales." In M. Giugni and M. Hunyadi (eds.). *Sphères d'exclusion*. Paris: L'Harmattan, 2003.

Castañeda, Hector-Neri. *Thinking and Doing*. Dordrecht: Reidel, 1975.

Daniels, Norman. *Just Health Care*. Cambridge, Mass.: Cambridge University Press, 1985.

Doyal, Len and Gough, Ian. *A Theory of Human Need*. London: MacMillan, 1991.

Hare, Richard M. *The Language of Morals*. London: Oxford University Press, 1952.

Hare, Richard M. *Moral Thinking*. Oxford: Clarendon Press, 1981.

Hare, Richard M. "Ontology in Ethics." In T. Honderich (ed.). *Morality and Objectivity*. London: Routledge & Kegan Paul, 1985.

Jackson, Frank. *From Metaphysics to Ethics*. Oxford: Clarendon Press, 1998.

Kim, Jaegwon. *Supervenience and the Mind*. Cambridge, Mass.: Cambridge University Press, 1993.

Kim, Jaegwon. *Mind in a Physical World*. Cambridge, Mass.: MIT Press, 1998.

Kramer, Matthew H. "Rights Without Trimmings." In M. H. Kramer, N. E. Simmonds & H. Steiner (eds.). *A Debate over Rights*. Oxford: Oxford University Press, 1998.
Lomasky, Loren. *Persons, Rights, and the Moral Community*. Oxford: Oxford University Press, 1987.
Nozick, Robert. *Anarchy, State and Utopia*. New York: Basic Books, 1974.

The Authority of Desire in Medicine

Matthias Kettner

1 Introduction

Operative within the medical profession as we know is an overlapping consensus that helps to draw the lines separating good and bad practice, proper and improper goals, adequate and poor competence, established and contested paradigms, valid and mock membership. Besides narrative components regarding the heroes, origins and turning points in the history of medicine, this overlapping consensus contains elements of a normative self-understanding of the profession which shapes the collective identity of the profession and nourishes a sense of identity among its individual members. This normative self-understanding, I will argue, has presently come under massive transformative pressures. In particular, observations of three analytically distinguishable trends indicate these transformative pressures. These three trends may be designated as post-conventionalism, medical utopianism, and commodification.[1]

2 Three Transformative Pressures

(1) *Postconventionalism*. Indications abound that so called alternative or complementary medicine is becoming increasingly attractive for health-care recipients as well as health-care providers. Interestingly, "alternative medicine" is less a sortal concept than a summative label that captures a great range of labels united only by a notably programmatic intention to deviate from recognizable, conventional standards of mainstream medicine. The variety of labels that fall under the heading of alternative or complementary medicine is impressive. Some examples are: Alternative healing, alternative healing therapies, alternative health, alternative medicine, alternative therapeutics, alternative therapies, complementary health care, complementary medicine, extended therapeutics, "fringe medicine," holistic healing, holistic health, holistic medicine, innovative medicine, mind body medicine, natural healing,

M. Kettner
e-mail: Matthias.Kettner@uni-wh.de

M. Düwell et al. (eds.), *The Contingent Nature of Life*,

natural health, natural medicine, "New Age medicine," "New Medicine," planet medicine, unconventional medicine, unconventional therapies, unconventional therapy, unorthodox healing, and unorthodox therapies.[2] This amorphous group of "therapeutic" and "diagnostic" methods is chiefly distinguished from establishment (science-based) healthcare by its acceptance of one or another enriched notion of health (e.g., "spiritual health") as a proper medical concern. Common to many though not all brands of alternative medicine are notions of empowerment, leadership, or both; e.g., the idea that a good practitioner is a teacher who can "empower" one. Its purported goal is not simply to cure, in the sense that mainstream medicine purports to cure, but to bring about "healing," often described as an experience of physical, mental, and spiritual "wholeness."

(2) *Medical Utopianism.* For an observer not bewitched by the spell of the post-modern credo of "the end of the great narratives" there is impressive evidence of a return to visionary thinking in contemporary medicine and its public discussion. It is certainly not unfair to say that over the last decade we have witnessed enormous bouts of a spirit of therapeutic over-assertiveness combined with ample public attention driven by the mass media. Stem-cell research provides a good case in point. Public attention to stem-cell research exhibits a dynamic of exaggerated hopes followed by exaggerated disappointment.

Another field of medical research combined with public attention driven by mass media can be pinpointed by the now popularized term *enhancement.* "The term enhancement is usually used in bioethics to characterize interventions designed to improve human form or functioning beyond what is necessary to sustain or restore good health" (Juengst 1998: 29). Medically crafted strategies of enhancement not only serve promethean aspirations that may characterize the mind-set of a minority of researchers in the life sciences. However, such strategies have given rise to a new genre of medical science fiction that in almost every instance unfailingly seizes the attention of a majority of laypeople interested in medicine. Despite vast differences in what passes as enhancement, the rhetoric of enhancement has opened a tremendous range of wishful thinking about medical utility along side, or perhaps over and above, the time-worn medical promise of "fighting disease." Partly at least, talk of enhancement is becoming medicine's second strategic theatre, whereas prospects of winning at the first battle line against major diseases (e.g., fighting cancer) seem utterly dull.

Any sketch of medical utopianism today would be incomplete without mentioning medically assisted reproduction and its associated diagnostic techniques. Interestingly, assisted reproduction has given rise to extremely positive and extremely negative visions. Much negative utopianism about this complex of medical research and practice invokes the historically substantiated examples of eugenics, discrimination based on somatic or intellectual traits, and racist adaptations of Darwinian "natural selection" to "social Darwinism." Positive medical utopianism about assisted reproduction and its diagnostic techniques takes up, e.g., concerns of reproductive freedom, the desire to reconcile "natural" parenthood with the lifestyle and constraints of couples where both partners pursue demanding professional careers, and an increasing cultural bias against accepting contingency in procreation. For instance, by ingenious applications of genetic testing we can (it is hoped) prevent disease or get "better than natural" children.

Some further significant examples of concerns that resonate with positive medical utopianism are these. We can (it is hoped) soon boost muscle strength, memory and *élan vitale* in the elderly, and give athletes a competitive "edge." As many a medical promise goes, we will be able to control aging and correlative morbidity processes, not only in order to extend life as such but to extend the personally good life by reducing or eliminating limitations induced by aging. The research interest to "compress morbidity" into the very final and, many hope, brief final phase of one's life – and the corresponding desire of many potential clients to appropriate the fruits of such research and thus have their lives extended – is a powerful expression of positive medical utopianism today.

(3) *Commodification.* A third powerful trend that helps to reshape the common normative self-understanding of the medical enterprise is commodification in the following sense: The relevant standards that control the growth of medical knowledge and its diverse products (medical know-how, recommendations, state of the art treatments, appliances, and pharmaceuticals) shift from standards based on an ethics intrinsic to the clinical mission and vision of medicine to standards based on the rationality of markets. This is not the place to examine in what precise sense markets within a broadly capitalist economy behave in ways that warrant (or fail to warrant) ascriptions of rationality. Let it suffice to say that the commodification of a formerly non-market transaction x (e.g., donating eggs) first creates a market for x, thus (second) bringing x under "market mechanisms" (which supposedly align the supply of, and demand for x effectively) which assign a market price to x, and third, create the complementary role of a customer-supplier relation regarding x, all of which (fourth) brings x into the scope of moral norms pertaining to business ethics, rather than leaving x under norms pertaining to other areas of practice, such as, the ethos of the healing professions.

It is hard to deny that we are witnessing today in G8 countries a marked tendency to commodify health-care services and other products which can be delivered by the medical profession.

This tendency is linked with libertarian redescriptions of patients as "health-care clients," "customers," or "medical consumers," for example, as consumers of "cosmetic surgery." Such redescriptions carry in tow a radical extension of libertarian background assumptions, such as the assumption that our lives are ours to shape as we wish, and by whatever means we choose (as long as we do not violate the law and do not harm others illegitimately).

More important perhaps – though certainly harder to pin down by means of empirical research in interpretative social science – is a gradual alteration, engendered by processes of commodification, in the way we think about medical interventions in our somatic and mental states once such states become matters of personal preference even though they are not related to any medically acknowledged malady. The commodification of relevant medical skills and knowledge, I submit, encourage and support attitudes regarding a broad range of features of one's body and mind analogous to attitudes of consumers with regard to the choices they face among consumer goods.

Illustrations of this can be found in extant practices which enable people to alter moods and personality states or even traits (e.g., by using without medical indication psychoactive medical drugs like Prozac, Ritalin, and Viagra). Paradigmatically, the "beauty industry's"[3] promise of fulfilment of personal aspirations to bodily beauty and other body-based aspirations of perfectionism, and the "anti-aging industry's"[4] equally luring promise to prolong (if not "turn back") time can serve well to illustrate how commodification and consumerist objectification are intertwined.

3 Reconceptualising the Medical Enterprise

A common pattern in post-conventionalism, medical utopianism, and commodification is a propensity of medicine to cater more than ever before to the desires, wishes, preferences, or demands of its recipients. There is a marked propensity within some branches of medicine to treat recipients' desires as authoritative over professional autonomy. Moreover, recipients' desires are often mobilized in order to harness them to big projects (e.g., stem-cell research, proteomics) that are ambitiously pursued within some parts of the healing professions but are hard to advance without widespread favourable public opinion.

What, if anything, is wrong with medicine opening up to peoples' desires and thus becoming more responsive to its clients? Are there any normative considerations from which medical professionals, bioethicists or public officials can derive justified limits on whatever authority clients' desires acquire and bring to bear on the evolving shape of the medical enterprise?

Apparently, the gist of normative opposition to the desire-driven transformation of the medical enterprise is a tangible resistance against the assimilation of the medical enterprise to a thoroughly business enterprise, i.e. a set of practices that are essentially and predominantly governed by norms of economic rationality.

Several strategies have been devised to argue against allowing medical care to become entirely commercialized. I will briefly characterise three prominent strategies whose common denominator consists in reconceptualising the medical enterprise in ways that provide grounds for limiting the authority of desire. I call these strategies detours, because they stop short of invoking a politically demanding position in bioethics. In the final section, I argue that adopting a politically demanding position in bioethics cannot, and need not, be avoided when bioethicists reflect on desire-driven medicine.

(1) *Technology as the essence of medicine.* The first detour construes the medical enterprise essentially as technology. If the path of evolving medical practices is at bottom a path of evolving technological practices, then it is appropriate to submit these practices to technology assessment in order to draw an authoritative line separating potentially desirable from potentially undesirable practices.

Construing medicine by and large as technology to be governed by the right kind of technology assessment will not do, however. Consider the standard notion of technology assessment and how it translates into the medical context. The "assessment of medical technology combines information about safety and efficacy with social values, costs, side effects, acceptability, and legal issues to reach conclusions about the value of the technology under study" (Bailar and Mosteller 1992). The background assumption of this first argument is obvious: The proper locus of good governance of medicine within a liberal state is not bioethics, but technology assessment, more precisely, biopolitics informed by expert technology assessment. Expert technology assessment in turn *presupposes* value commitments that by itself it can neither legitimately generate nor justify.

Non-expert, "participatory" forms of technology assessment such as, e.g., citizen consensus-conferences, generate or at least refine internally the "social values" and "acceptability" standards which are required in any evaluation that seeks to go beyond considerations of feasibility and profitability. Their "grass roots" consensus orientation gives participatory forms of technology assessment a certain *prima facie* democratic appeal. However, participatory forms of technology assessment, no less than expert forms, at most construct a de facto group consensus whose purported (moral) universalizability and (political) projectability onto all persons who stand to be affected by the respective technological regime remain methodologically unwarranted.

In order to fill out the normative lacuna of medical technology assessment, we need a normatively grounded position in biomedical ethics, combined with an ethically qualified notion of consensus-building as developed in "discourse ethics."[5]

(2) *Science as the essence of medicine*. Any attempt to view the nature of the medical enterprise in basically the same terms as the development of empirical sciences promises to provide strong normative grounds against unfettered postconventionalism, utopianism, and commodification of medicine.

The background assumption operative in this attempt to make medicine look like proper empirical science is a normative one: Science (and its ethos) is not and should not be assimilated to business (and its ethos). Neither should medicine, to the extent that medicine is like science.

This strategy of reconceptualisation, though *prima facie* more promising than modelling medicine as technology, is an unsatisfactory detour for at least three reasons.

It purchases normative grounds for the governance of medicine at the cost of seriously distorting the nature of medical practice as we know it. Medical practice as we know is methodologically most akin to science where it is strongly evidence-based. But evidence-base medicine is hardly representative of the practical knowledge and craftsmanship that are constitutive of large parts of the medical domain.

Moreover, if we idealize science proper, including medical science, as a pursuit of truth in which evidence is countenanced only to the extent that it its objectivity can be ascertained, then we lose sight of the prominent role of subjective evidence in medicine. Let me briefly illustrate this point: In Germany, health-care legislation no longer requires medical insurers to pay only for treatments and medical products whose medical value has been objectively proven by scientific standards. Health-care legislation has adapted a more lenient stance by allowing subjective utility to count to an extent as evidence in the assessment of the overall medical value of treatments and medical products, even when objective proofs by scientific standards are not forthcoming, providing that no good case can be made to discredit the respective treatment or product as based on downright irrational convictions. On this basis, for instance, acupuncture is counted in while "acuscope therapy" would be counted out.[6]

Finally, to the extent that the case for viewing medicine and empirical science as methodologically isomorphic can be substantiated, the problem of specific normative standards returns, now with regard to the norms that should govern the *application* of medicine-as-science.

(3) *The medical profession as the essence of medicine.* Any profession, in the sociological sense of the term as canonically defined by Talcott Parsons, characteristically provides some specific benefits that are externally valued in society at large and is internally shaped by particular goals and values that are considered proper ways and appropriate values by the members of the profession. Professions are entrusted with externally legitimate and internally effective powers of self-governance ("professional autonomy"), e.g. concerning inclusion and exclusion of their membership.

Construing the medical enterprise essentially as a tradition of professionals has definite advantages over the two possible strategies considered so far. This move makes available some powerful normative resources for (medico)-moral objections to a desire-driven transformation of medical knowledge and medical skills. First, by appealing to a normatively rich concept of health it is possible to bring issues of safety and bodily harm into moral discourse. For instance, some of the "risks" of somatic stem-cell treatments can now be interpreted as consisting in morally dubious pressures on standards of informed consent and other moral requirements of a recognizably medical nature. Second, some (though arguably not all) normatively rich concepts of health maintain conceptual links to normative concepts of *equity*, thereby paving the way to bring in issues of fairness and distributive justice. For instance, a case can be made that a commercialized health care system that would systematically condition the provision of highly valued medical benefits (e.g., organ transplants or other disease-related therapeutic benefits of vital importance) by ability to pay would be intolerably unjust. However, the force of such considerations as these does not transfer to the governance of medical benefits that are neither disease-related nor of vital importance, e.g., purely cosmetic surgery.

Third, by tapping into the values that have developed within the tradition of the medical professions, we get purchase on notions of *liberty*, thus permitting us to bring issues of freedom and coercion, both overt and covert, into moral discourse. For instance, it is not altogether implausible to redescribe influences of media-amplified popular culture (for instance, a fashion industry whose power can be tyrannical) on the body-based norms of self-esteem and aesthetic recognition as morally questionable forms of mind control.

Despite its evident appeal, basing the derivation of limits to desire-driven medicine on the normative grounds provided by construing medicine as essentially a profession with a specific tradition of care leaves open two crucial questions: How far do such grounds cover applications of medical knowledge that are unrelated to disease? (Again, think of purely aesthetic surgery as a case in point.) Second, how robust is the normative authority of the medical tradition when it conflicts with the emerging authority of desire; how robust *should* it be?

A natural move at this point is to embark on articulating what could pass as "the" goals of medicine, i.e, goals that are demonstrably intrinsic to the identifiable tradition of "the" medical profession. I use scare quotes for the singular article in order to highlight the problems of pluralism and contingency besetting this approach. The search for the goals of medicine (Cassell 1991; Hanson and Callahan 2000) has resulted in a short, widely accepted list of goals:

- Prevention of *disease* and promotion of *health*
- Relief of *pain and suffering* caused by maladies
- *Cure* of maladies; *care* when cure is not possible
- Avoidance of premature death and the pursuit of peaceful death

To the extent that this set, or some similar set, of goals can be identified and defended, such goals certainly help us to *specify* and also to *justify* the range of clinical treatment-indications. Such indications, in turn, help us to justify the profile of constraints that we think should govern the desires that drive the development of the medical enterprise.

However, owing to the fact that in most of its history the medical profession was dominated by a concern with classifiable disease, the search for the intrinsic goals of medicine, if authentic, will reveal little else than goals that express this historically dominant preoccupation with maladies. Therapy (if it is still to be called therapy) beyond clinical indications is off the mark from the point of view of the goals of medicine. Yet such therapy in an extended sense is exactly the problem. This problem cannot be addressed satisfactorily by an approach that is oriented to the past and is thus structurally conservative, as is the appeal to the values enshrined in the dominant tradition of the profession.

Observe furthermore, that whatever goals of medicine we can reconstruct, the result does not necessarily determine the normative nature of medical practices *as* medical practices.

Observe also that there is more than one goal in any plausible list of "the" goals of medicine. Where there is more than one goal, there will be conflict. Where conflicting goals are *prima facie* legitimate goals, competition for fixing the appropriate pattern and proportion of multiple goals will also be *prima facie* legitimate. Hence, the profile of constraints that we think should govern the desires that drive the development of the medical enterprise will as such not be harmonious; rather, whatever overlapping consensus there is will envelop a lot of dissent and contested distinctions.

The considerations advanced in this paragraph motivate a simple conclusion: Once we settle for a normatively rich concept of medicine we get a number of contested distinctions, i.e. distinctions we use with more or less justification in order to distinguish good and bad, welcome and unwelcome, desirable and undesirable transformations of the medical enterprise.

Take for instance the contested distinction between treatment and enhancement. Consider the human growth hormone. If the goals of medicine *should* be restricted to treatment of diseases, only children with a clinically relevant hormone deficiency *should* be given human growth hormone. Giving the hormone to children with normal hormone production, no matter what their height, would be a form of enhancement and thus go beyond the appropriate goals of medicine, thereby perhaps subjecting them to *risks* that could only be justified if they were suffering from a malady whose indicated treatment would presuppose acceptance of putting oneself at such risks. Depending on whether we focus on treatment of disease as a non-negotiable goal definitive of the medical enterprise as such, giving growth hormone to children with normal hormone production may either be suspect of violating the important bioethical principle of non-maleficence; or it may be a case of a morally laudable exploratory extension of the equally important principle of beneficence.

Imagine now that certain forms of physical or neuropsychological enhancement might give competitive advantages to some people, namely those who can afford the therapies. As Dan W. Brock has pointed out in a recent Hastings Center volume about the ethical and social implications of enhancing human traits: If beauty helps people get better jobs, and the rich can afford cosmetic surgery, then the rich and beautiful will become even richer. In that possible world, medicine would be co-opted in the promotion of *injustice* in society. Still another medioethical concern – respect for autonomy or, inversely, the due minimisation of coercion and heteronomy – crops up once we take account of the mounting evidence that medically crafted enhancement options can *pressure* people into using them who would otherwise not want to use them. Athletes using dangerous performance-enhancing drugs exemplify this pressure.

4 What sets the Goals of Medicine?

Why not step out of the game of arguing about notoriously contested distinctions that allegedly capture the normative essence of medicine as a historically extended concrete field of practices? Why not give up on the very idea of a normative essence of medicine? Why not adopt the – perhaps quite salutary – stance that any such distinction reflects nothing over and above the contingent evolution of medical practices? After all, if history comes with no transcendental guarantees then neither are such guarantees to be found in the history of professional medicine.

There is at least one powerful reply to this deflationary temptation: Such normative laissez-faire would loftily decide an issue that remains deeply controversial within the healing profession itself.

Consider, for example, the controversial issue of medical futility judgments. A medical futility judgment is a patient-centered determination of no benefit. Like all judgments, futility judgments carry a defeasible validity claim. The claim they make is backed by reasons which one expects can be evaluated as medically sufficient reasons from any suitably informed and competent member of the medical profession, and by the same token, reasons which one expects can be evaluated as morally sufficient reasons from the moral point of view of anyone, including the patient (Kettner 1999). Nearly all physicians, even those who oppose futility policies, will agree *that there are limits* to physicians' professional obligations to provide requested treatment, no matter how strongly patients desire such treatment. The important point is this: Limits and a refusal to go beyond them, or at least a reasoned recognition that going beyond them would be morally wrong, reflect normative distinctions that are inherent in the very practice at hand. Moreover, limits and a reasoned refusal to cross them reflect a determination to uphold these distinctions across contingent changes in the respective body of practices (cf. Will 1997). It would appear unwise to ignore the rational potential that is embodied in the tradition of medical practice.

But what if anything sets "the" goals of medicine? Where do "goals" come from? What ensures their perseverance? Who is to judge where and whether perseverance or change is called for? After all, medicine is no natural kind like H_20 and there is no equivalent to the science of chemistry for fixing its essence. At bottom, "society" has empowered the medical community to define "the" goals of medicine, which is thus to grant that professions and their normative essences are socio-historically malleable. Yet this does not mean that they are completely autonomous in setting their goals, because they work within a broader institutional context of morals, social purposes, and legal or political constraints.[7]

Here we reach a point where the authority of desire reveals itself as a constitutive ground of the medical tradition in its historical concreteness. Medicine is always already a response to certain powerful human desires, some of which are perhaps empirically universal or, for that matter, "anthropologically basic" (as purported desires to rid oneself of maladies and other evils with the help of designated healers) while probably others are not (for instance a desire to control body weight).

If this way of looking at the unfolding history of the body of practices we call medicine is not altogether wrong, the resulting picture of the unruly forces that have been shaping medicine is not very inviting for friends of essences. Consequently, it would be surprising if any true-to-the-facts list of the goals of medicine were to reveal that these goals formed a prestabilized harmonious whole.

Friends of (normative) essences might want to restore order by shifting the focus of analysis from medicine's goals to medicine's good. If medicine has an essence, this essence might reveal itself in the *values* that guide medical practice rather than in the criss-cross of goals that are operative in these practices. There is a tempting analogy here between the rational pursuit of truth (logic) and the rational pursuit of the good (ethics): If we understand belief as aiming at the true we could likewise understand desire as aiming at the good. If our values are rooted in our desires, so in turn are norms since our norms articulate our value-commitments. But if we can rationally judge norms and values (which we certainly can, because norms and values are good-dependent) then we can also rationally judge the underlying desires since they are good-dependent too, albeit indirectly.

This analogy, however, does not lead very far. Though desires *can* be good-dependent, they need not be, and where they are so dependent the question arises how to distinguish between genuine and merely apparent good(s). Desires can be unwelcome (e.g. the desire of an anti-aging clinic to maximise its profits even at the sacrifice of sound treatment). Desires can originate in processes far removed from processes of rational governance (e.g. a desire to turn back time originating in unconscious narcissistic fantasies). Finally, it is possible to have a desire without a rationalizing belief in the value of its object (as can be seen in animals and small children).

5 Bioethics and Biopolitics – No more Detours

If desires have been and continue to be formative forces that shape the sets of practices we recognize as the tradition of professional medicine, what normative resources do we have for specifying and justifying certain limits that we think ought to constrain the authority of desire in medicine?

In the preceding paragraphs I have sketched three strategies of reconceptualizing the nature of the medical enterprise. I have argued that they are detours because analyzing medicine respectively as technology, as science, and as a profession with a tradition cannot bring us much closer to solving this problem. These strategies of reconceptualisation, for all the insights they bring in other respects, are detours because they make it harder for us to realize the inescapably political edge in normative arguments (of whatever kind) by which we mean to rationally contribute to governing the authority of desire in medicine. What are the implications of this for bioethics? Is bioethics yet one more detour?

Not necessarily. Consider the *interventionist* nature of bioethics. Speaking pragmatically, applying ethics is an activity that aims at ameliorating, in some morally qualified sense of better and worse, practices that are beset by moral uncertainties serious enough to attract and merit the attention of experts, thus becoming the target of rational scrutiny and reconstruction.

Where bioethics does not strive for contemplation but for application, the practice itself of bioethics, "pure" bioethical argument included, has a political edge. Bioethics and biopolitics are Siamese twins.

Evidently, bioethics as biopolitical intervention must involve a moral stance of some kind or other because it aims at ameliorating target practices in some *morally* qualified sense of better and worse. Whence this stance? How are we to justify any particular moral commitment we make as bioethicists (arguing, e.g., for a ban on formats of televised beauty surgery because they harm young people by unduly extending the coercive grip of consumerist norms into our sense of self)? According to a hermeneutical model of application we take the moral commitments for criticising and perhaps refashioning some target practices from the normative textures and contexts of those very practices. According to a rationalist model of application we should invoke only moral commitments that are available to anyone capable and willing to engage in rational thought and action ("the moral point of view" in a sense long since made popular by Kurt Baier). According to a discourse model of application (see Habermas 1990, 1993) it is an unobjectionable moral policy to begin by invoking certain abstract moral commitments which are already operative in argumentative discourse and to import as much moral substance from the target practice (following the hermeneutical model) as will pass through moral discourse.

Within the confines of this paper I cannot compare the merits of these and other ways of construing the kind of governance to which bioethics as applied ethics can aspire. Instead, after having devoted much effort to bringing out the contingency of medicine owing to the desires that fuel the medical enterprise, I will end with a note on the contingency of bioethics.

It is appropriate to distinguish between contingency (C1) in the sense of the uncontrollable, the unpredictable, chance and surprise, and contingency (C2) as an awareness of alternatives or options that are or have been open; of "possible practices" if you will. Genetic mutations and *tsunamis* illustrate the former meaning of contingency, positive freedom and deliberation the latter. With this distinction in mind, there is no need to think of contingency as the great underminer and foe of normative justification in ethics. While contingency in the first sense (C1) must be accepted as an existential of real practice, contingency in the second sense (C2) is a precondition for normative justification in ethics. Bioethics, and applied ethics generally, should cherish C2 as a precondition for justifying its own consequential recommendations and interventions into practices that are the target of application. Leaving aside the notoriously contested question of rationally definitive foundations of moral thinking, clearly justification in bioethics, and in applied ethics generally, can be comparative, i.e. couched in terms of persuasive *comparisons* of *alternative* normative commitments and their foreseeable consequences for everyone who stands to be affected on either side of the alternatives. This relatively robust form of comparative justification requires reflexive awareness of possible practices.

To conclude: I have argued that bioethics cannot escape having a bio-political edge. It is therefore reasonable to advocate that bioethics, even in its theoretical core of rational moral discourse, should acknowledge and develop, rather than suppress, its bio-political edge. To those who are uncomfortable with this conclusion I would add as consolation that the inverse also holds, i.e. whatever passes as bio-politics inescapably has a bio-ethical edge. If biomedical ethics purports to have a bearing on the authority of desire in medicine then it must speak up in a bio-political voice, because this is the plane on which medical practice is now urged to address the demands of desire.

Notes

[1] With reference to the German health-care system I have backed the phenomenological description of these trends by empirical data (Kettner 2006b: 81–91). For the present purposes I have confined myself to qualitative generalizations and examples, while assuming that the description of these trends can be extended to all countries with similarly advanced and similarly priced health-care systems.

[2] This concatenation is based on the internet resources for alternative medicine (http://www.pitt.edu/~cbw/syst.html).

[3] A recent article in *Time* ("Face Facts," March 13, 2006, Vol. 167, No. 11, 40–47) ventures to explain a purportedly emerging European "obsession with external beauty" by the democratization of medical

knowledge, a search for beauty powered by TV programs designed to convince viewers that a makeover is something they need feel no guilt in desiring, and the needs of the "new male."

[4] As to popular media, cf. *Business Week*'s (March 20, 2006, 64–74) special report "Forever young." For a balanced assessment of claims related to anti-aging, see Post and Binstock 2004.

[5] Habermas 1990, 1993. For a critical account of Habermasian discourse ethics, see Kettner and Matthias 2003, 2006a.

[6] The example of acuscope therapy is drawn from Jack Raso's *Expanded Dictionary of Metaphysical Healthcare, Alternative Medicine, Paranormal Healing, and Related Methods* (1998), where acuscope therapy is explained as a form of energy medicine (vibrational medicine) that allegedly speeds healing of virtually any injury; its centerpiece is the Acuscope (also called the Electro-Acuscope), a computerized device that purportedly balances the body's electrical current.

[7] I wish to thank Kenneth Westphal for this point and for many other valuable comments.

References

Bailar III, John C. and Mosteller, Frederick. "Medical Technology Assessment." In *Medical Uses of Statistics* (2nd ed.). John Christian Bailar and Frederick Mosteller (eds.). Waltham, MA: Massachusetts Medical Society, 393–412, 1992.

Cassell, Eric J. *The Nature of Suffering and the Goals of Medicine*. Oxford: Oxford University Press, 1991.

Habermas, Jürgen "Discourse Ethics: Notes on a Program of Philosophical Justification." In *Moral Consciousness and Communicative Action*. Cambridge, MA: MIT Press, 43–115, 1990.

Habermas, Jürgen *Justification and Application: Remarks on Discourse Ethics*. Cambridge, MA: MIT Press, 1993.

Hanson, Marc J. and Callahan, David. *The Goals of Medicine: The Forgotten Issue in Health Care Reform*. Washington, DC: Georgetown University Press, 2000.

Juengst, Eric T. "What Does Enhancement Mean?" In *Enhancing Human Traits: Ethical and Social Implications*. Erik Parens (ed.). Washington, DC: Georgetown University Press, 29–47, 1998.

Kettner, Matthias. "Argumentative Discourse, Good Reasons, and Communicative Rationality." In *Rationality, Realism, Revision: Proceedings of the 3rd International Conference of the German Society for Analytic Philosophy September 15–18, 1997 in Munich*. Julian Nida-Rümelin (ed.). Berlin: de Gruyter, 331–338, 1999.

Kettner, Matthias. "Gert's Moral Theory and Discourse Ethics." *Rationality, Rules, and Ideals. Critical Essays on Bernard Gert's Moral Theory*. Walter Sinnott-Armstrong and Robert Audi (eds.). Lanham, MD: Rowman and Littlefield, 31–50, 2003.

Kettner, Matthias. "Discourse Ethics. Apel, Habermas, and Beyond." In *Bioethics in Cultural Contexts. Reflections on Methods and Finitude*. Christoph Rehmann-Sutter, Marcus Düwell and Dietmar Mieth (eds.). Dordrecht: Springer, 299–318, 2006a.

Kettner, Matthias. "Wunscherfüllende Medizin zwischen Kommerz und Patientendienlichkeit." *Ethik in der Medizin*, 18, 1, 81–91, 2006b.

Post, Stephen G. and Binstock, Robert H. *The Fountain of Youth. Cultural, Scientific, and Ethical Perspectives on a Biomedical Goal*. New York: Oxford University Press, 2004.

Will, Frederick L. "Reason and Tradition." In *Pragmatism and Realism*. Kenneth Westphal (ed.). Lanham, MD: Rowman and Littlefield, 105–120, 1997.

Procreative Needs and Rights

Norbert Campagna

1 Procreation: From Duty to Right

Not so long ago, most people believed in the existence of a duty to procreate. The justification of this duty was either religious – based for example on the Old Testament injunction: "Go and multiply" – or political – often based on the need of the State to have many soldiers for its army. For most people, procreation was an objective, economic need: having many children gave them a guarantee that there would be someone to look after them in their old days. Having children was, so to say, an investment in the future, and the more children one had, the greater were the chances that one would not be too worse off in one's old days, when one was unable to work anymore – it being presupposed, of course, that children had a duty to care for their parents. In many Third World countries this situation persists and it certainly at least partly accounts for the problem of overpopulation – and I don't wish to condemn people in those countries for thinking and acting as they do, given the situation they are left in by their national authorities and the world community.

In our present-day modern Western societies, the economical need for a child has taken a collective form: society as a whole needs many people to make possible the functioning of the welfare system, and especially the system of pensions. But it should be clear that the reason for which people procreate has nothing to do with a possible breakdown of the pension system. Their reasons for procreating are much more personal and they are also linked to personal rather than to social needs. For example, there are couples who procreate because they hope that a child will constitute new cement for their increasingly fragile marital relationship – a hope that often turns out to be an illusion. Or someone may feel something like an existential need for a child. In this case, the child is a means to give sense to a life that appears to have none. Once the child is there, questions about the meaning of life disappear. There may also be people who see in their child something like an ersatz for immortality, i.e. they think that they can live on through their child or at least through the child's memory – or think about the cult of the ancients to be found in many

N. Campagna
e-mail: norbertcampagna@hotmail.com

M. Düwell et al. (eds.), *The Contingent Nature of Life*,

religions and cultures. The possibility of cloning human beings has given a new drift to this hope for immortality, though the latter is conceived in crudely naturalistic terms – survival of a specific genetic code – rather than in symbolical terms. Finally, though the list is far from being complete, one could mention an emotional need: the child is someone one can love and cherish and by whom one can be loved and cherished in return. And many people feel a strong need to love someone and to be loved in return.

Today, the idea of a duty to procreate has more or less disappeared. Duty-talk has been replaced by what Mary Ann Glendon has called rights-talk: people have or at least claim to have a *right* to procreate. Of course, such a right to procreate existed also in the past, but it was grounded on or even implied by the duty to procreate: because God commanded procreation, human beings necessarily also had a right to procreate. The *raison d'être* of the right was to make possible the accomplishment of the duty. Today's right to procreate is grounded or claimed to be grounded not on a duty – not even a duty to oneself (by the way an incomprehensible notion for today's predominating rights-talk) – but on a personal need, be it emotional, existential or whatever. It is often because someone feels a strong need for a child that he or she claims to have a right to a child.

2 Procreation: From a Liberty-Right to a Claim-Right

Another change has also taken place within the last couple of years. The right to procreate was initially conceived as a right not to be hindered in one's procreative choices, that is to say as a negative right. It is this negative right which is embedded in article 16 of the Universal Declaration of Human Rights and in article 23 of the International Covenant on Civil and Political Rights, to mention only two important documents. If a couple wanted to have children, neither the State nor any other third party or person had a right to hinder the couple, materially or legally, from materializing their want. The decision to procreate, or family planning, as it is often called, was considered as a purely personal decision. People were to be left alone and also wanted to be left alone. Of course, this did not hinder the State from passing laws introducing different regimes of taxation, either privileging couples with many children – in most European countries – or couples with as few children as possible – for example in China.

Today, the right to procreate is more and more conceived as a positive or claim-right. The duty of the State is not only seen as a mere duty of abstention – leave people free to procreate – but it comes to be seen as a duty of positive intervention, generally of a financial, but also infrastructural nature – make it possible for anybody to procreate. This evolution is to a great extent due to medical developments, especially in the field of procreative medicine. Formerly, the incapacity to bear or generate children was seen as something one had to accept – even though it must be admitted that in the past people already tried many rather less than more efficacious devices to enhance their reproductive capacity. Nowadays, IVF makes it possible for many infertile couples to have a child. Moreover, preimplantation diagnosis makes it possible to choose among possible children the one you want to bear and bring to life. And cloning would even make it possible to have a child which is genetically quasi-identical with one of the parents.

This being so, the right to have a child can today be spelled in multiple terms. IVF makes it possible to have a child of one's own – rather than adopt a child. Preimplantation diagnosis and prenatal chirurgical interventions make it possible to have a healthy child – rather than a child with genetic defects or an eye-colour one doesn't like. And cloning makes it possible to have a child genetically almost like oneself. The relation between these new technologies and people's procreative needs is not a one-way relation, with needs unilaterally determining the evolution of technologies. There certainly was a preexistent need to have a child, but the existence of the technologies and the possibilities they opened up probably played an important part in fostering more specific needs, as for example the need for a child genetically almost like oneself, and also in making the frustration of having one's need for a child not satisfied more strongly felt. These needs have given rise to the claim that they be also met and they have put on the clothes of today's predominating rights-talk. Though the move from the initial "Leave us alone in our decision to have a child" to "Give us the money to have a clone" has not yet reached its end-point, the idea that the State should democratize cloning by subsidizing it in individual cases may one day become normal – as normal as the idea that the State should take charge of the costs of giving life to a child in the usual way.

As far as well-off people are concerned, the right to have more specific procreative needs met can be conceived as a pure liberty-right: a billionaire can be content with asking nothing else but to be left alone in his decision to pay a medical team so as to have himself cloned. Yet most people are not billionaires and for them the use of modern procreative technologies is only possible if the health insurance system takes over the costs of the medical interventions. Hence the question arises of whether the democratization of the need to have a child of one's own or a healthy child or a child genetically almost like oneself should also be accompanied by the democratization of the universal economic availability of the technologies allowing the creation of such a child. In order to answer this question, one has first of all to reflect on the moral relevance of the need to have a child *tout court* and on the moral legitimacy of the claim (a) to satisfy the need for child (the liberty-right) and (b) to have the need for a child satisfied (the claim-right).

3 Needs and Rights

When does the need of an individual create an obligation in other people, be it an obligation not to intervene in the actions the individual undertakes in order to satisfy his need or an obligation to intervene so that the individual may satisfy his needs or have them satisfied? One could also ask: When does the need generate a right in the person who has the need and a duty in the persons surrounding the person who has the need?

At least two things seem to be initially plausible. First point: not every need can serve as a legitimate basis for a rights-claim. If some people feel a need to drive very fast with their car, this does not automatically give them a right to do so. And even if it did, this right would not necessarily be an absolute right, trumping all the rights other people have. Second point: some needs can serve as a legitimate basis for a rights-claim. If I meet a starving person and can give him something to eat without an unbearable cost to myself, I am at the very least morally obliged to do so. Here one could also say that the starving person's need to have something to eat trumps my need to keep my food to myself. This being so, we need a criterion allowing us to distinguish the needs which create rights and obligations from those which don't. What is it about a need that can serve as a legitimate ground for creating such rights and obligations?

One could first of all suggest a qualitative criterion: the moral relevance of a need is determined by its strength. The stronger the need makes itself felt, the greater the obligation it creates. This purely qualitative criterion is problematic for a couple of reasons. It is first of all subjective in a very strong sense. I do not know how one can measure the strength of a need in an objective way – I do not believe that this strength can be put in relation with the amount of a certain hormone or whatever present in the blood. If someone says that he feels a very strong need, all we can rely on is what he says. Of course, if the need really is very strong, its non-satisfaction will probably have negative physical and psychical consequences for the person feeling the need – whereas the satisfaction will have positive consequences. The problem is that one cannot know these consequences in advance of the satisfaction or non-satisfaction of the need in question – and this is so because we cannot know whether the person really feels the need as strongly as he or she says. But even if one could know the consequences in advance, recourse to the purely qualitative criterion just mentioned makes us fall into a naturalistic fallacy, and this independently of whether the strength is defined in physiological or psychological terms. How strongly someone feels a need is a natural fact about that person and taking this natural fact as a sufficient basis for establishing a normative conclusion is more than problematic.

What then about the following, quantitative criterion: the moral relevance of a need is determined by the number of people feeling it? The more people feel the need, the greater the obligation to allow them its satisfaction or to help them in satisfying it. This criterion is objective, since it is possible to count people. The need to have something to eat is a need felt by something over 6 billion people, whereas as the need to have a clone of oneself is a need felt by maybe some thousands of persons – or so I think. Of course, what one counts is the number of people who *say* they feel the need and not the number of people who actually feel the need. But there is an even more serious problem: Whereas the former, qualitative criterion gave rise to a naturalistic fallacy, this criterion gives rise to a sociological fallacy and is therefore also very problematic, as numbers do not necessarily establish good normative conclusions. The fact that one hundred million people with a white skin feel a strong need to get rid of one thousand people with a black skin does not give the former a right to eliminate the latter.

What characterizes the two criteria just mentioned is their purely formal character: they try to discriminate needs independently of their content and of the reasons one may have for satisfying them or for having them satisfied. This formal character makes them in a certain sense attractive, because they allow us to bypass fundamental metaphysical questions. Though I do not deny that bypassing metaphysical discussions may sometimes have very real pragmatic advantages – for example if you have to find a legislative *modus vivendi* – I don't think that metaphysical discussions should be shunned by philosophers. Philosophers should of course not simply cast aside as *completely* irrelevant such factual elements as the strength of a need or the number of people feeling it. Strongly felt needs or widely spread needs should be considered seriously, but one should not adduce a moral relevance from these elements only. Strength and spreading are not *per se* morally relevant reasons. They can only be clues helping us to identify needs which might be backed by morally relevant reasons.

One should clearly distinguish between the strength of the need or the feeling accompanying the need and the strength of the reasons one has for wanting the need to be satisfied. Someone may strongly feel a need without anyone being in the least obliged to let him satisfy his need, let alone to help him satisfying it. Good reasons for doing this may be lacking. If someone strongly feels the need to kill his lover's husband so as to be able to marry her, nobody is under an obligation to let him do this, even if the husband's need to continue living is weaker than the potential killer's need to get rid of him. On the other hand, someone may only weakly feel a need – or maybe even not feel it at all – and there can nevertheless be an obligation to let him satisfy the need or help him satisfy it. A child usually doesn't feel a need to go to school and be educated, and I doubt that anyone of us will question the existence of that need and the pertinacy of the reasons for satisfying it. Needs must be distinguished from feelings of need and therefore information about the feelings doesn't yet tell us something conclusive about the need itself or about the legitimacy and strength of the reasons for satisfying it or for having it satisfied.

The fundamental question one has to ask is thus not "How strong is the need felt?", nor "How widespread is the need?", but "How good are the reasons for our wanting the need to be satisfied?". And the goodness of the reasons is not purely dependant on the strength of the feeling nor on its statistical presence in the general population. For normative purposes, needs should not be evaluated by subjective or statistical criteria, but by normative criteria, and these normative criteria should make reference to the intentions of the person who has the need.

4 The Problem of Instrumentalisation

After these general remarks concerning needs and the way to evaluate them, let us return to the specific need under discussion in this contribution, i.e. the need for a child. And let us first of all make the point that if a child is the object of a need, i.e. if people need or feel the need to have a child, the child is conceived as a good, as something satisfying the need. Hence, the child is instrumentalised. This could seem to be the end of the story – *Kant locutus, causa finita*. But let us remember that Kant's categorical imperative does not condemn instrumentalisation as such, but only a wholesale instrumentalisation, i.e. the reduction of a human being to the status of a pure instrument (Kant 1985). Whether we want it or not, we can't help treating other people *also* as instruments – you are not merely reading this contribution of mine to please me, but also and primarily because you hope to find in it something that can help you in your own meditations on the normative aspects of human procreation. What the categorical imperative forbids is treating other persons *only* as instruments. Thus, if I go to the baker's to buy some bread, I treat the backer as an instrument to satisfy my need for food. But as long as I pay him for the bread and treat him as minimally civilized people treat each other – I greet him, thank him when he hands over the bread etc. – I do not reduce him to a mere instrument. When I enter the baker's shop, my intention is to buy bread, not to instrumentalise the baker. If I stole the bread, I would be instrumentalising the baker in a wholesale way. Though buying the bread, I still treat him as a human being (on the notion of humanity in Kant's categorical imperative, see Joerden 2005).

If people engender a child because they feel a need to engender it, one has to look for the reasons underpinning that need. Take for example a couple who want to have a child because their marital relation is at a critical point and they believe that a child will help to save it. In this case, it is clear that the child is instrumentalised and the need for a child is equivalent to the need for something that will save the marital relationship. If something other than the child was – or was thought to be – more efficient, the couple would, *ceteris paribus*, probably use that other means. Thus, if a trip to the Seychelles would at least be as efficient as the begetting of a child, there is no doubt that the couple would pack its suit-cases and get on the next plane to the South Sea islands.

Since a well-functioning marital relationship is a very important human good – for the individuals directly involved but also for society in general – one cannot condemn the couple for wanting to save that relationship. Nor can one *prima facie* condemn them if they want to save it by having a child. What is condemnable, however, is the reduction of the child to a pure instrument, that is to something one casts away if it doesn't properly fulfill its function or something one does only care for inasmuch as such care is necessary for the child's properly fulfilling its function.

As a general principle to guide us in the matter under discussion in this contribution, I would state the following: The need to have a child may be legitimately satisfied if and only if one accepts the duty to care after the child according to its own needs, whether the child satisfies his or her instrumental function or not. The decision to procreate may thus not be grounded on the needs of the child to be born but on those of the parents, but as soon as the child is born, its own needs topple those of the parents who have engendered him or her. One can thus postulate a right of a couple to have a child, yet with the proviso that a legitimate use of that right presupposes that the couple will care for the child according to its own needs, even if the child does not contribute to saving the marriage.

The same holds true for a person who needs a child in order to give sense to his or her life. Giving sense to one's life is certainly something important and one shouldn't condemn *prima facie* someone who needs a child as a means to put an end to this quest for sense. But here again we should add a proviso: the person must accept to care for the child according to its own needs even if the child does not help to put an end to the quest for sense. After all, the child has not given its consent to being born (Kant 1982: especially p. 394).

The approach defended here makes a distinction between objects morally worthy of pursuit and objects not or at least less morally worthy of pursuit. You can't put on the same moral level someone who wants a child because the child will save a failing marital relationship and someone who wants a child because all of one's friends have children and one does not wish to be considered as an outsider. But the worth of the object of pursuit is at most – if at all – a necessary and not yet a sufficient condition of legitimacy. Unless one accepts the duty to care for the child according to its own needs – which also means looking after the world this child will grow in – one cannot claim a right to have one's need for a child satisfied. In other words: the right to have a child is inseparable from the duty to care for that child according to its own needs, and someone who claims a right to have a child must be conscious of the fact that in doing so he implicitly accepts the duty. Not accepting the duty amounts to reducing the child to the status of an instrument. But there is no right to reducing another human being to the status of a mere instrument (see also Campagna 2005b).

The need for a child may not be considered exclusively as the need for an instrument that will serve to fulfill a certain purpose – even though the child may in fact also serve to fulfill that purpose. The need for a child is the need for a being who has needs of its own, with at the very least some of these needs trumping the needs of the persons who decide to have the child. When potential parents put forward their need for a child, they should not forget that they are putting forward the need for a being who will need them. The present need of the parents should be seen as inextricably linked with the future needs of the child. And the child's need for the parents' caring for him or her is stronger than the parents' need for a child.

Those who claim a right to have a child should always be conscious of the fact that they are in a certain sense claiming a right to have duties. The child may well satisfy their needs, but its presence automatically creates duties that will frustrate many other needs of the parents. In the case of procreative needs, the future duties towards the child should always prime the satisfaction of the parents' present needs (see also Campagna 2005).

5 The Case of Cloning

What consequences does this analysis have for the moral evaluation of a practice like cloning? Two things should be said.

First point: a necessary condition for the moral legitimacy of cloning is that the satisfaction of the need for a child also serves a worthwhile end, like saving a marital relationship, giving sense to one's life, etc. Having recourse to cloning just to be "in" or to make the headlines of the newspapers does not make recourse to cloning morally legitimate. This first point entails a normative evaluation of needs and thus also a criterion that will permit us to say that the need to give sense to one's life is much more important than the need to be spoken of for a couple of weeks. Such an evaluation cannot do without substantial elements.

Second point: if cloning is not the only possible means to satisfy the need for a child, recourse to cloning is morally legitimate if and only if *it is in the child's interest to be cloned rather than engendered in a more traditional way*. In other words, even if one acknowledges a right to have a child, this right entails the duty to engender the child in that way which best corresponds to its long-term interests, always bearing in mind that as parents of the child, one has a special duty to make sure that these interests – and hence also the needs giving rise to them – be satisfied. I have no *principled* moral objection to reproductive cloning, but I do have principled moral objections to some types of intentions underlying the wish to have one's need for a child genetically almost like oneself satisfied. And I also have *pragmatic* objections to lifting the implicit or explicit legislative ban on reproductive cloning.

It must be specified that the analysis given here does not tell us how the matter should be regulated at the legislative level. A distinction should be made between what is morally legitimate or illegitimate and what should be permitted or forbidden or subsidized by the State. Given the fact that my moral approach implies an evaluation of needs and also lays great stress on intentions, it cannot easily be translated into a generally acceptable legal text. Nor should it, by the way, because the task of moral philosophy is not to prepare blueprints for legislative texts, but to bring to the mind of the legislator certain reference-points which should not be lost sight of.

6 Conclusion

To sum up the gist of the argument. We may acknowledge the existence of a procreative need without having to accept the idea that the sheer existence of this need, its felt strength or the mere number of people claiming to feel or have it automatically gives rise to a right to procreate and a corresponding duty of the State to make procreation possible. From a purely moral point of view, a right to procreate can only be granted where the decision to procreate refers (a) to a worthwhile need of the procreators that is to be satisfied by the child and (b) to the duties towards the future child one assumes through the act of procreation, it being said that among these duties we also find the duty to procreate the child in such a way that being procreated in this way rather than in another way is beneficial to the child. Though I deny the existence of a duty to procreate, as it was assumed to exist in the past, I affirm the existence of a duty to care for the child one has engendered and thus also to engender it if and only if one can and is prepared to care for it according to its own needs. The right to procreate, if it exists, does not rest on a duty to procreate, but the duty to care adequately for the child rests on the right to procreate – if, again, such a right exists.

References

Campagna, Norbert. "Gezeugtwerden und Naturwüchsigkeit. Überlegungen zu einem unreflektierten Begriff in der Habermasschen Argumentation." In *Gattungsethik – Schutz für das Menschengeschlecht?* Kaufmann, Matthias and Sosoe, Lukas (eds.). Frankfurt etc.: Peter Lang, 2005a.

Campagna, Norbert. "Le bébé-médicament et l'instrumentalisation de l'être humain." In *La bioéthique au carrefour des disciplines*. Haldemann, Frank; Poltier, Hugues and Romagnoli, Simone (eds.). Bern etc.: Peter Lang, 2005b.

Joerden, Jan C. "Der Begriff 'Menschheit' in Kants Zweckformel des kategorischen Imperativs für die Begriffe 'Menschenwürde' und 'Gattungswürde'." In *Gattungsethik – Schutz für das Menschengeschlecht?* Kaufmann, Matthias and Sosoe, Lukas (eds.). Frankfurt etc.: Peter Lang, 2005.

Kant, Immanuel. *Die Metaphysik der Sitten. Werkausgabe Band VIII*. Frankfurt am Main: Suhrkamp, 1982.

Kant, Immanuel. *Kritik der praktischen Vernunft*. Hamburg: Felix Meiner, 1985.

Needs, Capacities and Morality

On Problems of the Liberal in Dealing with the Life Sciences

Marcus Düwell

1 Introduction

In this paper I discuss some basic assumptions of a modern, liberal ethos against the background of the moral evaluation of the life sciences. It seems to me that moral statements concerning the life sciences presuppose evaluative assumptions that are partly in conflict with the scope of moral convictions that are covered by the classical liberal ethos or political liberalism. I will especially focus on the *ethical neutrality* concerning the moral evaluation of human needs and capacities. Therefore I will argue bioethics has to reflect on a substantial criterion, not merely formal, to weigh human needs and capacities. In the current debate Nussbaum and others have proposed the 'capabilities approach' as an Aristotelian framework for such a moral evaluation. I will examine whether the notion of 'human capabilities' can provide us with an evaluation criterion that is not arbitrary, yet substantial enough to justify moral judgments in our current bioethical discussions. Afterwards I will briefly propose a Kantian alternative to such a framework.

2 The Liberal Ethos

There is an important conviction we have about the treatment of the human body. Treatment of the human body has to be legitimated by the decision making and the self-understanding of the autonomous person. Hence, a person can decide what others may do to his or her body, and can set limits by saying 'you may not do that to me.' That the individual person is the only one who can decide what will happen to him or her it is morally important. Yet, if one's *capacity to control* one's own action is an important and morally justified limitation of the freedom of others, then we are evaluating this capacity to control our own action in itself as valuable, regardless of the aims we want to reach with our actions. In a specific way, the idea that one can control what happens to one's body belongs to the core of an

M. Düwell
e-mail: m.duewell@uu.nl

M. Düwell et al. (eds.), *The Contingent Nature of Life*,

ethos of the modern world. Social contract theories discuss the legitimization of institutions that can protect this control. Utilitarianism's aim is to protect our liberty and to coordinate the possibilities of preference fulfilment. As well in the centre of the moral philosophy of Kant we find the autonomous, free and rational being that gives laws to him/herself. Of course, the coordination of our actions in a social contract as well as Kant's idea of the lawgiver limits the scope of legitimate aims we can want to reach with our actions. Kant even thinks that the individual has some *duties to him/herself* and that not everything one wants to do to one's body is morally permissible but Kant defends these limitations of a person's self-determination by arguing that those prohibited actions would contradict important preconditions of individual autonomy. Furthermore, he argues that these duties towards ourselves are relevant for the way the individual treats him/herself (virtue ethics) and not for the moral regulation of action in the public or legal sphere, so even if there are reasons to assume that there are moral limits concerning the way we may treat our own body, the individual is the only one who may set those limitations for him/herself. No external institution is morally allowed to decide what happens with a persons own body. For the liberal ethos it is further important to note that this value of our ability of self-determination is not only one value next to others, but that it forms the *core of a liberal morality*. The protection of human dignity and human rights is seen as an explication of such an ethos and this ethos is seen as indispensable and universally binding.

This liberal perspective involves the conviction that we should *evaluate neither the ways of dealing with our own body* nor our *needs and desires* as such from a *moral perspective*. If someone wants to realize his homosexual desires, this is fine as long as his partner agrees. The needs and desires are not morally bad or good as such and whether he/she wants to develop specific capacities or not is something one has to decide. In the liberal ethos a direct evaluation of our needs and capacities is only made in certain specific instances such as, for example, if our needs affect the way we treat others. Sadistic desires, aggressive habits or the like are morally problematic due to their impact on our behaviour towards others. Similarly, if our desires affect our basic abilities for being an autonomous person (drugs, addictions) or if young people are influenced in a way that interferes with the development of their basic abilities, then we will evaluate those needs and attitudes towards ourselves directly. All these exceptions are related to the protection of the capability of autonomous decision making of the individual. Needs and desires *as such* are seen as morally neutral; they are like factual circumstances since we take them as they are. That the autonomous individual is the only one who deals with his or her own needs and capacities, and that others may do so only insofar as they are given permission by the individual is morally relevant.

The development of regulations of (bio-)medicine after World War II was strongly committed to this liberal ethos. The whole regulatory framework is very much concentrated on the idea of protecting the self-determination of the patient concerning the treatment of his or her body. The central role of patient autonomy and the establishment of the idea of informed consent is an indicator of that development (Dworkin 1988). The historical reasons are well known, with the abuse of medical authority in Nazi concentration camps the need for more regulations in medicine due to the development of biomedicine and the establishment of liberal societies being some factors worth mentioning.[1] In any case the ethos of informed consent replaced the Hippocratic ethos. Therefore bioethics was mainly concerned with problems that arise when the protection of the autonomy of the patient is of central importance, e.g. how to deal with those who are not able to consent. Furthermore, attention was drawn to the issue of how to protect the most vulnerable groups of society from the unintended side-effects of informed consent, e.g. disabled people who are in danger of being the victim of new discrimination by the developments of biomedicine. A significant number of debates in bioethics are concerned with regulatory problems regarding the implementation of a liberal ethos under complex circumstances. We find the moral conviction concerning our self-determination with regard to our own body at the core of a lot of bioethical declarations. And in general the protection of human dignity and human rights is interpreted in terms of individual self-determination and informed consent.[2] This implies that our body and our needs and desires, are in a distinct sense morally neutral.

Of course these observations concerning informed consent have to be described in much more detail. One relevant distinction shall be mentioned here: In recent years many critical remarks have been made with regard to informed consent. These criticisms are related to problems regarding the implementation of informed consent in practice (a lot of bureaucracy etc.), the problem of how people can be made competent enough to be able to give informed consent, and the overwhelming possibilities of medical choices to the problem of how to deal with informed consent from a public health perspective. In this context it is important to distinguish between the *value of patient autonomy* that forms the evaluative basis of what has to be morally protected and *informed consent as an instrument* of this protection. Informed consent is implemented as a mode of protection and not as a moral value as such. Morally there is the value of the self-determination of the patient that deserves protection. Under regulatory circumstances 'informed consent' became so important because it seemed to be the most appropriate way of implementing this protection, but if the practice of informed consent is criticized nowadays, a distinction has to be made between the criticism directed against the moral values behind the idea of informed consent and the criticism of informed consent as an appropriate instrument of protection.

3 The Liberal Ethos and the Developments in the Life Sciences

This liberal ethos came under pressure for several reasons. It is important to note that the liberal perspective is criticized by many authors beyond just those who have fundamental objections to the liberal ideas, such as Hegelians, communitarians and feminists. In the last decades the ethical debate has been accompanied by a criticism of the whole modern project. In old Europe the whole project of liberal ethics was especially criticized in relation to its metaphysical presumptions which became suspect in the context of the 'dialectics of enlightenment'[3] and in Heidegger's criticism of metaphysics (Heidegger 1953). Another prominent example is Elisabeth Anscombe's criticism of 'modern moral philosophy' (Anscombe 1958). Anscombe's criticism is precisely related to the problem of moral neutrality concerning human needs, desires and the human body. Anscombe argues that Kant, Hume, Hobbes and Mill, despite all their differences, commonly assume that our ideas of justice or moral obligations have to be legitimated by an idea of the independent individual and that for this legitimization no recourse to a substantial psychology or anthropology is necessary or possible. The birth of the modern moral subject, then, was accompanied by a destruction of the historically grown sources of traditional morality.[4] This kind of criticism does not focus on specific developments of the liberal world, but criticizes the whole idea of autonomous self-determination of the individual as a starting point of moral consideration. In consequence, the whole idea of human dignity and human rights came under pressure.

In this context we are not focussing on fundamental anti-modernist thinking or the internal dialectics of modernity. I focus on problems concerning the liberal framework by considering the development of the life sciences. So, even if we accept the concept of freedom of decision-making about our own body and about ourselves as the core of our moral convictions, there are several reasons why we can doubt whether this is enough to provide answers to crucial questions in the actual bioethical debates. I want to mention here some examples of discussions in which I think a liberal position seems to be insufficient.

First, the liberal ethos seems only to be concerned with the application of new technologies for the individual, but there are a lot of ethical discussions about the question of *whether certain biotechnological methods should be developed at all*, e.g. cloning. Since the existence of such a technology will affect the life of everybody, we cannot simply delegate such questions to the decision making of the individual. In several cases, the very existence of technologies in the life sciences is of crucial importance to our self-perception, as well as to the scope of possible actions. It seems that the liberal ethos is blind to the development process of new technologies and only wants to protect the individual against technologies that already exist, but given the impact of these technologies on our lives one could expect some kind of ethical standard that shows us why the development of those technologies should be accepted by everybody. Moral evaluations of the development of new technologies often seem to presuppose an evaluation of those needs that are the reason for the development of those technologies. If reproductive medicine is legitimated to create possibilities of assisted fertilization, the desire to have a child of one's own seems to be a legitimate reason to develop such a technology. Another judgment of this kind would be to say that the desire of a couple to clone a baby to replace a lost child is a problematic desire. So, it seems that for the moral evaluation of the development of new technologies, a value judgment concerning human desires and capacities is necessary.

Second we have to evaluate desires and needs when deciding whether or not a *technology should be offered*. Is the desire for a smaller nose a sufficient reason to offer *cosmetic surgery*? If we reject that this desire is a legitimate reason, we will perhaps argue that it is not a therapeutic measure against an illness. Since the distinction between health and illness is not only a biological one, we will need a stronger means of evaluation in order to make use of this distinction. If, in the case of cosmetic surgery, we adopt a liberal solution, then we need a *value judgment to justify whether or not insurance companies* must pay for the treatment. This issue became more important in the context of the new debate about enhancement technologies (Parens 2000; Presidents Council on Bioethics 2003). The scope of technologies where the medical need is in doubt is continuously expanding. Drawing distinctions regarding the legitimacy of different needs or desires seems to be unavoidable.

Third we have to make some general evaluations about the human body if we want to prevent modern technologies from even being used to produce disabilities. A very striking example is that of a *deaf couple that have the desire to have a child that is also deaf*. Can their desire justify the use of pre-implantation genetic diagnosis (PGD)? The deaf couple can refer to the fact that deafness is accepted as a specific way of living, as a 'culture of the deaf' in its own right. If we accept the use of PGD for that purpose, we accept that medical technologies are used for the fulfilment of the wishes of parents and are used to produce a disability. If not, we have to answer the question of what makes the desire of the deaf couple so different from the desire to have a so-called healthy child. How is such a judgment possible without a substantial evaluation of the relevant desires and wishes?

Last but not least, the discussions about *priorities in the healthcare system* force us to extend the scope of necessary valuations in moral judgments. If we argue for priorities in the health-care system we seem to presuppose a *non-arbitrary and substantial notion of human flourishing* that guides the formation of a hierarchy of important goods. We can, of course, delegate this solely to the market, but this means that we make no attempt at all to offer a moral legitimization for these priorities.

4 Bioethical Approaches and a Substantial Criterion

There are seemingly several discussions in bioethics where both the ethos of autonomous decision making and the delegation of decision processes to the individual patient are insufficient for an ethical evaluation. This 'insufficient' is not meant in a kind of intuitionist sense that the consequences are not meeting our moral intuitions. That would not be a strong argument since there may be reasons to reconsider those intuitions: they may be wrong. 'Insufficient' here means that this ethos is unable to formulate moral judgments on the basis of its own resources. It will have to refer to other kinds of belief, such as the presupposition that scientific progress is a value in itself or a belief that modern institutions will have the capacity to deal with the side effects of new technologies. These convictions may be true, but they go beyond the argumentative capacity of the liberal ethos itself.

Up to now we have not been given a reason to think that the liberal ethos is wrong. We have only seen that we are not able to give a moral judgment in several bioethical discussions without making value judgments concerning our bodies and our needs and without making some anthropological assumptions. If we take this tension between the modern ethos and the need for those evaluations into account, it seems to be unavoidable that we must find an evaluation criterion that allows for moral evaluations of needs and desires to a degree such that we can hope to find answers in these bioethical discussions. Also, the criterion must be (justifiably) universal or general. I will call this a '*substantial criterion*' because it must have enough content for a moral evaluation of needs and capacities. This does not mean that the criterion has to determine the value of specific needs and capacities from *all* perspectives. It is possible that some capacities are of value to me only insofar as I want to reach a specific goal in life. Such a capacity would be valuable in the context of a specific idea of a good life. A substantial criterion does not have to, or perhaps even should not, imply a sufficient determination of the value of needs and capacities from the perspective of, say, a perfectionist ethics (Hurka 1993). A 'substantial criterion' will only help us to determine whether such needs and capacities are morally valuable in the sense that it is morally obligatory, permissible or prohibited to develop specific capacities or to fulfil specific needs. Whether such a moral criterion is available is not obvious and is not self-evident. It is not self-evident that such a criterion can be formulated to an extent that is concrete enough for our bioethical debates, but these are the questions that should be discussed in detail.

Before we have a look at concrete proposals to formulate such a criterion, some remarks concerning the need for it are necessary. The mainstream of bioethical approaches will try to avoid formulating a substantial criterion. This will be the case, for example, in a *procedural approach* like discourse ethics. In concrete bioethical discourses there are only two options open to such an approach: Either the evaluation of our needs and capacities will be an issue discussed in the discourse, in which case we necessarily have to ask on what the search for an evaluation criterion can be based, or a substantial criterion with regard to the preconditions of the procedure as such will have to be found, e.g. by arguing that only those capacities that are compatible with the preconditions of the discourse procedure are morally valuable. In the first option the whole question of the legitimization of a substantial criterion would reappear in the context of the discourse, in the second option the approach would propose a substantial criterion of its own.[5]

Similar things could be said concerning the theories of *prima-facie duties or mid-level principles, like those of Beauchamp/Childress* (Beauchamp and Childress 2001; Düwell 2006a). We would need a substantial criterion in the application of such an approach. Judging the relative relevance of the four principles regarding the urgency and importance of prima-facie duties already presupposes that we have some value standards in the background. The justification of this evaluative standard is precisely the issue in question. Theories like *casuistry, particularistic or contextualistic approaches* are in a similar situation.[6] In their attempt to stick to concrete, practical debates, systematic explication and justification of criteria and principles is avoided. The evaluative assumptions implicitly play a role here, but in a completely unclear and uncontrolled way.

The urgency for a justification of a substantial criterion has even *increased in the development of bioethical debate.* Casuistry and principlism are methods for bioethical debates that could be successfully applied in contexts where established practices exist and where standards of good practice are generally accepted. If we are in a clinical setting where we have no doubt about what a good doctor has to do, we can try to transfer the characteristics of accepted evaluations in one case to other more complex cases. The characteristics of the bioethical debate, however, are increasingly changing with regard to the kind of questions that are at stake. We are confronted with discussions that do not concern extreme medical cases. Bioethics is confronted with the challenge that the *creation of the relevant knowledge and the development and the perspectives of the technologies as such require moral evaluation*. Bioethics has to face the fact that newly developed technologies can be implemented in very different cultural contexts and that the development of the life sciences is accompanied by a lot of unforeseen implications. The *challenge of biomedicine* is not primarily that bad people can use technologies for bad aims or that some technologies are inherently bad; rather, the central challenge is that we are faced with substantial research activities with increasingly global importance and a strong impact on our lives while we do not know what the possibilities of technological applications may be in the future. We do know that the applications will be in very different cultural, political and institutional contexts. Taking this into consideration, it seems inadequate to use methods for evaluating such developments in the life sciences that depend on moral convictions that are shared only in specific cultural and political circumstances. To meet this challenge, bioethics has to become moral philosophy in a much more fundamental sense and that implies what I have called the discussion about a substantial criterion.

5 Capabilities Approach

A promising proposal is offered by the '*capabilities approach*,' a notion that is used in very different contexts. There is some intuitive plausibility to the claim that capabilities necessary for human flourishing should be protected and supported. This notion seems especially fruitful for bioethics, since it seems to be intrinsically linked to the *idea of medicine*. Medicine should help people in cases where fundamental capabilities are endangered. This notion is furthermore closely linked to the idea of '*human rights*.' Rights protect fundamental capacities or empower us to develop and enlarge the spectrum of relevant capabilities. Determining which capacities are relevant for human beings to be able to discuss the content of human rights is necessary. Each concept of rights presupposes an idea about what kind of capacities are endangered. If we assume that we should have freedom of speech, for example, we presuppose that free communication and articulation are valuable and can be in danger. If we enlarge the spectrum of assumed rights, we suppose that the potential threats are changing. This has happened in recent decades in the debate on the environment. We realized that the destruction of nature could destroy the preconditions of our existence and that the scope of moral obligations had to be extended. Furthermore, we are in general presupposing that others, society and political institutions, should respect our fundamental freedoms, but we also assume that the community should support us, at least to some degree, in the development of our capacities. We think that children worldwide have a right to education and we think that people with disabilities or people in difficult social situations should be supported, at least to some degree. The entire *propaganda for the development of biomedicine* makes the assumption that governments have a moral obligation to use a significant amount of their available budgets for the development of biomedicine. We presuppose that there is a strong obligation to make large efforts to make people's lives easier, to provide the possibilities for medical treatments against Alzheimer's, etc. *There is some intuitive plausibility to the idea that we are morally obligated to protect and support the capabilities of human beings* and there is some plausibility in the assumption that this protection of human capabilities should be the central criterion for deciding which moral obligations we have – a criterion that could be used to solve moral conflicts.

Amartya Sen has used the notion of 'human capabilities' as a criterion to measure and compare the welfare level of different societies.[7] He seems to presuppose that we are morally committed to support equal life standards and that the notion of capabilities should function as an instrument for this measurement of the standard of equality. However, he does not tell us very much about why we have an obligation to protect human capabilities or why that is morally significant at all. *Martha Nussbaum* has in several books presented a list of human capabilities that are, she argues, essential for human existence. She further argues that the preconditions to develop those capabilities should be guaranteed in the entire world.[8] She internally links this notion of capabilities to the notion of rights. Everybody has a right to the preconditions for the development of his or her capabilities. What is, in several respects, unclear in Nussbaum's approach is the internal structure of the list: *First of all*, she does not tell us why we have an obligation to act according to these capabilities. In her Aristotelian view, the essential importance of some capacities for a 'human' existence is sufficient for her to assume their moral importance, but since she defends the claim that the capabilities ground the notion of rights, much more argumentation is needed. She has to defend why, in a strong sense, it is immoral not to act according to those capabilities and why everybody can expect others to act in this way. *Secondly*, there is no clear structure to her list of capabilities. Nussbaum mentions that the list should not be seen as a description of necessary preconditions of a good life but as a survey of the relevant aspects of a flourishing life. We cannot imagine a good life in which these aspects are not fulfilled, but are all of these capabilities of the same importance? If they are, then the list is so broad that it is unhelpful as a guide to resolving conflicts between different capabilities. So, if such a capabilities approach should have more than merely heuristic importance there should be more than a checklist of things that are – more or less – morally relevant and if the list really should provide us with a moral criterion, there should be an argument about a kind of non-arbitrary hierarchy in the list. Within an Aristotelian framework, however, the basis of such a hierarchy is not clear.

Besides the problem of the normative force of the capabilities and the internal hierarchy of the list, there should be some answer to the question of whether the protection of human capabilities is an exclusively moral criterion. In many bioethical discussions it is important to know whether we are only obliged to protect human capabilities or whether the protection of animals and the environment is of moral importance in itself. Nussbaum's new book *Frontiers of Justice* is concerned with questions such as these. Nussbaum wants to show that the capabilities approach has some strength in comparison to contract theories, especially when we are dealing with the question of to what extent vulnerable people in poor countries, disabled people and animals have to be taken into moral consideration. In *Frontiers of Justice* she holds that the capabilities approach is to be interpreted in the line of Rawls's political liberalism without the limitations of a contract theory. She defends a concept of dignity without relating it to Kant's concept of personhood. This concept of dignity is, however, an Aristotelian concept of dignity (Nussbaum 2006: 159–160) that defines the dignified life in terms of fulfilment of the necessary capabilities. Here Nussbaum defends the claim that *animal capabilities* must be protected analogously to human capabilities. She even defends the position that human dignity is only our species-specific form of dignity, whereas for horses a kind of horse dignity has to be supposed. That this concept has very much to do with Aristotle is not very likely, but how far one can use the concept of dignity in the Kantian sense of prohibition of treating entities that have dignity as means only, without referring of notion of personhood is also unclear. For Kant the fact that the rational being is capable of setting goals for him/herself is the central reason why we should not treat those rational beings as means only. It is furthermore unclear what the normative implications are of being an entity with dignity. In the chapter about animals we learn that sterilizing or killing animals is allowable, if this seems to be appropriate (ibid.: 371). Thus a systematic explication of what moral consequences flow from being a dignified being is missing. Furthermore an explication of why animals, and not plants and landscapes, should have such a moral status is also lacking.[9] Concerning moral status Nussbaum writes: 'Instead, we should adopt a disjunctive approach: if a creature has *either* the capacity for pleasure and pain *or* the capacity to movement from place to place *or* the capacity for emotion and affiliation *or* the capacity for reasoning, and so forth (we might add play, tool use, and others), then that creature has moral standing.' (ibid.: 362). Why we should give moral consideration to all those creatures, what the normative implications are if a being has such moral status and whether or not all of these moral statuses are of the same importance is very unclear. *Frontiers of Justice* has made unclear what seemed to be clear in the capabilities approach up to now.

6 Kantian Perspectives

My objections are not motivated by as negative an attitude against the capabilities approach as it may seem. My aim was to make a brief inventory of the problems that have to be answered before the notion of human capabilities can function as a moral criterion; and answers – I think – will not be found within an Aristotelian framework. If human capacities have to be protected because they are necessary for human flourishing, then it is not clear from where the normative force comes. How can the fact that one needs some goods in order to live a flourishing life stand as a reason to expect that others have the obligation to give these goods to me? We have to think of the fundamental interrelation among (a) the relevance of some capacities to us (b) our vulnerability and dependency on others for protection, and (c) moral obligation. We can only claim that everybody should act in accordance with the protection and the support of our fundamental capacities if we can show that there is a fundamental interrelationship between those capacities and the sources of moral normativity. Our agency is the reason that there can be moral obligations at all. Without the ability of human beings to take the interests of others into account there would be no moral obligation whatever. Human capacities are of central moral importance only because human beings, as bodily existing beings, are vulnerable and depend on the protection of others and, at the same time, are capable of action and capable of morality. Only if we are able to keep in mind the relation between moral obligations, our vulnerability and our agency as a *necessary* interrelation, can we have a reason to think that the protection of human capabilities has normative force.

In this line of argument one should expect that the Kantian notion of human dignity can provide us with a perspective from which to develop a substantial criterion.[10] If we relate human capabilities to human agency as the source of moral obligations, we find an evaluative standard that allows for a hierarchical perspective on our capacities. Not all human capacities that are important for a good life are on the same level, but the necessity of them for the possibility of being a moral agent will be the angle from which the importance of capacities has to be judged. The evaluative standard will then not be in concurrence with the basic liberal idea that self-determination with regard to our body is the starting point of a moral ethos. On the contrary, it will show that our existence as an autonomous agent is the standard that has to be protected. That approach, however, will deny that the consequence of such an ethos of the autonomous agent is that all bioethical debates can be solved by referring to the decision of the agent and that the freedom of informed consent is the core of the goods that have to be protected through bio-political measures. If we have to protect the conditions of agency for everybody first of all, we have to realize that there are some measures that are more urgent than others and this will provide us with some hierarchy of relevant goods. In several respects, the application of such a measure will not differ very much from Nussbaum's version of a capabilities approach. For example, I agree with Nussbaum that having the freedom to play, having the possibility to have aesthetic experiences and so on is important for our existence as moral agents, but I think we have to argue for the normative force of these judgments and we have to establish a hierarchy of morally relevant goods if such a concept is to play a relevant role in bioethics.

Notes

1 Jonsen 1998; Düwell and Neumann 2005.

2 See Beyleveld and Brownsword 2001.

3 It is worthwhile to mention that at least Adorno's criticism of the modern subject criticizes the raping of the subject in modern society through the idea of abstract rationality, but the idea of the emancipation

of the individual is always the normative basis of his criticism. Especially in his *Minima Moralia* one can see that he even criticizes the modern myth of the individual by referring to the value of individualism (Adorno 1951).

[4] Also see the discussion of Anscombe's approach in O'Hear (2004). A similar criticism can be found in MacIntyre (1981).

[5] In the debate between Habermas and Apel an discourse ethics it seems that Habermas chose the first strategy while Apel initially argued for the second strategy (Apel 1976). In his new book on bioethics however, Habermas's idea concerning the dignity of humankind includes something like a substantial criterion (Habermas 2001).

[6] Arras 1999; Dancy 2004; Jonsen and Toulmin 1988; Jonsen 2005; Steigleder 2003; Willigenburg 2005.

[7] He first introduced this notion in Sen (1980).

[8] Nussbaum 1988, 2000a. The most recent version of the list can be found in: Nussbaum 2006:76–78.

[9] Balzer et al. (1998); Düwell (2006a); Taylor (1986); Warren (1997).

[10] In that line: Gewirth 1978, 1996; Steigleder 2002.

References

Adorno, Theodor W. *Minima Moralia: Reflexionen aus dem beschädigten Leben*. Frankfurt a.M.: Suhrkamp, 1951.

Anscombe, Elisabeth. "Modern Moral Philosophy." In *Philosophy*, 33, 1–19, 1958.

Apel, Karl-Otto. *Transformation der Philosophie*. 2 Vols. Frankfurt a.M.: Suhrkamp, 1976.

Arras, John D. "A Case Approach." In *A Companion to Bioethics*. Helga Kuhse and Peter Singer (eds.). Oxford: Blackwell, 106–114, 1999.

Balzer, Philipp et al. *Menschenwürde vs. Würde der Kreatur: Begrifsbestimmung, Gentechnik, Ethikkommissionen*. Freiburg i.Br.; München: Alber, 1998.

Beauchamp, Tom L. and Childress, James F. *Principles of Biomedical Ethics*. 5th ed. New York; Oxford: Oxford University Press, 2001.

Beyleveld, Deryck and Brownsword, Roger. *Human Dignity in Bioethics and Biolaw*. Oxford: Oxford University Press, 2001.

Dancy, Jonathan. *Ethics Without Principles*. Oxford: Clarendon Press, 2004.

Düwell, Marcus and Neumann, Josef N. "Medizin- und Bioethik: Geschichte und Profile." In *Wieviel Ethik verträgt die Medizin*? Marcus Düwell and Josef N. Neumann (eds.). Paderborn: Mentis, 13–49, 2005.

Düwell, Marcus. "One Principle or Many?" In *Bioethics in Cultural Contexts: Reflections on Methods and Finitude*. Christoph Rehmann-Sutter, Marcus Düwell and Dietmar Mieth (eds.). Dordrecht: Springer, 93–108, 2006a.

Dworkin, Gerald. *The Theory and Practice of Autonomy*. Cambridge: Cambridge University Press, 1988.

Gewirth, Alan. *Reason and Morality*, Chicago; London: University of Chicago Press, 1978.

Gewirth, Alan. *The Community of Rights*. Chicago; London: University of Chicago Press, 1996.

Habermas, Jürgen. *Die Zukunft der menschlichen Natur: Auf dem Weg zu einer liberalen Eugenik*? Frankfurt a.M.: Suhrkamp, 2001.

Heidegger, Martin, *Einführung in die Metaphysik*. Tübingen: Niemeyer, 1953.

Hurka, Thomas. *Perfectionism*. New York; Oxford: Oxford University Press, 1993.

Jonsen, Albert and Toulmin, Stephen. *The Abuse of Casuistry: A History of Moral Reasoning*. Berkeley: University of California Press, 1988.

Jonsen, Albert. *The Birth of Bioethics*. New York; Oxford: Oxford University Press, 1998.

Jonsen, Albert. "Casuistical Reasoning in Medical Ethics." In *Wieviel Ethik verträgt die Medizin*? Marcus Düwell and Josef N. Neumann (eds.). Paderborn: Mentis, 147–164, 2005.

MacIntyre, Alasdair. *After Virtue*: *A Study in Moral Theory*. Notre Dame: University of Notre Dame Press, 1981.

Nussbaum, Martha C. "Nature, Function, and Capability: Aristotle on Political Distribution." In *Oxford Studies in Ancient Philosophy*. Supplementary Volume, 145–183, 1988.
Nussbaum, Martha C. "Aristotelian Social Democracy." In *Liberalism and the Good*. Bruce Douglas and Gerald Mara (eds.). New York: Routledge, 203–252, 1990.
Nussbaum, Martha C. *Women and Human Development: The Capabilities Approach*. Cambridge: Cambridge University Press, 2000a.
Nussbaum, Martha C. "Aristotle, Politics, and Human Capabilities: A Response to Antony, Arneson, Charlesworth, and Mulgan." In *Ethics*, 111, 102–140, 2000b.
Nussbaum, Martha C. *Frontiers of Justice: Disability, Nationality, Species Membership*. Cambridge Mass.; London: The Belknap Press of Harvard University Press, 2006.
O'Hear, Anthony (ed.). *Modern Moral Philosophy*. Cambridge: Cambridge University Press, 2004.
Parens, Erik. *Enhancing Human Traits: Ethical and Social Implications*. Washington DC: Georgetown University Press, 2000.
Presidents Council on Bioethics. *Beyond Therapy: Biotechnology and the Pursuit of Happiness*. New York; Washington DC: Harper Collins, 2003.
Sen, Amartya. "Equality of What?" In: *The Tanner Lectures on Human Values*. Vol. 1. Sterling McMurrin (ed.). Salt Lake City: University of Utah Press, 195–220, 1980.
Steigleder, Klaus. *Kants Moralphilosophie: Die Selbstbezüglichkeit reiner praktischer Vernunft*. Stuttgart: JB Metzler, 2002.
Steigleder, Klaus. "Kasuistische Ansätze in der Bioethik." In *Bioethik – Eine Einführung*. Marcus Düwell and Klaus Steigleder (eds.). Frankfurt a.M.: Suhrkamp, 152–167, 2003.
Taylor, Paul W. *Respect for Nature: A Theory of Environmental Ethics*. Princeton: Princeton University Press, 1986.
Warren, Mary Anne. *Moral Status: Obligations to Persons and Other Living Things*. Oxford: Oxford University Press, 1997.
Willigenburg, Theo van. "Casuistry in Medical Ethics." In *Wieviel Ethik verträgt die Medizin*? Marcus Düwell and Josef N. Neumann (eds.). Paderborn: Mentis, 165–178, 2005.

Further Literature

Düwell, Marcus. "Moralischer Status". In *Handbuch Ethik*. 2nd ed. Marcus Düwell, Christoph Hübenthal and Micha H. Werner (eds.). Stuttgart: JB Metzler, 434–439, 2006b.
Höffe, Otfried. *Demokratie im Zeitalter der Globalisierung*. München: CH Beck Verlag, 1999.
Kant, Immanuel, *Fundering voor de metafysica van de zeden*. Thomas Mertens (transl.). Amsterdam: Boom, 1997.
Krämer, Hans. *Integrative Ethik*. Frankfurt a.M.: Suhrkamp, 1992.
Nussbaum, Martha C. and Sen, Amartya (eds.). *The Quality of Life*. Oxford: Clarendon Press, 1993.
Pogge, Thomas. "Can the Capability Approach be Justified?" In *Philosophical Topics*, 167–228, 2002a.
Pogge, Thomas. *World Poverty and Human Rights*. Cambridge: Polity Press, 2002b.

Moral Judgement and Moral Reasoning

A Critique of Jonathan Haidt

Albert W. Musschenga

> *[. . .] what readers can hope to achieve after working through the book.*
>
> - *They should have improved their reasoning skills (such as identifying and evaluating reasons, conclusions, assumptions, analogies, concepts and principles), and their ability to use these skills in assessing other people's arguments, making decisions and constructing their own reasoning. [. . .]*
> - *They may have strengthened certain valuable tendencies in themselves – to reason, to question their own reasoning and to be fair minded*
>
> *(Thomson 1999: 3).*

One of the courses I offer to Ba-students in philosophy is on ethical theory and moral reasoning. The goal of this course is that students learn what moral reasoning is and what the relevance of ethical theories is for moral reasoning. They also need to make the exercises from Anne Thomson's *Critical Reasoning in Ethics*. At the end of the course they have to write, as a test, a paper in which they set up a moral argument on a certain subject. I hope, of course, that they not only learn some reasoning skills but will also improve their moral judgements. I assume that reasoning skills contribute to taking better and more justifiable standpoints. I also assume that reasoned judgements will be translated into action. Am I right?

Most psychologists agree that there are two types of cognitive processes or 'reasoning systems'. Roughly, one system is associative and its computations reflect similarity and temporal structure; the other system is symbolic, and its computations reflect a rule structure (Sloman 1996). Stanovich and West labelled these systems or types of processes 'System I' and 'System II' (Stanovich and West 2000). There is now considerable agreement on the characteristics that distinguish the two systems. The operations of System I are fast, automatic, effortless, associative, and difficult to control or to modify. The operations of System II are slower, serial, effortful, and deliberately controlled; they are also relatively flexible and potentially rule-governed. The perceptual system and the intuitive operations of System I generate *impressions*

A.W. Musschenga
e-mail: aw.musschenga@mdw.vu.nl

M. Düwell et al. (eds.), *The Contingent Nature of Life*,

of the attribute of objects of perception and thought. These impressions are not voluntary and need not be verbally explicit. In contrast, *judgements* are always explicit and intentional, whether or not they are overtly expressed. The label 'intuitive' is applied to those judgements that directly reflect impressions. As in several other dual-process models, one of the functions of System II is to monitor the quality of both mental operations and overt behaviour (Kahneman 2003: 1450–52).

Recent studies show that most of our judgements are not simply the outcome of conscious – System II – reasoning. To a large extent, they are intuitive and automatic – System I – responses to challenges, elicited without awareness of underlying mental processes (Bargh 1996; Bargh and Chartrand 1999). Moreover, people are often not very adept at describing how they actually reached a particular judgement (Nisbett and Wilson 1977). In his by now famous article 'The Emotional Dog and Its Rational Tail: A Social Intuitionist Approach to Moral Judgment', Jonathan Haidt extends these findings to the area of moral judgements (Haidt 2001). Haidt thinks that especially philosophers and moral psychologists working within the rationalist (Kantian) tradition overestimate the causal role of formal reasoning – of System II – in moral judgement (815). He argues that moral judgements, in addition to being largely intuitive, typically amount to post hoc reasoning with a defensive character after a judgement has been made (818f.). Moral reasoning is similar to the reasoning of lawyers who construct justifications for antecedent intuitive judgements (820f.). People may at times reason their way to a judgement by sheer force of logic, overriding their initial intuition. However, such reasoning is rare (819). Moreover, referring to sources such as Blasi's review on the literature on moral cognition and moral action (Blasi 1980), Haidt states that the relation between moral reasoning and moral action is much weaker than that between moral emotion and moral action (823f.).

Haidt's approach of moral judgements is descriptive and explanatory. He is well aware of the limits of this approach. That is why he stresses that his view on how moral judgements are made is not a claim about how they should be made (815). He also knows that we cannot always trust our intuitions. Haidt does not deny that deliberate reasoning takes place and is necessary. He says it is rare. If the only point of disagreement between Haidt and moral philosophers would be the frequency of deliberative reasoning, more and refined empirical research should prove who is right. However, Haidt is also sceptical about the power of deliberative reasoning. First, its power to correct intuitive judgements is limited, since moral reasoning is largely post hoc, biased and not objective (821ff.). Second, its impact on moral action is weak, much weaker than that of moral emotion (823f.). If Haidt is right in his view on moral reasoning, many moral philosophers, especially those standing in the Kantian tradition, should revise their belief in the role of deliberative reasoning in producing moral judgement and in its power to influence action. Haidt is not only an empirical social psychologist, he is also, as many psychologists who study morality, a Humean in his view of the role of reason. It is not clear to what extent his views on the role and the power of reasoning are corroborated by his empirical work.

My objective in this paper is to critically examine Haidt's view that moral reasoning is post hoc, biased and unable to actually motivate action. In Section 1 I start with summarising Haidt's views. To clarify how much they are in need of correction, I chart in Section 2 the reliability of intuitive judgements. I argue, in Section 3, that Haidt does not do justice to the actual role of moral reasoning. In Section 4 I discuss how we can counteract the weaknesses in the reasoning process signalled by Haidt. In Section 5 I examine Haidt's views on the relation between moral reasoning and moral action.

1 Haidt's Social Intuitionist Model

Before giving a quick sketch of Haidt's theory of moral judgement and moral reasoning which he calls 'the social intuitionist model' (SIM), I summarise the definitions of his central concepts. Moral intuition he defines as '... the sudden appearance in consciousness of a moral judgment, including an affective valence (good–bad, like–dislike), without any awareness of having gone through steps of searching, weighing evidence, or inferring a moral conclusion. Moral intuition is therefore the psychological process that the Scottish philosophers talked about, a process akin to aesthetic judgment. One sees or hears about an event and one instantly feels approval or disapproval' (818). His definition of moral judgements is: '... evaluations (good vs bad) of the actions or character of a person that are made with respect to a set of virtues held to be obligatory by a culture or a subculture' (817). Moral reasoning is defined as '... conscious mental activity that consists of transforming given information about people in order to reach a moral judgement. To say that moral reasoning is a conscious process means that the process is intentional, effortful, and controllable and that the reasoner is aware that it is going on' (818). Note that Haidt does not mention states of affair as objects of moral judgement and that in his view moral judgements are always relative. Note also that for Haidt moral reasoning is always a conscious mental activity.

Haidt's SIM is composed of four principal links or processes the existence of which, according to him, is '... well established by prior research in some domains of judgment, although not necessarily in the domain of moral judgment'. (Haidt 2001: 818f.). The first is the *intuitive judgement link*, already sufficiently described above. The intuitive process is, according to Haidt, the default process. The second one is *the post hoc reasoning link*. Moral reasoning is an effortful process, in which a person searches for an argument that will support an already-made judgement. The third link is *the reasoned persuasion link*. Moral reasoning is produced and sent forth verbally to justify one's already-made judgements to others. Such reasoning, Haidt says, can sometimes influence other people, although moral discussions and arguments are notorious for the rarity with which persuasion takes place. Haidt hypothesises that reasoned persuasion works not by providing logically compelling arguments but by triggering new affectively valenced intuitions in the listener. The *social persuasion link* is the fourth link. The mere fact that friends, allies and acquaintances have made a moral judgement exerts, according to Haidt, a direct influence on others, even if no reasoned persuasion takes place. These four links constitute the *core* of SIM. The core of this model gives moral reasoning a causal role in moral judgement *but only when reasoning runs through other people*. SIM posits that moral reasoning is usually done interpersonally rather than privately. The *full* model includes two other links. The fifth one is *the reasoned judgement link*. Moral reasoning occurs when intuitions conflict or when the social situation calls for thorough examination of all the facets of a scenario (820). People may at times reason their way to a judgement by sheer force of logic, overriding their initial intuition. However, such reasoning is, according to Haidt, rare. The sixth link is *the private reflection link*. Thinking about a situation a person may, e.g. by role taking, spontaneously activate a new intuition that contradicts the initial intuitive judgement. Private reflection is also rare.

2 The Reliability of Intuitions

Intuitive reasoning makes use of heuristics – mental short-cuts or rules of thumb. Heuristic operate through a process of attribute substitution (Kahneman and Frederick 2002). They are used when people are interested in assessing a 'target attribute' and when they substitute a 'heuristic attribute' of the object, which is easier to handle. The use of heuristics gives rise to intuitions about what is true or right. Well-known are the heuristics that reduce the complex tasks of assessing probabilities and predicting values to simpler judgemental operations (Kahneman and Tversky 1974: 1124). These heuristics usually work well, but may sometimes lead to severe and systematic errors. Kahneman and Tversky investigated the biases and the errors in probability assessments.

Only recently researchers have started to investigate the possible errors caused by intuitions in the moral and political domain. Baron conducted extensive research on intuition and error in public decision-making (Baron 1994a, 1995). He showed that people following their moral intuitions may generate nonoptimal or even disastrous consequences. There is still little work done on the errors produced by moral heuristics. Moral heuristics often represent generalisations from a range of problems for which they are well-suited (Baron 1994b). According to Sunstein, moral heuristics become a problem when they are wrenched out of context and treated as freestanding or universal principles, applicable to situations in which their justifications no longer operate (Sunstein 2005). Sunstein points to a number of heuristics that lead to errors. He distinguishes four categories of heuristics: those that involve morality and risk regulation, those that involve punishment, those that involve 'playing God' – particularly in the domains of reproduction and sex, and those that involve the act-omission distinction. For reasons of space, I will only mention two heuristics. We condemn people who knowingly engage in acts that may or will result in human deaths. At the same time, we do not disapprove of people who, e.g., fail to improve safety measures while believing that there is a risk but appearing not to know for certain that deaths will ensue. Sunstein suggests that a moral heuristic makes us condemn the former people. Another heuristic which Sunstein suggests is 'Do not play God' or, in secular terms 'Do no tamper with nature'. This heuristic might explain the wide-spread repugnance against e.g. cloning.

Let us assume that we do use moral heuristics in our everyday moral judgements. Contrary to heuristics in probability assessments, there is no neutral, theory-independent standard for determining when the use of moral heuristics leads to an judgemental error. Baron derives his standards from utilitarianism. Sunstein does not really discuss this question. His main point is that the presence of moral heuristics can be accepted by people of diverse general (ethical) theories.

3 The Role and Nature of Moral Reasoning

I think that most moral philosophers have no problem in accepting Haidt's findings on the automaticity of moral judgements and on the prevalence of biases in everyday moral reasoning. More controversial is his downplaying of the role of reasoning. Haidt's view on moral reasoning can be unpacked into four related statements: (1) moral reasoning is usually post hoc and defensive ('lawyers' reasoning'), (2) only when reasoning runs through other people, moral reasoning has a causal role in moral judgement, (3) deliberative reasoning and reasoned judgements are rare, and (4) moral reasoning is biased. I consider these points in this order.

(ad 1) According to Haidt's SIM, the intuitive process is the default process, which regulates everyday moral judgements in a rapid, easy and holistic way. Referring to Nisbett and Wilson (1977), Haidt states that moral reasoning is generally a post hoc construction intended to justify automatic intuitions (823). Is he right? In my view this is at best the case to a limited extent. Usually we only take a defensive stance – and 'reason like a lawyer' – when we have a firm intuition about an issue. If a pro-lifer says, at a meeting of consociates, that the destruction of superfluous in vitro fertilised eggs is murder, no one there will ask him for supporting reasons. Moral discussions generally start within a group of people who are not all like-minded. Someone who is passing a judgement, is challenged to provide supporting reasons. After exchanging arguments, this person might be able to convince his discussion partners. Or he himself might become convinced by his partners' counterarguments. However, when we engage in a moral discussion, we often start not with a firm but with a rather weak intuition. We use the exchange of opinions and reasons to test the plausibility of this intuition. If the intuition is not defendable, we drop it. Even if we do not immediately and publicly concede that our initial opinion is no longer tenable, we may change our view afterwards. Thus, contrary to Haidt, I argue that people who participate in a moral discussion not always behave like lawyers. They are often willing to become convinced of the opposite of their initial intuition.

There is still another argument against the idea that moral reasoning is post hoc. Haidt believes that many moral intuitions, e.g. sympathy, reciprocity and loyalty, are partially built in by evolution (826). Moral development then is primarily a matter of the maturation and cultural shaping of these intuitions (828). Because of its cautious formulation ('many moral intuitions', 'partially built in by evolution'), it is hard to argue against this view. Haidt does not mention that at least some of our intuitions are the product of discussion and deliberation. Reasoned judgements can become automatic. Let me give an example. Many people who are vegetarians, have been raised within families where the eating of meat was normal and morally unproblematic. These people once must have come to the conclusion that eating meat is morally wrong. In the course of time not eating meat has become a habit for them. If they are asked for their view on the slaughtering of animals, they won't have any problem in giving their judgement. Their judgement has become intuitive and automatic; it does not require deliberative reasoning. However, it is a post-reflective and not a pre-reflective intuition.

(ad 2) As we have seen, the core of Haidt's SIM gives moral reasoning a causal role in moral judgement, but only when reasoning runs through other people. He hypothesises that reasoned persuasion works not by providing logically compelling arguments, but by triggering new affectively valenced intuitions in the listener. Many people will know the experience that sometimes new intuitions pop up during a discussion. These intuitions may be triggered by what the conversation partner said, but sometimes they just emerge from the processes going on in one's mind. They are just a by-product of the discussion. Discussions often are merely exchanges of beliefs and convictions, not specifically aimed at determining whose judgements are supported by the best reasons. However, sometimes, usually when opinions conflict, we do have argumentative discussions. Imagine that you have a friend who is speciesist. She believes that we do not owe animals much moral consideration since they are amoral beings lacking (self-)consciousness, intelligence and sophisticated forms of communication. You happen to have got Frans De Waal's *Our inner ape* (2005) as a present for your birthday. This book contains a wealth of evidence which you use in the discussion with your friend. In the light of all the evidence your friend comes to the conclusion that her intuitions about animals are no longer tenable. The discussion causes her to see chimpanzees, gorillas and bonobos in a different light. She now realises that they are much more alike humans than she thought before. So conversion induced by reasoning is, in my view, possible (see also: Saltzstein and Kasachkoff 2004: 275).

(ad 3) In Haidt's view moral reasoning occurs when intuitions conflict or when the social situation calls for thorough examination of all the facets of a scenario (820). First, reasoning does not only occur when *intuitions* conflict. Reasoning also occurs when moral intuitions conflict with moral *convictions*. Moral judgements often result from a process in which stereotypic attributes are used for identification, categorising and inference. In her discussion of Haidt, Fine (2006) refers to research by Monteith and colleagues showing that people do not apply activated stereotypes to stereotyped groups if they believe that it is unacceptable (Monteith et al. 2002). She concludes from these findings that controlled cognitive processes can intervene prior to social judgement formation (2006: 91). Fine also thinks that the model proposed by Monteith et al. suggests that an individual's conscious reflection on their automatic responses may eventually lead to the successful 'automatisation' of prejudice control (2006: 91f.).

Second, it is not entirely clear what kind of situation Haidt has in mind when saying that moral reasoning also occurs when the social situation calls for thorough examination of all the facets of a scenario. The following example may clarify that. When a political party proposes to lower the income tax, a lot of people will not immediately have an intuitive judgement about it. It is always nice to have the" command over a larger part of one's income, but I would like to know how the government should compensate the lower revenues. It might lead to lowering the social benefits or an increase of tuition fees, both of which I would deplore. Lowering the income tax is a complex issue that affects the interests of many parties. However, complex multi-party issues are not the only ones that cannot be expected to be already covered by intuitions. There are also many novel issues. For example, should we try to control the processes that are responsible for ageing? Is it justifiable to invest public money in that kind of biomedical research? There are many complex and novel situations which are probably not covered by intuitions. So it is not clear why Haidt thinks that deliberative reasoning is rare. In his reply to similar critique by Pizarro and Bloom (2003), Haidt repeats that people can agonise over a discussion and can have conflicting intuitions. He recognises that deliberative reasoning takes place, but holds on to the view that it is rare. Haidt nor his critics provide evidence for their belief on the frequency of deliberative reasoning. So this issue remains undecided.

(ad 4) If a dual process model is appropriate for a theory of moral judgement, one has, according to Haidt, to specify the relationship between the intuitive processes and the reasoning processes (2001: 820). He raises the possibility that the reasoning process is the 'smarter' but cognitively expensive process that is called in whenever the intuitive process is unable to solve a problem cheaply. According to him evidence does not corroborate that deliberative reasoning is smarter. This is because two major classes of motives have been shown to bias and to direct reasoning. The first class he calls *relatedness motives*. People tend to agree with their friends and allies. Their judgements highly influence their own judgements. Desire for harmony and agreement have strong biasing effects on judgements. The second class of motives he calls *coherence motives*. The existence of these motives was established in the research on cognitive dissonance (Festinger 1957). This research showed that people try to keep their attitudes and beliefs congruent within the beliefs and attitudes that are central to their identity. They tend to ward off evidence that threatens these identity-constitutive attitudes and beliefs. This tendency leads to accepting evidence supporting their prior beliefs uncritically, while subjecting opposing evidence to much greater scrutiny (Lord et al. 1979).

According to Haidt, these two classes of motives explain why people often behave like 'intuitive lawyers' and not like 'intuitive scientists'. Let me assume, for the sake of argument, that the evidence for the existence of the relatedness and the coherence motives is conclusive. If present, these motives threaten the objectivity and openness of reasoning. Such moral reasoning can indeed, as Haidt argues, be compared to the reasoning of lawyers. It is not the prime duty of lawyers to find the truth, but to defend the interests of their clients. Lawyers indeed examine the evidence for the allegations against their clients more closely than exempting evidence.

They search for the best arguments they can find to prove that at least reasonable doubt is possible. I have no ground to deny that we often do reason like lawyers. Rather than engaging in an open discussion, we often try as long as possible to defend our antecedent standpoints. My point is that we do not and cannot always reason like lawyers. As I argued above, in many situations we do not have clear intuitions. And if we do have an intuition, we are often prepared to drop it if there are too many reasons pleading against it. Moreover, as I will argue in the next section, we need not always be the non-voluntary victims of biases.

4 Can the Biases of Moral Judgements be Corrected?

Moral intuitions are not always reliable. However, for several reasons deliberative reasoning appears, in Haidt's view, to be inappropriate for counteracting the biases and errors of intuitive judgements. One reason is that reasoned judgements rarely are able to completely replace the initial intuitions. This might be connected to the fact that reasoned judgements are seldom translated into actions. Another reason is that deliberative reasoning itself is not objective and free from biases. Haidt recommends '... to try to treat moral judgment style as an aspect of culture and try to create a culture that fosters a more balanced, reflective and fair-minded style of judgment' (2001: 829). Here he thinks of Kohlberg's 'just community schools'. Last but not least, people should get other people to help them improve their judgement. Discussions with wise and open-minded persons, and exchanges of reasons and evidence can help one trigger a variety of intuitions. Haidt thinks that, as more conflicting intuitions are triggered, the final judgement might be more nuanced and ultimately more reasonable (829). I am not sure what Haidt means by 'reasonable', but, apart from that, these suggestions seem sensible. Discussions, especially with wise, experienced and open-minded people, may help us see an issue from diverse points of view. Further research is required to examine whether these suggestions really work.

Surprisingly, Haidt also suggests that it may be possible to use SIM to get reasoning and intuition working more effectively together in real moral judgements. He even speaks of directly teaching moral thinking and reasoning skills (829). What kind of reasoning skills are required to avoid the errors and the lack of objectivity that Haidt himself signalled? Can we think of more reliable methods of reasoning than that of everyday deliberative reasoning? Is it possible to build in procedures that counteract the biases of deliberative reasoning?

Before going into these questions, we need to take a closer look into the deficiencies of everyday reasoning. As we have seen, in explaining the biases and errors of reasoning Haidt points to two classes of motives. First, the relatedness motives that induce us to agree with our friends and allies. Second, the coherence motives that make us accept evidence supporting our prior beliefs uncritically, while subjecting opposing evidence to much greater scrutiny. In their discussion of the foibles of human prediction, Bishop and Trout mention a number of other deficiencies and frailties of human reasoning, part of which are also relevant to the subject of moral reasoning (2005: 37–45). Humans are bad in reliably detecting correlations. Certain cognitive limits, including limits on memory, attention, and computation could well be implicated in the relative unreliability of social judgements. Another problem is that we tend to be overconfident about the power of our reasoning and our predictions. These problems are exacerbated because we often do not receive sufficient and accurate feedback for learning from our mistakes. On the basis of his study of relevant literature, Horton (2004) mentions still other biases. First, the vivid/pallid dimension, the tendency to be much more influenced by vivid, concrete data than by the same data, or even much more probative data, presented in a pallid or abstract way (Nisbett and Ross 1980). Second, wishful thinking, the tendency to be differentially inclined to believe what we want to be true (Trope and Liberman 1996).

If we have already so much insight into the biases and errors of human judgements, we should be able to design strategies for improving human reasoning. Drawing from research in psychology, statistics, machine learning, and Artificial Intelligence, Bishop and Trout offer prescriptions for how we ought to reason about certain sorts of problems. These prescriptions include e.g. making statistical judgements in terms of frequencies rather than probabilities, considering explanations for propositions one does not believe, ignoring certain kinds of evidence (e.g. certain selected cues that improve accuracy only very moderately, and certain kinds of impressionistic information, such as opinions stemming from unstructured personal interviews) (2005: 12, see also Bishop 2000). The research Bishop and Trout refer to has resulted in Statistical Prediction Rules (SPR) that enable us to make successful judgements in a wide range of real-life reasoning problems, e.g. the prediction of violent recidivism, and the diagnosis and prognosis of prostate cancer. Strikingly, the often very simple SPRs are often more reliable than people having large experience and training at making certain sorts of predictions (2005: 25).

Some, perhaps most, of Bishop and Trout's recommendations are only relevant for the area of predictive judgements. Since issues of evidence also play a role in moral arguments, these recommendations may be indirectly relevant for moral reasoning. Interesting is what they say about the reliability of expert judgements. Neo-Aristotelian thinkers often suggest that, in deliberating about what to think or what to do, we should ask ourselves what the 'moral expert', the phronimos, the wise and experienced person would think or decide. One could object that experts in predictive judgements cannot be compared to those who have wisdom and experience in moral judgement. However, I am not convinced that this objection is valid. Neo-Aristotelians often compare moral judgements to clinical judgements. Well-trained and experienced doctors are supposed to make better diagnoses because they are better in discerning what is relevant. Similarly, wise and experienced persons are thought to pass better moral judgements because they are better in perceiving what is morally relevant in a given situation. The empirical findings Bishop and Trout refer to, do not show that expertise does not make a difference for the reliability of predictions. They only show that using SPRs leads to judgements that are more reliable than those of experts. Predictions of well-trained and experienced people might still be more reliable than those of non-experts.

Is there anything comparable to the SPRs in moral reasoning? Baron thinks there is (Baron 1994b). He makes use of Hare's distinction between two levels of reasoning, the intuitive level and the critical level (Hare 1981). Hare argues that, in case of a conflict between intuitions (conventional rules of thumb), we should switch over to the higher, critical level of reasoning and directly apply to the critical principle of utility. Baron's solution presupposes that utilitarianism is the correct ethical theory. This view is not generally shared among moral philosophers. Others might turn for a solution to Rawls' theory of reflective equilibrium (Rawls 1971). A narrow reflective equilibrium consists in a good fit between a person's intuitions (his well-considered judgements) and a set of principles. This equilibrium is reached in a process of mutually adjusting judgements and principles. Thus, Rawls considers intuitions as revisable. A problem for this solution is that one cannot simply substitute Rawls' intuitions-as-well-considered-judgements for the psychological notion of intuitive judgement. For reasons of space, I will not further explore whether there are higher levels of critical reasoning that can correct the biases and errors of everyday moral reasoning. I will focus upon countervailing strategies that can be build into everyday moral reasoning.

Psychologists conducted research on debiasing strategies that may provide us with useful insights (Arkes 1991; Wilson and Brekke 1994). Wilson and Brekke review studies that have attempted to reduce biases in information processing and judgement. The results of these attempts are ambiguous. Some of studies have shown that awareness of biases – Wilson and Brekke speak of 'mental contaminations' – leads to their elimination, some have shown that awareness leads to undercorrection because people adjust insufficiently, some have indicated that awareness causes people to adjust too much, resulting in overcorrection, and some have shown that awareness does not cause people to adjust their responses (1994: 130). Analysing these studies, Wilson and Brekke conclude that three steps are necessary for successful debiasing. First, increasing awareness of biases. The success of attempts to increase people's awareness of biases depends in part on the extent to which researchers succeed in convincing the research participants that their judgements are indeed open to bias. Second, the studies reveal that awareness of potential bias is not sufficient. People must also be motivated to correct it. Third, some of the studies indicate that even when people are aware that information can bias them and are motivated to resist that bias, they adjust their response either too much or too little. Wilson and Brekke suggest that the reason is that they are unaware of how much they are biased and thus do not know how to alter their responses (1994: 130f.).

Horton (2004) is the only philosopher I know of who explicitly reflects upon the implications of research on bias and error in moral judgements for what he calls 'moral methodology'. He has attempted to translate insights from debiasing studies into countervailing strategies that can be deployed in the moral reasoning process. Interestingly, he notes that many of the debiasing strategies that have been shown to work, are rather obvious given the nature of the biases in question, and are already part of standard philosophical practice. The first strategy consists of taking steps to elicit thought processes that work in a direction contrary to the bias in question.

One technique is that of role-reversal, of imagining what one should think, feel or do when in the place of someone affected by one's actions and decisions. Another technique is trying to find the strongest case for claims other than one is inclined to defend. These techniques may help to counter Haidt's coherence motives and the errors associated with wishful thinking. The second strategy consists of two stages. One begins by taking a step back and surveying one's thinking about an issue with a critical eye, searching for symptoms of the biases. If symptoms are found, one then re-examines the substantive issues related to them, taking account of the possibility that one's thinking about them so far may have been affected by bias. The first strategy and the second strategy are based on the hope that, by redirecting one's thinking in certain ways, one may be able to correct any distortion caused by biases. The third strategy dispenses with this hope and is therefore a last resort, according to Horton. Instead of trying to counteract the influence of biases, one opts for a conclusion that takes account of the bias. 'One says to oneself, I find myself inclined to believe x, but I know that this is likely to be due to the influence of one or more biases. Taking this influence into account, y is more plausible' (2004: 556).

The strategies Haidt advises for counteracting biases fall under the category of 'getting other people to help you improving your judgement'. Discussions with others, especially wise and experienced others, will help us become aware of our biases and will force us to take account of views different from our own and the reasons supporting these views. Horton's strategies seem to be more monological. They can even be deployed in private deliberation. The two sets of strategies can easily be combined. Research has shown that there is no guarantee that counteracting strategies work. However, the studies reviewed by Wilson and Brekke provide us with some insights concerning the conditions under which these strategies may be effective.

5 The Relation Between Moral Reasoning and Moral Action

In the previous sections I examined Haidt's view that moral reasoning is usually post hoc, and almost unavoidably biased. I do not think that these two characteristics of moral reasoning fully explain why Haidt downplays the role of moral reasoning in moral judgement. I suspect that part of the explanation is also that, in his view, reasoned judgements have less motivating force than intuitive judgements or that reasoned judgements only influence actions through triggering (new) intuitions.

As we have seen, Haidt argues that the statistical relationship between moral reasoning and moral action found by Blasi (1980), turns out to be weak once the factor of intelligence is partialed out (823f.). In referring to Mischel and Mischel (1976) Blasi states that emotional and self-regulatory factors seem to be more powerful determinants of actual behaviour. Haidt's article is above all an attack on the work of Kohlberg (1981, 1984) and the contemporary approaches stemming from that tradition (Turiel 2002; Rest et al. 1999). They emphasise the role of cognition in moral functioning. Seen from the perspective of (meta-)ethical theory, Kohlberg was not only a *cognitivist*, but also an *internalist*. Cognitivist internalists believe that moral beliefs can directly motivate into action, thus without the help of an antecedent desire. While Kohlberg is highly influenced by the Kantian moral philosopher John Rawls, Haidt feels more affinity with the philosophy of David Hume. In what is known as 'the Humean theory of motivation', an action can only result from a combination of a belief and a desire. Thus, reasoned judgements can only lead to action if the beliefs grounding the judgements are connected to certain desires. Contrary to reasoned judgements, intuitions are, in Haidt's view, not just cognitions, they also have an 'affective valence'. With regard to intuitive judgements Haidt seems to be a *noncognitivist internalist*. According to noncognitivists, moral judgements express attitudes, e.g. of (dis)approval or of (dis)like. In the noncognitivist view the motivational force of moral judgements is constituted by their affective aspect. Haidt appears to take sides with another approach in moral psychology, that of Hoffman. Hoffman sees moral emotion as the primary source of motivation. It is the role of empathy to transform moral principles, learned in 'cool' didactic contexts, into 'hot cognitions', thus giving them motive force (Hoffman 2002: 239). Other theorists provide a more integrated perspective in which moral cognition and moral emotion are interlinked and can both function as primary sources of moral motivation (e.g. Gibbs 2003). If I am right in suggesting that Haidt is a noncognitivist internalist with regard to intuitive moral judgements and that he regards reasoned judgements as merely cognitive, his view that reasoned judgements only influence actions by triggering new intuitions becomes understandable. He thinks that reasoned judgements as such have no motivational force.

I am not going to defend cognitive internalism. I will assume that the noncognitive internalists are right in saying that reasoned judgements have no inherent motivational force. If so, reasoned judgements can only lead to action if they tie in with a desire. Thus, the gap between reasoned judgement and action has to be bridged by a person's motivational profile. What kinds of motivations are required for bridging this gap? In the literature we find a host of proposals. I will just mention two of them. In his attempt to reconcile the Kantian view on reasons and judgements with the Humean theory of motivation, Smith says that insofar as we are rational, we will desire to do what we think we have most reason to do. Rational persons desire to act in accordance with their evaluative judgements (Smith 1994: Ch. 5). The motives that bridge the gap between judgement and action are motives of rationality. Zangwill's solution is the introduction of a generic desire to do the morally preferable thing (Zangwill 2003: 144). A similar idea can be found in Wren (1991).

Haidt suggests that the statistical relation between deliberate moral judgement and action is weaker than the relation between intuitive judgements and actions. This phenomenon can be explained by assuming that some of the persons who do act in accordance with their *intuitive* judgements, lack the appropriate motivational profile to translate their *reasoned* judgements into action. If we also assume that morally mature persons act in accordance with both their intuitive judgements and their reasoned judgements, the conclusion must be that cognitive moral development – the process of learning to reason morally – and the development of the motivational profile required for translating reasoned judgements into action, do not necessarily coincide. Struck by the numerous findings that people differ in fairly regular ways as to their willingness to translate moral judgements into action, Kohlberg responded in his later work by starting to assess judgement and conduct at first separately, and then to ask about the relation between the two (Kohlberg and Candee 1984: 508). In this view reasoned moral judgement is necessary, but not sufficient for moral conduct. Kohlberg introduced a distinction between first-order judgement of rightness ('deontic judgements') and second-order affirmations of the will to act in terms of that judgement ('responsibility judgements') (1984: 518). He suggests that deontic judgements are in the higher, post-conventional stages of moral reasoning always accompanied by responsibility judgements, whereas this is not the case in the conventional stages. In my view, August Blasi offers the most promising theory that explains how in the course of moral development the gap between judgement and action is bridged (Blasi 1984, 1993, 1995). In his theory which aims to integrate moral cognition with moral personality, self-identity is the central explanatory concept in moral functioning. The self-identity model distinguishes three major components of moral functioning. The first one focuses on the significance and salience of moral values in one's identity. For some people moral values and principles permeate their perception and reasoning because they are rooted at the core of their identity, where for other people these values and principles are not particularly salient in their self-concept and in their daily activities. Morality has different degrees of centrality in people's perception, thinking and acting. The second component refers to an individual's sense of personal responsibility for moral action – an element, as we have seen, already present in Kohlberg's later work. The third component is self-consistency. Self-consistency is a fundamental motive that can only be satisfied by congruence between judgement and action. These three components are present in the functioning of morally mature persons. Morality and identity/self-concept are according to Blasi separate psychological systems which only slowly, and sometimes imperfectly, come together and become integrated.

The empirical evidence on the relationship between moral identity and moral action is reviewed by Hardy and Carlo. They conclude that, although the research on moral identity and moral action is sparse, results thus far validate Blasi's conception of moral identity as a source of moral motivation (Hardy and Carlo 2005: 242). Blasi's view on the role of moral identity seems to be superior to theories of moral development that regard cognition and/or affect as the only sources of moral motivation.

6 Conclusion

It is important that moral philosophers take notice of social psychologists' findings on the automaticity of judgements and on the role and the nature of everyday moral reasoning. Moral reasoning is indeed often post hoc, biased and unable to actually motivate action. These are the lessons we should learn from Haidt. I argued in this paper that Haidt's theory is rather one-sided than wrong. In the context of discussions on complex and novel situations, moral reasoning is not post hoc. And if moral reasoners are aware of biases, they can counteract them when using the right strategies. Haidt's distinction between intuitive and reasoned judgements is important and useful. Depending on their meta-ethical assumptions, many philosophers and psychologists of moral development see either intuitive judgements or reasoned judgements as 'the' moral judgements. I think that we should accept that there are two different types of moral judgements, intuitive judgements and reasoned judgements, which have a different meta-ethical profile. I assumed that morally mature persons are able to act in accordance with both their intuitive judgements and their reasoned judgements. If I am right, cognitive moral development and development of the motivational profile required for translating reasoned judgements into action, do not coincide. This view finds support in the work of Blasi who says that morality and identity/self-concept are separate psychological systems which only slowly, and sometimes imperfectly, come together and become integrated. Much of what I said are assumptions that require further philosophical and empirical research.

References

Arkes, H.R. "Cost and benefit of judgement errors: Implications for debiasing." *Psychological Bulletin*, 110, 486–498, 1991.

Bargh, J.A. "Automaticity in social psychology." In *Social psychology: Handbook of basic principles*. E.T. Higgins & A.W. Krugalski (eds.). New York: Guilford, 169–183, 1996.

Bargh, J.A. and Chartrand, T.L. "The unbearable automaticity of being." *American Psychologist*, 54, 462–479, 1999.

Baron, J. *Judgment Misguided: Intuition and Error in Public Decision Making*. New York: Oxford University Press, 1994a.

Baron, J. "Nonconsequentialist decisions." *Behavioral and Brain Sciences*, 17, 1–42, 1994b.

Baron, J. "A psychological view of moral intuition." *The Harvard Review of Philosophy*, 36–40, Spring 1995.

Bishop, M.A. "In praise of epistemic irresponsibility: How lazy and ignorant can you be?" *Synthese*, 122, 179–208, 2000.

Bishop, M.A. and Trout, J.D. *Epistemology and the Psychology of Human Judgment*. New York: Oxford University Press, 2005.

Blasi, A. "Bridging moral cognition and moral action: A critical review of the literature." *Psychological Bulletin*, 88, 1–45, 1980.

Blasi, A. "Moral identity: Its role in moral functioning." In *Morality, Moral Behavior and Moral Development*. W. Kurtines & J. Gewirtz (eds.). New York: Wiley, 128–139, 1984.

Blasi, A. "The development of identity: Some implications for moral functioning." In *The Moral Self*. G.G. Naom & T. Wren (eds.). Cambridge: MIT Press, 99–122, 1993.

Blasi, A. "Moral understanding and the moral personality." In *Moral Development*. W.M. Kurtines & J.L. Gewirtz (eds.). Boston: Allyn & Bacon, 229–253, 1995.

Festinger, L. *A Theory of Cognitive Dissonance*. Stanford: Stanford University Press, 1957.

Fine, C. "Is the emotional dog wagging its rational tail, or chasing it? Unleashing reason in Haidt's social intuitionist model of moral judgment." *Philosophical Explorations*, 9, 1, 83–98, 2006.

Gibbs, A. *Moral Development and Reality: Beyond the Theories of Kohlberg and Hoffman*. Thousand Oaks: Sage Publications, 2003.

Haidt, J. "The emotional dog and its rational tail: A social intuitionist approach to moral judgement." *Psychological Review*, 108, 814–834, 2001.
Hardy, S.A. and Carlo, G. "Identity as a source of moral motivation." *Human Development*, 48, 232–256, 2005.
Hare, R.M. *Moral Thinking: Its Level, Method and Point*. Oxford: Clarendon Press, 1981.
Hoffman, M.L. *Empathy and Moral Development: Implications for Caring and Justice*. New York: John Wiley, 2002.
Horton, K. "Aid and bias." *Inquiry*, 47, 545–561, 2004.
Kahneman, D. "Maps of bounded rationality: Psychology for behavioural economics." *The American Economic Review*, 93, 1449–1475, 2003.
Kahneman, D. and Tversky, A. "Choices, values, and frames." *American Psychologist*, 39, 341–350, 1974.
Kahneman, D. and Frederick, S. "Representativeness revisited: Attribute substitution in intuitive judgement." In *Heuristics and Biases: The Psychology of Intuitive Judgment*. T. Gilovich, D. Griffin & D. Kahnemann (eds.). Cambridge: Cambridge University Press, 1–18, 2002.
Kohlberg, L. *Essays on Moral Development, Vol. I: The Philosophy of Moral Development*. Cambridge: Harper & Row, 1981.
Kohlberg, L. *Essays on Moral Development, Vol. II: The Psychology of Moral Development*, Cambridge: Harper & Row, 1984.
Kohlberg, L. and Candee, D. "The relationship of moral judgment to moral action." In Kohlberg, L. *Essays on Moral Development, Vol. II: The Psychology of Moral Development*, Cambridge, Mass.: Harper & Row, 498–581, 1984.
Lord, C.G., Ross, L. and Lepper, M.R. "Biased assimilation and attitude polarisation: The effects of prior theories on subsequently considered evidence." *Journal of Personality and Social Psychology*, 37, 2098–2109, 1979.
Mischel, W. and Mischel, H.N. "A cognitive social-learning approach to morality and self-regulation." In *Moral Development and Behavior: Theory, Research and Social Issues*. T. Lickona (ed.). New York: Holt, Rinehart & Winston, 84–107, 1976.
Monteith, M.J., Ashburn-Nardo, L., Voils, C.I. and Czopp, A.M. "Putting the brakes on prejudice: on the development and operation of cues for control." *Journal of Personality and Social Psychology*, 83, 1029–1050, 2002.
Nisbett, N. and Wilson, T.D. "Telling more than we know: Verbal responses to mental processes." *Psychological Review*, 84, 231–259, 1977.
Nisbett, N. and Ross, L. *Human Inference: Strategies and Shortcomings of Social Judgment*. Englewood Cliffs: Prentice Hall, 1980.
Pizarro, D.A. and Bloom, P. "The intelligence of the moral intuitions: A reply to Haidt (2001)." *Psychological Review*, 110, 193–196, 2003.
Rawls, J. *A Theory of Justice*. Oxford: Oxford University Press, 1971.
Rest, J., Narvaez, D., Bebeau, M.J. and Thoma, S.J. *Postconventional Moral Thinking: A Neo-Kohlbergian Approach*. Mahwah: Erlbaum, 1999.
Saltzstein, H.D. and Kasachkoff, T. "Haidt's moral intuitionist theory: A psychological and philosophical critique." *Review of General Psychology*, 8, 273–282, 2004.
Sloman, S.A. "The empirical case for two systems of reasoning." *Psychological Bulletin*, 119, 3–22, 1996.
Smith, M. *The Moral Problem*. Oxford: Blackwell, 1994.
Sunstein, C.R. "Moral heuristics." *Behavioral and Brain Sciences*, 28, 531–542, 2005.
Stanovich, K.E. and West, R.F. "Individual differences in reasoning: Implications for the rationality debate?"*Behavioral and Brain Sciences*, 23, 645–665, 2000.
Thomson, A. *Critical Reasoning in Ethics: A Practical Introduction*. London: Routledge, 1999.
Trope, Y. and Liberman, A. "Social hypothesis testing: Cognitive and motivational mechanisms." In *Social Psychology: Handbook of Basic Principles*. E.T. Higgins & A.W. Krugalski (eds.). New York: Guilford, 239–270, 1996.
Turiel, E. *The Culture of Morality*. Cambridge: Cambridge University Press, 2002.

Waal, F. de. *Our Inner Ape*. New York: Riverhead Books, 2005.
Wilson, T.D. and Brekke, N. "Mental contamination and mental correction: Unwanted influences on judgments and evaluations." *Psychological Bulletin*, 116, 117–142, 1994.
Wren, T.E. *Caring About Morality*. London: Routledge, 1991.
Zangwill, N. "Externalist moral motivation." *American Philosophical Quarterly*, 40, 143–154, 2003.

Philosophical Reflection on Bioethics and Limits

Theo van Willigenburg

I see three themes that are at the core of current philosophical reflection on bioethics and limits:

1. The source and power of values and of meaningfulness in light of the abhorrent contingency of human life.
2. The meaning of naturalness in light of the rapidly growing possibilities of technical intervention.
3. The nature of reasonable reflection in the light of new insights into the role of emotion and intuition in processes of reasoning and judgement.

In this paper, I will concentrate on the first theme, but I will also briefly sketch the problems that are being discussed with regard to the second and third theme.

1 The Source and Power of Values and of Meaningfulness in Light of the Abhorrent Contingency of Human Life

There is a dominant tendency to think that the possibility to uphold values and to give meaning to our lives is seriously threatened by the awareness of the vulnerability and sheer contingency of our existence. Subjectively we experience ourselves as the middle point of the universe and we cannot but think of our own lives as overwhelmingly important. But seen from a more detached standpoint, we become aware that our existence is a sheer accident and that all of our lives just take up a split second of eternity. The idea is that this awareness explains the drive to develop technical means in order to extend the range of control that we have over the beginning and ending of human lives and over the natural world that we live in. We want to develop means to transcend biological limits and we see the life sciences as sources of possibilities to become master of our fate.

The thought that the awareness of our vulnerability and contingency poses a problem to us is defended by many philosophers. Thomas Nagel, author of *The View from Nowhere*, says: "to see myself objectively as a small, contingent, and

T. van Willigenburg
e-mail: vanwilligenburg@kantacademy.nl

M. Düwell et al. (eds.), *The Contingent Nature of Life*,

exceedingly temporary organic bubble in the universal soup produces an attitude approaching indifference" (Nagel 1986: 210). Nagel contrast a so-called first-personal, subjective perspective with a third-personal, more objective perspective. From the more detached third-personal perspective "my birth seems accidental, my life pointless, and my death insignificant". But as seen from inside "my never having been born seems nearly unimaginable, my life monstrously important, and my death catastrophic" (ibid.: 209). Our life has subjective importance, but objectively it is not important at all. Subjectively, we take our existence for granted, but objectively seen our existence is a sheer contingency. For we must admit that "[A]lmost every possible person has not been born and never will be, and it is a sheer accident that I am one of the few who actually made it" (ibid.: 211). Subjectively we are committed to a personal life in all its rich details. But seen from a more general vantage point, all our concerns, motives and justifications seem entirely gratuitous. From an external view our strivings and concerns are nothing but vanity. It wouldn't even have mattered, if we had never existed. According to Nagel, this "forces on us a kind of double vision and loss of confidence which is developed more fully in doubts about the meaning of life" (ibid.: 214). The same person who finds himself immersed in value and meaningfulness, "finds himself in another aspect simultaneously detached", thereby destroying the feeling of importance. These two viewpoints cannot be harmonized and the result is a divided self, a psyche torn apart. Our subjective self resists the reduction to meaninglessness and unimportance that the more objective view tries to force on it. We are "dragged along by a subjective seriousness" that we cannot even attempt to get rid of. But we also cannot ignore the more detached view on our lives. "[T]he objective standpoint, even at its limits, is too essential a part of us to be suppressed without dishonesty" (ibid.: 210). Objectively seen we are just part of contingent processes.

In order to escape this troubling tension, humans are prone to find means to escape the contingency of life. Religion has always been a universal source of structure, necessity and purpose. But humans also try to expand their own subjectivity. Nowadays, they do so by using evermore far-reaching technical means to intervene in what seems naturally and biologically given. Or, so this story goes.

I am not convinced, however, by this picture of the *condition humaine*. It hinges on Nagel's contention that questions about the meaning of life originate from the sketched tension between a first personal subjective and a third personal objective point of view. I believe that Nagel's description of the character and relation of these two viewpoints is incorrect.

We can see this by asking why taking a third personal detached view on our existence is important or even essential to us. Why would it be important to look upon our lives from a distance, when this might destroy the experience of importance and meaningfulness? The answer, of course, is, that we care about having a sincere and true view on our lives. We want to know whether what seems important to us *really* is important. Is what seems meaningful and valuable to us really meaningful and valuable? We ask this kind of question because we care about sincerity and truthfulness. But, of course, this concern itself can only be understood from a first personal perspective. We care about truth, and therefore we strife for a detached view on our existence, but the importance of this concern cannot be understood from such a detached stance. From a detached stance ultimately nothing really matters, also not our concern for truth. That truth is important to us is something that can only be grasped from the first personal point of view. We care about taking an objective standpoint, but this concern only matters to us from a first-personal non-detached perspective. This means that we can never detach from all our concerns. We can only detach and reflect on what is important to us, by taking at least one of our concerns as stepping-stone. We care about truthfulness. But we also care about fulfilling the projects that give meaning to our lives. We care about striving for important goals and upholding core values. We are concerned about our children, our family, and our friends. These are concerns that form the sources of what is meaningful and normative to us. I see compelling reason to help this child, because she is my daughter and I love her. I see compelling reason not to use my tickets for the opera house tonight, because I care about finishing my new book. *Concerns* are the source of reason giving considerations that are normative to us.

But how does our concern for truth influence those other core concerns? Taking a more detached view may lead us to ask whether our reason giving core concerns are really so important as we tend to think. What reason do I have for thinking that my concern to write a new book is really something worth striving for? In what way can I *ground* the importance of what is important to me? There are two ways to answer that question.

Rationalists will look for *reasonable justifications* of our core concerns. They look for reasons that explain why such concerns are genuinely important. Without such reasonable grounds our core concerns cannot function as the sources of rational necessitation, which is the kind of necessitation that is provided by our insight into the justifying force of reasons. The only way we can be necessitated is by the normativity of reasons.

The problem with this approach will be obvious. If our core concerns are the source of reason giving considerations that are normative for us, what source of reasons is left to provide for a reasonable justification of the concerns themselves? Core concerns encapsulate what matters to us. Is there any (other) source of reasons for thinking that some concerns are more worthwhile than others?

Rationalist are confronted with a serious problem here, and that is why *volitionalists* contend that we should not look for rational justifications of concerns, but that we should look upon them as *brute facts*. Our psyche, and more specifically our will, is pre-structured in such a way that some things simply count for us categorically. As Harry Frankfurt says, our core concerns should be understood as "contingent volitional necessities by which the will of the person is as a matter of fact constrained" (Frankfurt 1999: 138).

> ... the fact that a person cares about something (...) need not derive from or depend on any evaluations and judgements that that person makes or accepts. The fact that something is important to someone is a circumstance that naturally has its causes, but it may neither originate in, nor be at all supported by, reasons. It may simply be a brute fact
>
> (Frankfurt 2002: 161).

But how could we be necessitated by a brute fact? How might some concern we just happen to have become a source of what is rationally irresistible? The fact that one's core concerns have become part of one's own volitional structure does not entail that one is necessitated in a normative sense by what those concerns involve. Even a person's own concerns, strivings and worries are not, without further ado, authoritative for that person. Of course, concerns may push us. No parent can ignore her child crying in the middle of the night, even if she knows that for the child the time has come to learn that not every cry will be rewarded by parental consolation. You may know that, and still feel inescapably pushed to get up to comfort your child. Still, we should make a distinction between such inner pushes and the normative grip that concerns may have on us. We should distinguish the un-freedom involved in psychic forces we cannot escape, and the un-freedom involved in our being normatively necessitated by what is vitally important to us. Concerns do not just force or push us to take certain things as important. The "Here I stand, I can do no other" is not an expression of one's feeling captivated by inner psychic forces. "Here I stand, I can do no other" is an expression of one's feeling captivated by conviction. It is not an excuse (sorry, I am psychologically forced to stand her), it is a justification: I must do this, because I see that it is right and good to do so. I am normatively required to do so.

Volitionalists, like Frankfurt, cannot give a good explanation why we find ourselves rationally necessitated (not forced or pushed) by our concerns. They cannot explain why we find ourselves normatively required to take certain things as important. Our core concerns do not just force us, they have authority over us, even if we cannot find a reasonable justification for this authority. What explanation can we give for this seemingly paradoxical phenomenon?

I believe that there is a *third position*, in between rationalism and volitionalism that can provide a solution here (Willigenburg 2005b). According to this position we can explain the foundational and at the same time rational status of core concerns, by highlighting the way in which these concerns play a role at the background of rational reflection. If we think about the structure of rational reflection, it will strike us that reflection always takes place against the background of an innumerable amount of considerations that inform our reasoning without themselves figuring in or being subject to reflection. This is generally the case, in theoretical as well as in practical reasoning. Lewis Carroll (1895) taught us that for any valid argument, there must be a background rule of inference that is endorsed without having this background rule itself being made explicit in the argument as a premise. This means that in reasoning we have to make a distinction between the explicit reasons that figure on the foreground as the premises in the argument and background considerations that we could make explicit if pressed, but that we do not need to make explicit in order to be justified to draw the conclusion.

Background reasons, like those expressed by rules of inference, do not fulfil a supportive role in the inference. The inference is complete and the conclusion is justified, in spite of the fact that we have not cited considerations that seem to be presupposed as a matter of course. This does not mean that the rules of inference in the background are never taken in explicit consideration. Someone may opt for a deviant system of logic, which makes it necessary for me to make the inference rules that I take for granted explicit. The famous mathematician L.E.J. Brouwer has, for instance, rejected Aristotelian logic, defending an intuitionistic logic according to which the principle of the excluded middle (p v –p) is not a theorem. This has as a result that from (NOT (for all x) Fx) it does not follow ((there is an x)-Fx) (see van Dalen 1999, 2005). However, the situation in which basic logical principles are challenged is unusual. Usually, we may endorse rules of inference, necessary to make rational reflection possible, without having the background reasons expressed by such rules play a role on the foreground of deliberation.

The lesson that we may learn from this is that rational necessitation, i.e. necessitation by the insight that results from rational reflection, may be based on considerations that play no role in reflection itself, but that determine what counts as considerations that *do* play a reason-giving role in rational reflection. If we generalise and apply this lesson to evaluative reasoning, we see why we may be rationally driven to care about things and find conclusive reason to act in a certain way, without having *explicit* arguments for doing so. It may be that in the background of reflection, there are concerns that inform the process of reflection, without figuring as either explicit premises or implicit premises that we need to make explicit to complete the argument. These concerns are, in normal circumstances, not in the scope of deliberation, just like the rules of inference. They are not contesters in the process of rational reflection. They make rational reflection possible, by determining what counts as reasonable for us.

So, the idea is that core concerns are needed in the background, but they need not to be cited to complete the reasoning that is made possible by them. This explains why our deep concerns may issue in constraints by which we are normatively necessitated (as they are part of the structure of reason), without being themselves part of rational reflection. We simply need not think about them – for instance think about whether they are really so important as we take them to be – in order for them to provide a rational grounding of practical thought. Concerns need not be justified before they can play a role in reasoning, and this is due to the *essential structure* of rational reflection (in which we always need background considerations that inform reflection, without themselves being part of reflection).

Still, this does not mean that concerns may *never* be scrutinized. They function *default* as sources of what counts as reasons for us, but this role may be *challenged.* Just as the rules of inference in logic may be made explicit and scrutinized, one day, even a mother may come to question the obviousness of her unconditional maternal dedication (perhaps because of other conflicting concerns she has). And, all of us may come to question the obviousness of some of our concerns, driven by our concern to develop a sincere and correct view on what is really important in our lives. What can we say, then, against the argument that from a more detached view on what matters to us, many things are not as important as they seem to be? Does the awareness of the sheer contingency of our existence indeed threaten to destroy our convictions about value and meaning?

I believe, that on this profound level of reflection, second-order considerations are available that sustain the idea that certain concerns really have the import that we ascribe to them, just as there might be second-order reasons for adopting a certain system of logic. With regard to our core concerns, it is amazing that often the contingency of events is the source of their value and importance. The consciousness that human existence is a sheer accident and that our lives just take up a split second of eternity seems to ground – at least in our experience – the uniqueness of every human person. The deep unlikeliness of existence seems to be the source of its value. As a Dutch poet once wrote: "Al that is valuable is feeble."

This connects to the way we give meaning to our lives. In general, there can only be valuables against a background of scarcity of valuableness. Our life projects and endeavours only gain meaning against the background of the limitations inherent in living a human life (its finiteness, the limitations of every phase of life etc.). We want to go on living, but if there were no terminus to our lives, our activities would lack a shape and significance. What we simply cannot lose has less value than what can be lost or can deteriorate. If we lacked all experience of limitation, depression and despair, our lives would be shallow and empty, because happiness can only be experienced as a positive experience in contrast with experiences of unhappiness. Omnipotence would result in a loss of meaning and value. The choices we make and the goals we strife for would loose their import, if we were omnipotent without a need to make selections. If all goals were simply attainable, striving for this one important goal in your life would lose its value.

Humans have a capacity to turn what is at bottom contingent and conditional into something that is categorically important.[1] But this is only possible against a background of impossibilities and limitations. Our self-consciousness and our abilities of emotional "marking" and intellectual insight create the significances, the values and the ideals that necessitate us. But these values only emerge against a contrasting background of contingency, insignificance and vulnerability.

My contention is that life's sciences attempt to turn us into masters of our fate may do little to reduce the experience of vulnerability and helplessness. Life sciences will make us aware of new contingencies and new areas where significance may be found. This is, clearly visible in the way we give meaning to the idea of naturalness in light of the rapidly growing possibilities of technical intervention. Let me, therefore, shortly say something about our dealing with the idea of naturalness.

2 The Meaning of Naturalness in Light of the Rapidly Growing Possibilities of Technical Intervention

In spite of the fact that we have overwhelming control over the process of procreation, most parents consider the birth of their child as a precious gift. They do not present nor experience the newborn child as the result of careful planning and choice, they consider their child as something that is given to them by God or nature or whatever. They have received something, something that is beyond their planning, action and organisation.

The experience that lots of what happens in our lives – windfalls, disasters, gifts, luck – is not the result of our choices and doings is crucial to our self-understanding. Actions can only be understood as *actions* against a background of *events* that we feel confronted with. Actions are actions in contrast with happenings. This contrast is crucial. To understand ourselves and others as actors we need to ascribe to ourselves and others responsibility for certain "happenings" that are the product of our activity. This ascription of responsibility takes place, even if we are aware that we are susceptible to "moral luck". For instance, our *character*, source of many of our choices and actions, has been formed by factors most of which are not under our control. And we know that very often the *circumstances* determine the outcome of our choices and actions (think of Oedipus and his tragic fate). In spite of these factors which we cannot control, we take and assume responsibility. We need to presuppose a realm in which we are in charge. And just as we need to presupposes such a realm where we are in control and responsible for our actions, we need a realm of happenings, of things given – the "givenness" of which functions in the background of what we do. This is not just a psychological need. It is not that we wouldn't be able to live with the idea that everything is the result of our choices and actions. It is a metaphysical need: we cannot understand action and control, if not against a background of happenings beyond control. On the one hand, we cannot understand ourselves and what we do as completely determined by the circumstances (we are agents, not patients). On the other hand, we cannot understand ourselves as the cause of everything, as if everything that happens (health, illness, death, even climate changes) can be related to human choice and action. Just a determinism threatens agency, so complete voluntarism threatens to undermine our self-understanding as actors.

Such a metaphysical need is the reason for speaking about the "natural" and the "unnatural". *Nature* is understood as that which is not our creation. We need such a metaphysical understanding of the natural versus the artificial. We need to conceive of nature as the domain of what is not the result of human action and intervention. This conception as such is crucial to our self-understanding, even if there is a lot of discussion about what phenomena can be labelled "natural" and whether it is possible to sharply distinguish the natural from the artificial.

It is important that we do not mix up a metaphysical understanding of the distinction between the "natural" and the "artificial" or "unnatural" with a normative understanding. The concept of the "natural", as that which is not touched upon by human hands, can be understood in a metaphysical way as "that which is not our creation", but also in a normative way as "that which is unspoiled by human hands". In the bioethical literature, there is a recurrent discussion on the normative significance of calling something "natural" or "unnatural". It appears that it is very difficult to use such a distinction as a source moral evaluation. History shows that we adapt the distinction between natural and unnatural so as to suit our distinction between what is right or good and what is wrong or bad. So, it is not the case that something is bad because we consider it to be unnatural. No, we call something unnatural because we think it is bad. And we call something natural because we think it is good.

But even if the distinction between natural and unnatural or artificial as such has no normative significance, it is a distinction that is crucial for our self-understanding as agents and persons. This importance does not preclude the intervention in things that were up to now beyond human control. It means that we always need time and effort to reconstruct our self-understanding in terms of what we are responsible for and what not, given developing technological possibilities. Again and again, we need to reconstruct the border between actions and happenings. We need to delineate the domain of "givenness" in contrast with the domain in which we are in charge. We need to sort out what we can take responsibility for and what not. This may provide a basis for developing moral principles and rules that may have a say in regulating ground breaking developments in medicine and health care. Such principles may prescribe, for instance, a certain level of caution, while at the same time encouraging particular ways of the span of human control over life, death and procreation.

3 The Nature of Reasonable Reflection in the Light of New Insights into the Role of Emotion and Intuition in Processes of Reasoning and Judgement

Philosophical reflection tries to deepen our understanding of the conceptual worlds in which we are at home (the houses of concepts and ideas in which we live). It analyses the structures and presuppositions of our thinking, choosing and acting. Philosophical reflection is a discipline of rigorous argumentation and rational analysis. Still, philosophical reflection also hinges on intuition and non analysable insight. It is a typical philosophers' game to come up with all kinds of exotic examples to trigger and test our conceptual intuitions.

Similarly, moral judgement rests on more than rational argumentation. Especially in case of groundbreaking biotechnologies the moral judgment of people is influenced, and sometimes dominated by strong feelings of discomfort, worry, suspicion and fear, but also feelings of immediate enthusiasm and trust that strongly influence people's opinions. Often these "gut feelings" are the starting point of reflection and sometimes they fix a person's position in the moral debate.

In the philosophical literature there is a growing discussion about the normative status of these "gut feelings". Some think of these emotional and intuitive reactions as primarily primitive and not trustworthy. They argue that we need to rationalize en correct these gut feelings by giving people better information and by enhancing rational reflection. Other philosophers regard "gut feelings" as the most important indicators of a person's moral position, because they believe that in fact morality is nothing more than emotions and strong feelings.

My own position has always been that emotions and intuitions can be vehicles of important normative insights, but that these insights have to be unpacked and developed in a process of reflection and deliberation (Willigenburg 2003: 353–368, 2004a: 81–99, 2005a). Our intuitions about naturalness and unnaturalness are a case in point. However, in the last years, I have been doing conceptual and empirical research about the role of emotions in moral evaluation that has me made more suspicious about the influence of "gut feelings". More than 1000 students were included in an empirical research in which we asked half of the respondents to morally evaluate a sober description of a case of human cloning (of a "replacement" child) and the other half to evaluate the same case but now presented in the form of a documentary in which many emotion triggers were hidden. The documentary deliberately induced in respondents both a positive identification with the technological possibilities, but also negative, aversive feelings.

Our findings are that if emotions, in this way, influence the process of evaluation, students not only develop a more pronounced evaluative position, but that all students (also those who initially were rather positive about the cloning of the replacement child) come to evaluate the cloning case more negatively: they are more worried about the medical and psychological damage for the child, they consider the cloned child less as an individual, they are more afraid of slippery slopes, they are more worried about humans "playing God" etc. Strong emotional involvement induces in respondents a negative evaluation of groundbreaking biotechnology, even in those respondents who were initially positive in their assessment. This effect was overwhelming, even though the documentary was clearly evenly balanced in triggering both emotional reactions for and against having a replacement baby via cloning (van Willigenburg 2006).

A preliminary explanation of this phenomenon goes back to the idea that complex higher order social emotions are strongly imprinted by basic primitive emotions (Damasio 2000, 2004). Such primitive emotions are predominantly negative: fear, anger, disgust, and sorrow (only happiness is a positive emotion). If emotions are triggered, respondents come to be predisposed more negatively and appear to become more worried about groundbreaking biotechnologies. This is a disturbing conclusion, because research has also shown, that emotions ("gut feelings") play a crucial role in enhancing the efficiency of complex decision making and in the moral evaluation of complex developments and future scenarios (Prinz 2004). It seems that we cannot dispense of emotional reactions, but, at the same time, it may be that such emotions induce unwarranted negative and aversive judgements.

Note

[1] This is a complicated process, however, in which significance results from developed 'matches' between natural phenomena and typically human sensibilities. See my 2004b article: 91–104.

References

Carroll, Lewis (C.L. Dodgson). "What the Tortoise Said to Achilles." *Mind*, 4, 14, 278–280, 1895.

Dalen, Dirk van. *Mystic, Geometer, and Intuitionist: The Life of L.E.J. Brouwer, Volume 1: The Dawning Revolution*, 1999; *Volume 2: Hope and Disillusion*. Oxford: Oxford University Press, 2005.

Damasio, Antonio. *The Feeling of What Happens: Body, Emotions and the Making of Consciousness*. London: Vintage, 2000.
Damasio, Antonio. *Looking for Spinoza: Joy, Sorrow and the Feeling Brain*. Orlando: Harcourt, 2004.
Frankfurt, Harry G. *Necessity, Volition and Love*. Cambridge: Cambridge University Press, 1999.
Frankfurt, Harry G. "Replies" (to Michael E. Bratman, J. David Velleman, Gary Watson, T.M. Scanlon, Richard Moran, Susan Wolf, Barbara Herman and Jonathan Lear). In *Contours of Agency: Essays on Themes from Harry Frankfurt*. Sarah Buss and Lee Overton (eds.). Cambridge, Mass.: The MIT Press, 2002.
Nagel, Thomas. *The View from Nowhere*. Oxford: Oxford University Press, 1986.
Prinz, Jesse. *Gut Reactions: A Perceptual Theory of Emotion*. Oxford: Oxford University Press, 2004.
Willigenburg, Theo van. "Shaping the Arrow of the Will: Skorupski on Moral Feeling and Rationality." *Utilitas*, 15, 353–368, 2003.
Willigenburg, Theo van. "De status en rol van sterke morele gevoelens in morele oordeelsvorming." *Algemeen Nederlands Tijdschrift voor Wijsbegeerte*, 95, 2, 81–99, 2004a.
Willigenburg, Theo van. "Understanding Value as Knowing How to Value, and for What Reasons." *The Journal of Value Inquiry*, 38, 91–104, 2004b.
Willigenburg, Theo van. *Moral "Gut Feelings": Their Role and Trustworthiness in the Ethical Assessment of Ground-Breaking Technologies*. Manuscript (unpublished), 2005a.
Willigenburg, Theo van. "Reasons, Concerns and Necessity." *European Journal of Analytic Philosophy*, 1/1, 75–87, 2005b.
Willigenburg, Theo van. *The Influence of Emotions on Moral Judgements in Complex Cases*. Manuscript (unpublished), 2006.

Part III
Cases of Limits

Finite Lives and Unlimited Medical Aspirations

Daniel Callahan

Human beings have always struggled to find ways to deal with their finitude. We are not all we would like to be, and not able to have everything we might desire. Pleasure is fleeting and happiness elusive. It often seems as if it is easier for life to go wrong rather than to go right. Sadness and suffering are, though variable in their afflictions, something none of us can long avoid. The Old Testament story of Job never ages. The great religions of the world have all tried to find meaning in the face of suffering, pain, and death, and those who reject religion have always sought to find meaning as well, just in different ways.

In the modern world, medicine is one of the main instruments we use to combat the finitude of our body, that body which sickens, ages, and dies, and along the way is likely to bring assorted mental disorders. War, domestic violence, hurricanes, earthquakes show our vulnerability to lethal external forces. Sickness, by contrast, is a more intimate kind of harm, coming to us from the inside (even if, as with infectious disease, it comes to us from the outside, as an invader in our private bodily space). If illness and death in ancient times were often accepted in a fatalistic way, modern medicine – backed by science – has aggressively fought back. Francis Bacon and Rene Descartes held out the prospect that medical advances could change the human condition, and the famous American inventor and statesman, Benjamin Franklin, wrote in 1780 that "It is impossible to imagine the height to which may be carried, in a thousand years, the power of man over matter .... All diseases may be prevented or cured, not excepting even that of old age, and our lives lengthened at pleasure even beyond the antediluvian standard." (Franklin 1817: 57).

One can hardly doubt that, while "all diseases" have not been prevented or cured, and probably never will be, enormous progress has been made, with the medical progress superimposed on progress in most other areas of human life as well. We need only look briefly at the latter to see the ways in which improved nutrition, sanitation, housing, education, transportation and technological progress of a job-creating kind, have made life much easier to live and, at the same time, themselves make important contributions to improved health. At least 60% of the

D. Callahan
e-mail: callahand@thehastingscenter.org

M. Düwell et al. (eds.), *The Contingent Nature of Life*,

improvement in mortality rates in the twentieth century, close to 40 years longer average life expectancy, can be traced to socioeconomic improvements (Nolte and McKee 2004). The 40% attributable to organized medical care, blessed with innumerable new technologies, have left their mark as well (Newhouse 1992). Medical progress has, in short, been a great success.

Yet there are some paradoxes and problems. The greatest paradox is that, even as health improves and mortality rates decline – for all age groups and for most lethal diseases – the developed countries spend more, not less, money for health care. Research budgets, public and private, continue to rise and health care costs constantly rise. Moreover, the scope of medical care continues to widen, a tribute of a kind to the force of medicalization, the essence of which is to find a medical classification and solutions to human problems of a kind once taken to be "just life," as was once the case with erectile dysfunction and forgetfulness among the elderly. It seems clear, that is, that the medical struggle against bodily finitude has no limits or boundaries. It does not matter how well the struggle against bodily decay is going, new and ever-rising standards are constantly set about what counts as good health and no money seems enough money to fight all the corporeal evils that stand in our way.

Not only do we expect to live longer lives now, we expect to do so in much better health than our parents and grandparents – yet, even as we get those benefits, we can be more dissatisfied with our health than they were (Barsky 1988; Easterbrook 2003). The financially rich, it has long been noted, usually want to be even richer than they are, however rich that might be. There is always the new, advanced model of personal jets to be sought, just a little faster, a little more spacious, with a little longer flight range. So too, however healthy we might be, there is some higher stage we can aspire to, if only the prevention of future disease if we are already in perfect health (and there is much sense in the old observation of some anonymous wise physician that a "healthy" person is just someone who has not been well diagnosed).

The problems with the progress, for all of its benefits, are well known. The most obvious is its economic cost. In the United States, health care costs are now rising at a level of close to 7% a year, and 4% or so in Western Europe (and the latter, though not as bad as the U.S., is still above the rate of the rise in general inflation). Many schemes and ideas have been tried, or proposed, on both sides of the Atlantic Ocean over the past three decades to control costs, but – save for tight, often unpopular, government control of health care, no managerial or organization nostrum has met with sustained success. And, even in Europe, there are limits to taxation, already reached in many places, as well as aging societies and a demand for the constant stream of new, usually more expensive, technologies.

The other problem is what I call the great tradeoff, that of a longer life but with a heavier burden of disability. Two hundred years ago, most deaths were caused by infectious disease, killing both young and old. But death, when it came, was relatively quick, a matter of a few days or weeks at most. Now death is later but, because of chronic and degenerative disease, much slower, marked by a long decline, and sometimes so slow that it is hard to say just when someone is dying.

With those over 80 the fastest growing age group, and with an increased need for care by others in the years thereafter, long-term home and institutional needs and costs are now approaching the level of their medical costs. Moreover, the effective use of contraception and abortion, as well as a variety of social pressures, means that low birth rates, well below the population replacement rate of 2.1 children for every woman in every country, have become the norm (Grant et al. 2004). That means a change in the dependency ratio, with fewer young people to support more old people; not only will there be a shortage of caretakers for the old, the financial strain on young people to support the costs of care for the old could be considerable and painful.

In short, the present trajectory of medical progress and improved standards of living, together with increased demand for ever better health care, has the making of a social and financial disaster, or if not quite a disaster, a major problem for every developed country (and one also beginning to make an appearance in developing countries).

The characteristic response to these problems has been to treat them as issues of organization and management. It is assumed (1) that medical progress will and must go on, even if it creates problems along the way, and (2) that the proper response to them is to organize health care systems that can better manage them. There are three tacit assumptions behind those dual propositions, each thoroughly modern in its origins. One of them is that life is improved, and more fully human, when the range of choices of all kinds available to people are increased. The second is that fatalism and resignation in the face of evil, whether physical or social, is unacceptable. The third is that, when progress creates problems, more and better progress can solve them.

Finitude, that is, can with unremitting zeal be banished or, if not quite banished, then radically reduced in its power. Economic growth and prosperity, sought by means of technological innovation, can provide the necessary economic and social foundations for a good human life; and medical innovation can provide the good health necessary to make the most of that life. The key to overcoming finitude turns out, in the end, to require a firm commitment to overcoming it and a resolute unwillingness to accept it. Let me briefly look at the three assumptions noted above.

Choice. In one of the earliest books that appeared just as contemporary medicine was making some of its greatest strides after World War II – and one of the precursor books of the bioethics that was to blossom in the late 1960s – the Protestant theologian Joseph Fletcher struck a prophetic note. In Morals and Medicine he wrote in 1954 that "Choice and responsibility are the very heart of ethics and the sine qua non of a man's moral status . . .Just as helplessness is the bed-soil of fatalism, so control is the basis of freedom and responsibility . . . of truly human behavior . . . Technology not only changes culture, it adds to our moral stature." (Fletcher 1954: 10–11). The main obstacle to increased choice, and thus our moral stature, is nature itself. We will be free to the extent that we can dominate and subdue nature, giving to ourselves control over our fate.

Rejecting Fatalism. While writing in a different vein than Fletcher, less enamored with choice, the political scientist Michael Walzer wrote in his 1983 book

Spheres of Justice that "What has happened in the modern world is simply that disease itself, even when it is endemic rather epidemic, has come to be seen as a plague. And since the plague can be dealt with, it must be dealt with. People will not endure what they believe they no longer have to endure." (Walzer 1983: 8). If the expansion of choice is seen as one route in overcoming human finitude, the war against disease can be seen as another. The logic of Walzer's contention has behind it the fact of enormous success over the past 100 years in fighting disease. The conquest of most infectious diseases by the middle of the twentieth century was taken to serve as a guarantee that, sooner or later, the chronic and degenerative diseases of old age would be conquered as well. In 1970, the then-President Richard Nixon announced a "war against cancer," with every hope and full confidence that the war would soon be won.

Michael Ignatieff has added another important insight. "The modern world" he has written, "for very good reasons, does not have a vernacular of fate. Cultures that live by the values of self-realization and self-mastery are not especially good at dying, at submitting to those experiences where freedom ends and biological fate begins. Why should they be. Their strong side is Promethean ambition: the defiance and transcendence of fate, material, and social limit. Their weak side is submitting to the inevitable." (Ignatieff 1988: 32).

Progress as Self-Correcting. One of the long-running debates in western societies turns on the notion that problems created by technological progress can be cured by either more and better progress or progress that aims to rehabilitate what progress has injured. A central feature of environmentalism, for instance, has been a debate between those who believe that growth of almost all kinds must be limited if the environment is to be saved and those who believe continued growth is necessary for economic vitality and that, in any case, there are many technological ways that its harms can be avoided. The latter include more environmentally friendly ways of gaining energy (wind towerd, solar panels), cleaner ways of dealing with pollution (scrubbers on smokestacks, better emission controls on automobiles).

Health care has had its own versions of such debates. Half-way technologies such as dialysis machines, which keep people alive but not well, can and should be eventually replaced either by cures for diseases that bring kidney failure, such as diabetes, or by prevention strategies that avoid disease in the first place (exercise and a good diet to avoid heart disease). A much repeated claim is that the long-term key to expensive health care is more medical research, aiming to cure the chronic and degenerative diseases, and to better deal with frailty, deafness, arthritis and other debilitating conditions. To even slow progress at this point in history, it can be argued, would be to leave in place expensive systems of health care that simply perpetuate inferior medical care for diseases and conditions that could be better dealt with.

If claims of that kind are familiar enough in the ongoing wars against disease, a parallel version can be found in dealing with health care systems. The most widespread claims are that: inefficiency and waste are widespread and endemic, that too few of much used technologies have undergone evidence-based analysis, that better use of information technology would lead to more efficient medicine, that a greater(or lesser) use of market practices and mechanisms would make a decisive difference, and so on. The message is that, yes, too much money is spent on health care, but that is simply an accident of poor organization, too little health services research, and assorted vested interests that have a financial or other stake in the status quo.

The three assumptions I have described appear deeply imbedded in western culture, but particularly in the United States. Yet each of them is flawed, sometimes in ways not immediately evident. It is useful to understand those flaws because to do so puts us at least part of the way toward regaining a sense of our finitude, which I believe is not only necessary for sensible, affordable, and sustainable health care, but also for the living of a wise and sensible life.

We can look first at the notion that choice is the key to the humanly moral life, as if we can not be happy without choice and that, more broadly, an expanding range of choice as an outcome of overcoming nature is somehow the royal route to a good life. There are many reasons, moral and psychological, to doubt those views. Most obviously, a good part of the moral life consists of denying ourselves choices that would do harm to others and that might distort of own life as well. A major tenet of the environmental movement is to argue that many of the choices available in the past are now known to be harmful: the choice of throwing industrial or human wastes in our rivers, or using cheaper unleaded gasoline.

The environmental common good has led us (or most of us) to see that our freedom is sometimes best used to take from us our freedom to choose; freedom is a good but not the only good. The system of government approval of drugs for safety and efficacy takes from individuals, and their doctors, the right to choose any drug they might fancy, and takes from pharmaceutical companies the right to sell anything they want to see in the name of allowing patients an expanded range of choice. Human subject research requires that researchers limit the research strategies they might use in the name of the protection of human subjects. A stable, just society is one that, against a libertarian or anarchist philosophy, limits a wide range of choices that many people might like to make.

There is, moreover, a literature showing that simply expanding people's choices does not carry with it greater happiness or satisfaction. It may in fact overwhelm them, particularly if, in the name of unfettered choice, there are no recognized or time-tested ways of knowing which choices to make (Schwartz 2005). A life of poverty, mental illness, or a crippling disease can rob one of all meaningful choices in life. But that truth does not entail that an ever-expanding range of choices will add meaningfulness in any way proportional to the range of choices.

The enduring problem in combating fatalism is that, whatever progress medicine makes in lengthening our lives, and improving them along the way, we will still die. Now it may well be that the extra years we gain of life, and even good life, is a benefit, but – given the many years we will be dead (quite a few) – it seems to me a minor benefit. If average life expectancy brought us all to 125 years, we would soon enough reach that age, and all would be over.

Moreover, it is not unreasonable to guess that, if we all reached that age, we would not be too happy when that happened. If those years had been good, we would not want to stop; and, if they had been bad, we might still prefer years of that kind to being dead (as now happens with many people who want to hang on to life, even a life they might in earlier years have rejected out of hand). Almost all of those of us who have become old (I am 75) say that "it has all gone so fast," and they say that as much at 90 as at 70, and no doubt would say it at 125.

The obvious drawback with the notion of progress solving the problems of progress is that the historical record suggests that the solutions are almost always too little and too late. The phenomenon of global warming has generated a number of proposed technological solutions, including an increased use of non-fossil fuels, alternative energy sources, and taxation and other penalties on polluting factories. The Kyoto treaty aims at a comprehensive control of environmental hazards. But the technological solutions have been slow in coming, so far making little difference, and a variety of political obstacles (including the American refusal to sign the Kyoto treaty) have been further impediments.

In health care, with from 40%–50% of increased costs traceable to either new technologies or intensified use of older ones, more research correlates with higher costs. In most cases, the research does not save life but, instead, prolongs morbidity. Heart disease is a classic example of that phenomenon. Most forms of heart disease are incurable; its problems will last for a life time. A variety of new technologies have become available to cope with them, most of them highly expensive and many providing marginal benefits only. Now it is clear enough that many patients, and their doctors, are prepared to accept the chronic morbidity as the price of prolonging a life. But it is just as clear, from a societal perspective, that the added costs can make a great difference in the fiscal well being of health care systems.

Moreover, it is sometimes forgotten that, even if one's life is saved from one disease, this invariably means either that a later episode with the same disease will take our lives or that a substitute, successor illness will kill us – what I have called the "longitudinal" costs. A number of economists have tried, in the face of the correlation between technological innovation and health care costs, have tried to show that the costs are "worth it," in terms of the years of life gained by the innovations. This is too complex an area to be taken up adequately here, but that line of thought rests on the theory of "willingness-to-pay" theory, which calculates the value of life on the basis of what people say they would be willing to spend to extend their life for a set period or, alternatively, spend to reduce the risk of death. That theory, however, does not have a good way of reckoning the economic side-effects of extending the life, including the costs of treatment, work years lost to disability in the aftermath of life-saving treatment and the negative economic effect of innovative treatment costs on overall system costs. (Callahan 2003: especially Ch. 9, 233–234).

I have so far tried to show that, in its quest for infinite progress, without boundaries, limits, or even some clear final goal, contemporary medicine (with full public support) generates a host of problems. Not only do the technological innovations have various side-effects – true of almost all technologies – coping with those side-effects has generated a wide variety of tactics to manage and reduce them, and a huge effort to rationalize and minimize their importance. Put another way, modern people love progress, particularly medical progress, but they seem driven to evade facing up to its unfortunate features. The problems are either explained away, or if fully acknowledged, said to be worth the costs and undesirable side-effects. The rejection of fatalism, and of the very idea of finitude – however much daily life drums its reality into us – is by now a deeply ingrained feature of modern culture.

It is, in addition, a feature well supported and driven by the market. Modern industry, medical and non-medical, believes in and requires constant growth and innovation. Whether for reasons of time-limited patent protection (as in the pharmaceutical industry), or public demand for the new and the latest (cell phones, computers, automobiles), or industry generated public demand based on effective advertising (highly successful direct-to-consumer drug advertising in the U.S. and New Zealand).

No one who owns even lap top a computer can fail to notice the constant advertising for faster computing speeds and storage capacity. That advertising can be resisted, provided one is stalwart and phobic enough about change. What can not be avoided is the discovery, after owning a lap top for 4–5 years, that it can not perform all necessary functions any longer (too small a storage capacity, for instance, or too slow), that parts can no longer be bought to repair it, and that your children complain about its limited, outdated features. So, like it or not, a new one must be purchased. And who is likely to request of his physician that old-time massive invasive surgery, with much cutting and stitching, instead of laparoscopic surgery, with minor incisions and laser cutting?

If one concedes that there are indeed a variety of unattractive features that appear to come with an attempt to reject finitude, putting medical progress and innovation in its place, is there anything to be done about it? One response, seemingly the most common, is simply to accept it – in a more or less fatalistic way. One could say, yes, there are all kinds of problems, but that they are better than the opposite, which is not to have the progress at all, or at most in a thin stream. It is no doubt true, as argued by many economists, that the material success and high standard of living in the developed countries can be traced to scientific and technological progress, and that we are better off than without them. Turning my thesis about finitude on its head, it might be added that progress can do no more than push finitude around a bit, and that all progress will be tinged with evidence of our finitude. To recall the words of Joseph Fletcher, cited above, he is wrong to think, or at least imply, that we can ever have perfect control, and that we will instead, as our choices increase, have an increase of untoward results as a consequence. But one could say, troubles or not with control, we should keep trying, accepting the side effects.

Another response would be a variant on the first, but imbued with a sense that something, at least, should be done to better manage the downside of progress, even if it can not be eliminated. It would start with the presumption that progress must continue; it has now become part of human nature to pursue and, in any case, it is hard to imagine doing something about a harm-tinged progress without devising at least a few new technologies to cope with them. Most proposals, in environment and health care to deal with the problems are managerial and organizational in nature, and incremental in their desired effect. It is reform, not revolution, that is sought. Some of the reform measures have been mentioned above, such as evidence-based medicine and information technology. Expensive technologies can be measured in terms their cost-effectiveness ratios (are the new technologies better than the old in achieving the same ends); or cost-benefit balances (is the benefit worth the cost); or by making use of QALYs (measuring the value of a technology in terms of the relationship between quality of life and length of life it provides). Those methods all have significant limitations, however, both theoretical and practical.

The third possibility requires a cultural revolution. It would cease to struggle against our finitude but work, instead, to build into our culture a lively acceptance of it, using it to determine how best to live our lives. It would not give up the idea of progress, or put aside technological innovation. They would be aimed and designed to help us live better with our finitude, not to overcome it. It would take advantage of our present knowledge of the determinants of population health – education, jobs, good housing and nutrition – and focus scientific progress on enhancing them. It would in particular invest considerably more money in environmental research, in the quest for alternative forms of energy, and at the design of cities and towns to minimalize environmental harm.

In the medical arena, the essence of the revolution would be to recognize, accept, and build into health policy a perception that should by now be perfectly evident: both the costs and many of the side effects of trying to improve health by the development of new technologies have a diminishing return of health benefits combined with a chronic problem in paying for the health care on which that stategy is based. Most important, it would be necessary to accept some fundamental biological reasons why the campaign against disease can never fully succeed. As the great biologist Rene Dubos argued (with good scientific reasons) in his book The Mirage of Health, a "complete and lasting freedom from disease is but a dream remembered from imaginings of a Garden of Eden." (Dubos 1979: 2).

The most radical feature of the revolution would to work toward an agreement, even if of a rough and contentious sort, that the average life expectancy of those living in developed countries, now approaching 80 years, is a sufficiently long life to assure most (though not all) people time enough to live a decently long and full a life. Socioeconomic conditions, assuming a continuation of prosperity, will almost surely continue to add years to life even apart from medical care, but – given that likelihood – there is no need to actively pursue longer lives by research. The focus of research would be on behavioral research to enhance the socioeconomic contributions to health, a priority given to causes of premature death, on research to reduce disabilities and to facilitate independent living, and on research to minimize environmental harms to health. New technologies neither could nor should be stopped. But they should be subject to severe evaluation: not reimbursed unless they could be demonstrated prior to public release that they were cost effective, highly likely to provide a significant health benefit, and unless their likely cost and economic impact was publicly known.

Are changes of this magnitude, changing not just the practice of medicine and the delivery of health care, but the way we think about their nature and the contemporary culture of both of them possible? It may take a severe economic or other crisis to stimulate a willingness to think beyond the myriad proposals for managerial and organizational solutions, obviously not working in any effective way and fed only by hopes for incremental changes that have not worked in the past. And, even if they might work, hardly anyone who supports them claims they will deal effectively with the cost problem.

The harder work may be to persuade people to give up, or treat more lightly, the rejection of finitude that has been a mark of contemporary medicine and research. Longer lives and better health have improved the lives of all of us, but the fact that they improved it in the past does not prove they will do so in the future. The fact that considerable progress has already been made needs to be kept squarely in mind: where, given that progress, do we want to go now? There is already, I believe, a decline in health benefits that is in an inverse relationship between health research and expenditures. If nothing else, the fact that medicine now enables us to miss one bullet, and maybe a few, does not save us from being felled from the next ones that come along.

Finitude always catches up to us, and putting if off for still a few more years does not guarantee added happiness or satisfaction with life, and particularly if it comes at the cost of communal hypochondria, never satisfied, driving up costs and thus giving to the health sector money and resources that might better be spent on other social needs. What can be witnessed now is the equivalent of the trench warfare of World War I, with slow gains for the winning side but, when looked back at with the eye of history, appearing to be a huge waste. Medical progress has produced far more fruits than that war. But we are now in the later stages of the progress and the ground gets much harder to gain with each yard already gained.

References

Barsky, Arthur J. *Our Troubled Quest for Wellness*. Boston: Little, Brown, 1988.

Callahan, Daniel. *What Price Better Health: Hazards of the Research Imperative*. Berkeley: University of California Press, 2003.

Dubos, Rene. *Mirage of Health: Utopias, Progress, and Biological Change*. New York: Harper Colophon Books, 1979.

Easterbrook, Greg. *The Progress Paradox: How Life Gets Better While People Feel Worse*. New York: Random House, 2003.

Fletcher, Joseph. *Morals and Medicine*. Boston: Beacon Press, 1954.

Franklin, Benjamin. "Letter to Joseph Priestly, Passy, France, 8 February, 1780." In *The Private Correspondence of Benjamin Franklin*. William Temple Franklin (ed.). London: Henry Colburn, 57, 1817.

Grant, Jonathan et al. *Low Fertility and Population Ageing*. Santa Monica (CA.): The Rand Corporation, 2004.

Ignatieff, Michael. "Modern Dying." *New Republic*, 26, 28–33, December 1988.

Newhouse, Joseph P. "Medical Care Costs: How Much Welfare Loss?" *Journal of Economic Perspectives*, 6, 3, 3–21, 1992.

Nolte, Ellen and McKee, Martin. *Does Health Care Save Lives? Avoidable Mortality Revisited*. London: The Nuffield Trust, 2004.

Schwartz, Barry. *The Paradox of Choice: Why More is Less*. New York: Harper Perennial, 2005.

Walzer, Michael. *Spheres of Justice: A Defense of Pluralism and Equality*. New York: Basic Books, 1983.

Reproductive Choice: Whose Rights? Whose Freedom?*

Brenda Almond

1 Choice, Kinship and Society

Assisted reproductive rights are often seen in terms of the freedom and choices of mature individuals, and especially linked to women's rights and interests. In contrast to either of these approaches, I propose to look at some of the ethical issues involved from the point of view of those born by assisted reproduction (AR). In the United Kingdom, it was estimated that by the end of 2005, 57,000 people had been born following donor assisted conception, with current annual numbers estimated at 2,000. This is now similar to the numbers of children being adopted each year, the other main area in which professionals are involved in creating families where one or both parents are not biologically related to their offspring.

The United Kingdom is by no means unique in this. 1% of all births in the USA and 5% in Denmark involve assisted reproduction. Technology and medical intervention, then, now play a major part in reproduction in the Western democracies. But while it must be conceded that assisted reproduction has indeed ushered in an era of wider choice in the area of reproduction and while, in the context of a free society, it is hard to see choice as anything but a good, the question of whose choices are enlarged is often overlooked: those of adults or those of the children who result from those choices?

Human reproduction has until the very recent past been linked to two apparent immovables: one, the biological fundamentals of male and female; the other, the idea of the family as the basic building block of human society.

But each step in the new technologies of reproduction has brought new and unfamiliar ethical dilemmas. Previously inseparable aspects of parenthood have acquired new divisions: fathers can now be defined as genetic, social, or legal, while motherhood faces a further possible division between the mother who supplies the egg, or even just the nucleus of an egg, from which the child develops, and the mother

B. Almond
e-mail: brendalmond@yahoo.co.uk

*This paper is based upon Ch. 6 of my book *The Fragmenting Family*, Oxford and New York, Oxford University Press, 2006.

who carries the child through pregnancy to birth. Embryos can also be frozen, to be born years after conception and even after the death of their progenitors. Welcomed by some, regarded with concern or suspicion by others, these new possibilities have also opened the door to new kinds of life-style arrangements, as single parents, lesbians, gay men, and cooperating groups make use of assisted reproduction. Generous transfer of gametes has also sometimes produced unprecedented numbers of half-siblings, most unknown to the others, and in some cases this has been taken to surprising extremes, as in the case of a Danish sperm donor who, as the website of sperm-bank Cryos International once claimed, had unknowingly fathered 101 children worldwide.[1]

This wealth of choice seems to assume a narrow view of family as no more than parent and child. The genetic family is, however, a broader notion than this. And as some of those involved in fertility treatment increasingly lean towards the idea that, once donated, the source of a child's genetic material is irrelevant – that parenthood is a purely *social* concept – the science of genetics has begun to unveil its own secrets to a world in which individuals are increasingly concerned to understand their own complex genetic inheritance and to have access to the world of their genetic relations – a biological family that includes grandparents, aunts, uncles and cousins, as well as forebears and descendants. This fabric of connections has until now formed the webbing underpinning most known cultures and societies. But children born from donated gametes or embryos are deprived of the chance to locate themselves in this biological network – a network that has until now provided the individual's deepest conception of their identity and offered them the social space within which to find their earliest sense of self.

The argument that will be developed here, then, is that while many welcome the expansion of reproductive choice, and while moral and political theory assign an important role to freedom, there is a risk that not everybody's rights, welfare, and interests will receive equal attention. In particular, as increasing numbers of children who owe their origins to assisted reproduction reach adulthood, it is easier to see that their role as players in the AR game has not been adequately recognised and that choices have been made for them before they were in a position to choose for themselves or to consent to the choices of others.

Many of these donor children who are now reaching their twenties or thirties are asserting their own claims and also telling the medical world about their own feelings on the matter. One person who has made her voice heard in this way is Joanna Rose, who took her claim to the House of Lords in London. Born as a result of fertility treatment, the records of which could no longer be found, Joanna Rose claimed that the United Kingdom Government and the HFEA had a duty to assist her in finding out the identity of her natural father. Although she did not succeed in obtaining this, the Department of Health did later set up a voluntary register (UKDonorLink) to help people trace relatives.[2] Rose represents many who are concerned to know more about, or even to have access to, not only parents, but siblings, half-siblings, grandparents and others. This, though, may be one of the more straightforward of the legal and moral claims that are involved here, and other cases can be found that illustrate some yet more complex conflicts that can arise within the new world of reproductive medicine.

Increasingly, courts in jurisdictions as diverse as Canada and the USA, Australia and New Zealand, as well as European countries including the United Kingdom, Denmark and the Netherlands have been awarding visiting and custodial rights to individuals with no genetic connection to a child on the basis of either their original intention to participate in its care, or their actual participation in it. These are criteria that can be used, for example, in resolving disputes in surrogacy cases or in cases involving donor-assisted reproduction. In these circumstances, the question of who set out to have a child may be judged to outweigh that of who was physically instrumental in achieving it. Other accounts replace the biological or genetic connection with psychological criteria – for example, the emotional attachment between child and adult.[3] It is a short step from the view that strong attachments between child and carer are the central component of a child's well-being to the conclusion that they should take priority over the child's relationship with its biological parents (Skolnick 1998). In place of the older understanding of the primary relationships of human beings, then, family law, sociology and psychological theory seem no longer to see parenthood as rooted in the natural facts of procreation.

Some commentators are willing to take a more cautious approach to the issue and give *some* weight to biology. The British philosopher David Archard puts this in the form of a concession: 'There may be nothing wrong with a State permitting natural parents in the first instance to bring up their own children as they choose and within specified limits. What the State should not do is presume that natural parents have a right to rear which derives simply from biological parenthood' (Archard 1993: 109). Archard's own account draws a distinction between biological and *moral* parenthood. He writes: 'Moral parenthood is the giving to a child of continuous care, concern and affection with the purpose of helping to secure for it the best possible upbringing. "Parent" should only be understood as meaning one of several adult caregivers. Thus moral parenthood is not restricted to any particular familial form' (ibid.).

Children born, like those involved in these cases, by assisted reproduction using donated gametes or embryos are sometimes described as 'Children of choice'. This seems to give high prominence to children and certainly children rank high in most individuals' moral consciousness, just as they do in the verbal commitments of politicians and national leaders. But children are probably the most vulnerable of all human beings. Their rights are difficult to express and hard to enforce. In this area, where a new and unfamiliar gulf has opened up between genetic or biological relatedness and social relatedness, I believe that not nearly enough attention has been paid to the question of children's rights and welfare, looked at separately and apart from those of adults.

2 Protecting the Interests of the Child

Although some commentators, then, see the area of assisted reproduction primarily in terms of adult choice and argue for less regulation, or indeed for none at all, I would suggest that when the intention is to produce a child, this is a serious responsibility and the onus is on *all* those who play a part in it – health professionals, lawyers and politicians as well as those directly involved as would-be progenitors, parents or donors – to accept a requirement that is a current, but challenged, aspect of law in the United Kingdom, as expressed in the 1990 HFE Act, to have regard for the welfare of that child. In other words, children should not be viewed simply as commodities medically generated to satisfy the needs or desires of adults.

Some see this as asking for a judgement on parents. One commentator, Emily Jackson, has even suggested that a tacit eugenics policy is being applied to would-be parents, in the sense that their fitness to reproduce is being questioned (Jackson 2002). Doctors, too, because they may be asked to reply to a clinic's enquiry concerning would-be users of donor-assisted reproduction, may believe they are being asked to sit in judgement on their patients. But it is wrong to interpret the enquiry in this way. What doctors are actually being asked by the clinics, and what it is right and proper to ask, is not whether their patient *deserves* a child, but whether they know of any reason why their patient(s) should not be helped to have a child, especially a child which is the genetic offspring of another individual who has donated gametes or embryos for that purpose. For it has to be assumed – or at least hoped! – that those who donated their gametes did so in the belief that their biological offspring would not become, for example, the victim of abuse, or cruelty. As in the case of adoption, care is both necessary and justified when children, at no matter how early a stage of life, are being transferred between strangers.

This is considered objectionable by some philosophers who argue that, since these questions are not asked of people who procreate in the usual way, without medical assistance, there is no case for treating the users of donor-assisted reproduction differently. But this is a specious argument. What I am suggesting here, although I know that some will disagree, is that in providing treatment, the medical professionals involved take on some of the responsibilities of the more direct progenitors – they become, if you can put it this way, part-parents! Of course, they are not expected to become all-knowing, but it is not unreasonable to ask them to take the degree of care implied by the welfare clause in the Act.

Others, though, seem ready to abandon the 'welfare of the child' condition in the 1990 Act. John Harris, indeed, sees this requirement as problematic. He writes: 'This legislation constitutes prima-facie discrimination against infertile people since couples who can conceive naturally are not expected to demonstrate their potential fitness as parents' (Harris 1998b: 6; see also Harris 1985: Ch. 7). A further reason why this issue has become controversial in the British context is that the clause in the Act in which the welfare of the child is mentioned in so closely linked to another condition – 'the need of a child for a father'. It is therefore possible to argue that concern for the welfare of the child poses a challenge to the rights of single women and lesbians, thus raising questions about discrimination and gender equality. It seems to me important, therefore, to separate these two issues. The debate about non-standard families is a broader one that needs to be conducted on wide social and political grounds, taking into account empirical findings about such things as the likely length of some kinds of relationships relative to others, and the consequences this may have for their children. If the empirical findings show adverse consequences for children, this would, of course, *become* a welfare issue, but this could be set aside in the present context, since the welfare of the child, whether or not it was included in the Act with these aspects in mind, raises broader considerations. To address these, it is worth taking a closer look at some of the philosophical arguments that lie behind this debate.

3 To be or Not to be – the Problem of Non-existent Entities

Those who challenge the need to treat the child's interest as paramount sometimes talk as if there is a queue of children waiting in limbo for a chance to be born, so that the onus is on those who *stop* them being born to justify their decision. But this is, of course, nonsense. No-one is injured by not being born. The phrase used in the Act is 'child who may be born as a result of the treatment' and it is clear that this 'may' puts the matter in the same category as books that you or I might write, motorways that may be built, or laws that might be passed.

In the case of children born by donor-assisted reproduction, however, the assumption seems to be that they already have a kind of shadow existence in which they confront the alternatives of existing or not existing. It is assumed that existence is bound to be the better option, no matter what the circumstances. As John Harris puts it: '(To) choose to bring avoidable suffering or injury into the world is wrong. But unless the injury to the individual is so great as to make life intolerable, then this individual is not thereby wronged' (Harris 1998a: 91). A striking example of this is discussion surrounding a proposal to solve the shortage of donated human eggs for use in fertility treatment by taking eggs from cadavers or from aborted fetuses. While accepting that the interests of children born in this way might be involved here, John Harris considers the relevant question to be how such children might *feel* about their situation. His answer is another question: 'Will this knowledge be so terrible that it would be better that no such children had ever been or were even born? It is difficult to be certain how to answer this question, but it is surely unlikely that the consequences would be unacceptably terrible' (Harris 1998b: 14). As far as objective assessment of the child's interest is concerned, he continues: 'One question we should ask is whether the act of producing such a child is in the overall interests of the individual who is thereby produced, or is wrongful for some other reason. In the expectation that it will live a normal lifespan and have a reasonably favourable balance of happiness over misery in its life, it is overwhelmingly likely that the individual will have what would be objectively judged to be a worthwhile life' (ibid.). But while no-one is injured by not being conceived, and none of us can imagine regretting *not* having been born, most people can imagine what it would be like, unfortunately, to wish they had never been born. And it is worth noticing, just as a matter of interest, that the people who put forward the argument that it is better to exist than not to exist in connection with Assisted Reproduction never seem to entertain it for a moment when discussing abortion, where even minor disadvantages like being born into too large a family, or possible poverty, or suffering from minor handicaps, are taken as acceptable reasons for not bringing a fetus to term. Nor are they usually opposed to screening embryos for serious disease. But it is inconsistent to favour either abortion, or the selection of embryos, for reasons like these, and at the same time to be opposed to thinking in advance about the welfare of merely possible children. Indeed, now that PGD (Preimplantation Genetic Diagnosis) is available, the door is opened to 'wrongful life' cases brought either by parents of handicapped children who have not been offered testing, or by the children themselves against their parents for allowing them to be born.

Others opposed to making consideration of the welfare of the ART child a legal requirement do so because they do not want to take seriously the future claims of an embryo. They tend to take a blinkered view of what an embryo is. It has no thoughts, feelings or expectations, so is 'morally insignificant'. Its moral claims on us are nil. Susan Okin puts this point succinctly: 'a human infant originates from a minute quantity of abundantly available and otherwise useless resources' (Okin 1989: 83). But the idea of protecting future claims before their owner can assert them is well-established in both law and ethics. For example, an infant's inheritance can be protected, and a child can apply for compensation for an incapacitating injury it suffered at the fetal stage, providing it survives those injuries and passes the birth threshold. It is possible to make excessive claims about embryos; nevertheless, these are surely significant considerations.

4 The Right to Found a Family

There is, nevertheless, another side to the matter. For while children cannot be hurt by not being born, it is often argued that adults may be injured by being prevented from having children, and even by not being helped to have them when that would be possible. So is there a right to reproduce? And if so, how should it be interpreted? These questions have received much philosophical attention. Looking at the matter from a legal and political point of view, Ronald Dworkin argues that: 'the principle of procreative autonomy, in a broad sense, is embedded in any genuinely democratic culture' (Dworkin 1993: 166–167). Commenting from an ethical perspective, John Harris, agreeing with this judgement of Dworkin, has modified his own utilitarian position to discuss reproductive choice in terms of legally instituted rights. He explains: 'In so far as decisions to reproduce in particular ways or even using particular technologies constitute decisions concerning central issues of value, then arguably the freedom to make them is guaranteed not only by the United States Constitution but by the constitution (written or not) of any democratic society, unless the state has a compelling reason for denying that control' (Harris 1998b: 36). And John Robertson who, like Harris, defends a *prima facie* moral right to reproduce, argues that control over reproduction is 'central to personal identity, to dignity, and to the meaning of one's life' (Robertson 1994: 24). Robertson uses the principle of reproductive choice to defend various kinds of 'collaborative reproduction' as well as commissioned pregnancies, paid adoptions and similar contracts. He writes:

'An ethic of personal autonomy as well as ethics of community or family should ... recognize a presumption in favor of most personal reproductive choices' (ibid.: 145). Citing cases such as Romania under Ceausescu where contraception and abortion were forbidden, and China where forced abortion and sterilization have been used to enforce the one-child policy, he argues that the burden of proof lies on those who would limit freedom in this area.

But how strong is the case for extending the interpretation of reproductive choice in the way these commentators propose? This must depend, at least to some extent, on how far their case can indeed be supported by rights that have already secured international recognition. Two rights in particular are cited: the right to privacy, and the right to found a family. The first is expressed in Article 8 of the European Convention in these terms: 'Everyone has the right to respect for his private and family life, his home and his correspondence.'[4] The second is put in Article 16 of the UN Declaration: 'Men and women of full age, without any limitation due to race, nationality or religion, have the right to marry and to found a family.'[5] Here it is worth pointing out – and it is something that these commentators themselves are ready to concede – that what the authors of the international declarations had in mind was not technological assistance in child-bearing, but rather the possibility that a totalitarian state might attempt to *prevent* people having children by methods such as forced sterilisation or abortion, as in the cases Robertson himself mentions. A further and perhaps stronger consideration is one succinctly put by Maura A. Ryan: 'the success of Robertson's argument depends on accepting the view that persons can be the object of another's right ... he is asserting the right to acquire a human being.' Such a position, she argues, fails to respect offspring as autonomous beings (Ryan 1990: 7).

Three steps are conflated in the reasoning that has led so many moral commentators to stretch the case beyond this limited boundary. The first step, which the international declarations were certainly intended to support, is to defend the freedom of two individuals to marry and have children together. The second step is to advocate a right to receive assistance from medicine and science to do this – something that, even if not necessarily included in the first freedom, is not incompatible with it. The third step, though, has a more problematic status: this is the claimed right to be helped to have children using donated gametes, and it has come into the picture only because some of the resources and assistance that can be used to help people to have their own genetic offspring can also be used to help them have the genetic offspring of others. This immediately expands the circle of stake-holders or concerned individuals and changes the role and responsibilities of the decision-makers. A right of this sort, then, is far from being included in the right to found a family and indeed in some ways runs counter to it. So are the international covenants to be read as endorsing the progression of reasoning through each of the steps that lead from a right to marry and found a family to a right to medical intervention for this purpose using donated gametes? For, to begin with, the conventions apply only within a very specific framework – that of a couple who are in a position to marry. So neither a man nor a woman separately can find support in these conventions. As far as men are concerned, subsequent legal decisions have found that a man has no right to prevent his wife using contraception or having an abortion, and since he certainly has no right to compel any woman other than his wife to bear a child for him, it would seem that Article 8 of the European Convention cannot be read as giving a man *as an individual* an enforceable right to procreate. But unless a gender imbalance is accepted, this throws into doubt, too, a *woman's* right to reproduce. That conclusion seems to have been indirectly accepted in English law, though possibly not in some other European countries, in relation to assisted reproduction involving artificial insemination, since in situations where sperm has been stored or where sperm could be taken from a man who has died, the man's consent to the use of his sperm is judged to be essential to its use. Hence a woman's putative right, too, would seem to be qualified, even if marriage is discounted, by her need to find a willing procreative partner.

But some would argue that the advent of IVF has changed all this. In the case of a woman, a voluntary sperm donation service does, after all, give her a practical way of exercising her right to have a child with the help of a consenting male person and for men, too, a parallel might be claimed because of the possibility of obtaining the services of a surrogate and an egg-donor, whether the surrogate herself or another woman. This might seem to suggest that a right to procreate could, after all, be asserted and be enforceable outside the 'couple' framework. Of course, there is nothing to stop nations or jurisdictions creating a legal right of this sort. Nor does it prevent the assertion, informally, of a moral right. But for this, it is necessary to persuade others to accept the case. IVF is a collaborative procedure, so one person's right appears to entail another person's duty – in this case that of medical and technical personnel to assist in the procedure. Even if this were considered achievable, perhaps by following the practice in the case of abortion where only supportive medical staff would be involved, it would require some reinterpretation of the conventions so as to make them compatible with the possibilities opened up by new technologies of reproduction.

Again, though, this is to approach the issues involved from the perspective of the parents and would-be parents. But if we shift our perspective to look at matters the other way round, we might find ourselves asking a different question: are children losing something valuable, perhaps indeed a basic human right, if the reproductive choices of others, supported by medical and scientific expertise, necessarily deprive them of a genetic link to those who will have care and control of them till adulthood? In particular, how should we rate the loss (in the sense of deliberate deprivation) of either a father or a mother? The early loss of a mother by death or abandonment is viewed by most people as a particular tragedy. The same applies, though possibly not in the same way, to the early loss of a father. By 'early' loss, I mean loss before the child is old enough to understand, or indeed before the child is actually born. Of course, the later loss of either parent is likely to be deeply traumatic. But I would argue that the sense that a child has suffered an irreparable loss if circumstances have deprived it of either or both parents should dictate a much more cautious approach to the transfer of gametes and embryos and certainly support the view that any commerce in these materials for reproductive purposes is ethically wrong.

5 Does Human Genetic Material have Special Status? The Commodification Issue

Despite these arguments, many would say that donor-assisted reproduction does not raise any important human rights issues as far as the children involved are concerned. This may be because they discount the significance of original genetic material – eggs, sperm, or embryos. And since demand for the raw material of baby manufacture far outstrips supply, particularly when the requirement for eggs and embryos for stem-cell research are also taken into account, one solution that occurs to people who, for whatever reason, have an interest in increasing the supply is to introduce a financial incentive into the equation. But a special taboo has long been attached to the sale of human genetic material. There are a number of reasons for this. First, there is a general objection to the 'instrumentalisation' of the human body that applies, as well, to the sale of organs and tissue. Second, there is a well-founded fear that financially vulnerable individuals could be exploited. And thirdly, there is a general judgement that the sale of human eggs, sperm or embryos is contrary to human dignity. This aversion to the sale of genetic materials has been formally expressed in a number of declarations by international bodies.

The International Bioethics Committee of UNESCO (United Nations Educational, Scientific & Cultural Organization) has ruled that the transfer of human embryos can never be a commercial transaction and that measures should be taken to discourage any financial incentive. The Council of Europe has stated that the human body and its parts should not, as such, give rise to financial gain. And the European Union, too, has insisted that the prohibition on making the human body and its parts a source of financial gain must be respected.[6]

Individual countries, too, have adopted laws on the matter: Sweden threatens up to two years imprisonment for anyone who seeks to profit from the transfer of biological material from a living or a dead human or tissue from an aborted fetus. Switzerland prohibits the gift of embryos and any commercial transaction involving human germinal material and any resulting products from embryos. The case of Australia, though, may be more typical of what can happen in practice. While it is an offence there to intentionally give or receive value for the supply of human eggs, sperm, or embryos, and a 10-year jail sentence may be imposed for trading commercially in human eggs or embryos, Australians are bypassing the law by travelling to the USA to achieve what they cannot access in their home–country.[7]

Despite the international agreements which exist to ban such transactions, then, the fact is that they can be avoided, not least because they are not outlawed in the USA.[8] There are, of course, other countries with a permissive regime. But even countries which would prefer to appear to be adhering to international agreements on the matter are seeking ways to side-step the isssue. In the United Kingdom, proposals for payment described as compensation plus comprehensive expenses may, intentionally or not, be a de facto way of achieving this. So is compensation payment? In the United Kingdom, payment for supplying gametes and embryos is prohibited but it does allow reimbursement of donors' expenses and, in egg-sharing arrangements, women can be offered free IVF treatment in exchange for donating eggs.[9]

Given that there is a demand, it is not surprising that the whole matter may be viewed by entrepreneurs in business terms. My own view is that argument cannot settle these matters, which are deeply intuitive. They bring into question conceptions of family, social and legal conventions, and a judgement about the value of nature versus human artifice.[10]

6 Rights and the Adoption Analogy

Let me suggest, then, trite though it may sound, that a child has a special need for its mother, and that the separation of mother and child should not be taken lightly. But who *is* the mother? Except in certain special cases, egg donation does deprive a child of its genetic mother. But apart from when a surrogacy contract has been made to exclude this, it does provide someone else who has a close claim – and many would say a sufficiently close claim – to be a mother in another sense, i.e., she has invested her own body in supporting the essential stages from embryonic to infant life, creating another kind of intimate biological bond. In a number of countries, such as the United Kingdom, it is the birth mother who has legal recognition. I would suggest, though, that for reasons, both medical and social, that have already been widely discussed in relation to sperm donation, it is a step too far to expunge from all records the genetic origin and hence ancestry of the child. I would also suggest that any such arrangements are morally invalidated if the egg donation is not genuinely voluntary, and it is doubtful if it is so in egg-sharing arrangements where rich paying patients are the recipients and poor 'free' patients are the donors. As for the overt commercial sale of eggs for reproductive purposes, this is offensive to conceptions of the value of human life that it has taken millennia to establish and which we sometimes claim, even if a degree of self-deception is involved, as the foundation of our twenty-first century civilisation. But what of the loss of a father? Here some difference has to be acknowledged. For example, where sperm donors often prefer not to know if they have fathered children, women who donate eggs seem to have a much stronger sense that they are parting with their biological sons or daughters and a deeper sense of involvement and concern.

To sum up the judgements involved here, I believe the focus in Assisted Reproduction using donated gametes should be on the rights of the children actually born by assisted reproduction and only in a qualified way on those of adults who want to make use of it. Because children born from donated gametes are in one important sense – genetically and biologically, although not socially and legally – someone else's children, a situation is created that lies somewhere between natural procreation on the one hand and the adoption of an existing infant on the other. One important feature of adoption in the present day is that adopted children *are* acknowledged to have a right, not only to knowledge of their origins, but also to their own cultural and ethnic identity. Indeed, Article 8 of the United Nations 1989 Convention on the Rights of the Child specifies that: 'States Parties undertake to respect the right of the child to preserve his or her identity, including nationality, name and family relations as recognized by law without unlawful interference.'

7 The Case of the New 'Disappeared'

This principle has been conspicuously violated in several instances in modern times. In Argentina, babies were taken from their parents and given to childless supporters of the regime who then brought them up as their own. Some later discovered that they were children of the 'disappeared' – those who had died under torture or been murdered in the prisons of that regime. The reasons for their situation, of course, tragically intensified their situation, but their alienation constituted a loss and a rights deprivation in itself. These children were a new 'disappeared.' Those they believed to have been their parents were not in fact so, and the cultural and kin context in which they found themselves were often completely alien.[11]

There are also some historical cases which, while they fortunately lack those malevolent overtones, are in some ways comparable: for example, that of the 'Stolen Generation' – aboriginal children in Australia placed with incomer families – or that of the children shipped there from England around the time of the Second World War who never saw their original families again. But what if the crime is overt and no deception is involved? Between approximately 1619 and 1850, tens of millions of inhabitants of West and Central Africa were taken to the Americas, and so deprived of their genetic and cultural inheritance. Generations later, many black Americans are seeking their personal roots in Africa.

The analogy between these historical incidents and some of the current developments mentioned earlier may be limited. But, bearing in mind cases like that of the Danish student mentioned earlier whose hundred or more offspring will have unrecognised numbers of siblings and half-siblings worldwide, it is nevertheless worth using the analogy in order to argue that things could be arranged differently. As far as openness is concerned, there could be wider use of voluntary registers and the participation of third parties or mediators. It would also be possible to set things up differently from the outset as has recently happened, for example, in Sweden, New Zealand, and the state of Victoria, where donors of gametes do so agreeing to be ready to be contacted by their adult children if that is what those children want.

Whatever the situation, then, 'children of choice' have a right to at least one choice of their own: a right to choose *knowledge* of their parentage – not, that is, to be deliberately deceived about their origins by a medico-legal conspiracy. Without this, they are born as exiles from the kinship network and are orphans in a sense previously unknown to human beings. They may in fact have unknown half-siblings, cousins, aunts, grandparents, but they will never meet them. Of course, there is every chance that they will be provided by an alternative family network that will provide love and security, but the subtle similarities of genetic relationships may come to haunt them in the future, particularly when they have children of their own and start to look for such things as shared resemblances, attitudes, interests, tendencies, qualities of character and physical features in their own offspring. And, as one philosophical commentator has argued: 'Knowing one's relatives and especially one's parents, provides a kind of self-knowledge that is of irreplaceable value in the life-task of identity formation' (Velleman 2005).

Of course, there are secrets in many families and many people are deceived by family members about their origins for personal reasons and without the intervention of medical science; many may be mistaken about who their father is, but by no means as many as advocates of secrecy claim. A well-founded estimate is 2%.[12] Whatever the exact numbers, however, this is no reason for the state to refuse to divulge information it holds in official records to those for whom the matter is, whether for emotional or medical reasons, pressingly important. It would also suggest reconsideration of moves in a number of jurisdictions to record on birth-certificates as parents, without any qualification, people who are not genetically related to the child. It may also be a reason for a more cautionary approach in general to the new technologies, and for resisting the enthusiastic expansion of the possibilities they hold out.

8 Conclusions

I have argued, then, that the state is right to involve itself in protecting the interests of children born by assisted reproduction and right, in particular, if it acknowledges their right of access to their genetic identity. Is this, as some have claimed, to conflate the ethics of personal decision-making and the ethics of state interference with individual liberty? I do not think so. For while the state should not interfere in what is genuinely the province of the individual, assisted reproductive technology is a compound practice involving medical and legal professionals, taxpayers and most of all, future children who will become rights-bearing adults themselves. All of these, but especially the last, need to be brought into a picture too narrowly focused on individuals and couples.

Nevertheless, it is impossible to ignore the fact that this is an area of special concern to women. Up to now, a woman has been seen as the primary carer, even when not conceded any rights to her child, and, of course, to date, only a woman can gestate and bring to birth a child. There is, too, a common perception, supported by a certain amount of research, that while there are notable exceptions on both sides, women are particularly sensitive to matters of human relationships and perhaps also more ready than men to accept that children count in their own right, not merely as the appendages of adults. To what extent, then, is this a feminist issue? Feminism has stood for procreative liberty, but the emphasis has been on contraception and abortion – freedom from child-bearing. But it is also a matter for female, if not feminist, concern to think about the conditions that should govern freedom *to* procreate where this involves medical intervention and the participation of strangers.

Certainly these issues also concern men, but it is notable that in the famous story of Solomon's judgement, male justice saw the child as divisible, but the true mother was recognised by her willingness to sacrifice her own need to her child's interest. (Solomon himself is, of course, rightly famous for having recognised this.)

To sum up, then: in enabling intervention at the embryonic stage, the new technologies of reproduction have created both new risks and new rights. A commitment to freedom of choice in this area means not only respecting the choices of adults, but also recognising a responsibility to protect the welfare and rights of people at a stage when they are in no position to protect them for themselves. This may mean that some well-established rights, such as the right to found a family, must not be interpreted in a way that would single out the 'donor-conceived' as a class for unequal treatment by blocking their access to their genetic origins and their family relationships or, more controversially, by depriving them of a specific type of parental or other intimate relationship. But equity in the preservation of genetic identity – particularly the rights of the donor-conceived – has not so far not received as much attention as the rights of adults to fertility treatment. As the class of the donor-conceived has reached adult life, and as they have come forward with their own claims, the case they are pressing appears more compelling: not to be treated differently from those whose origins are more conventional and in particular not to be deprived of rights that are enjoyed by the naturally born majority.

Notes

[1] The Denmark-based Cryos International is the world's largest sperm bank. It markets its sperm throughout the world, shipping it to more than 40 countries, including Spain, Paraguay, Kenya, Hong Kong and the United States. Denmark's laws protect donor anonymity and the number of children a donor can father depends on where he lives and where his sperm is sent. In Denmark the limit is 25, a number that is supposed to guard against accidental incest between siblings. In Britain it is 10. In the United States the number is 25 births for each donor within a population of 800,000, according to guidelines issued by the American Society for Reproductive Medicine. Cryos claims a good track record. According to its web-site, (www.cryos.dk), since the company opened in 1987, its banked Danish sperm has led to 10,000 pregnancies around the world.

[2] Rose vs Sec. of State for Health and the HFEA [2002], EWHC 1593.

[3] The landmark work on attachment theory, originally a three volume trilogy, is Goldstein, Solni, Goldstein and Freud 1996.

[4] The European Convention for the Protection of Human Rights and Fundamental Freedoms, 1950, as amended by Protocol no. 11. 1998, Article 8 (Brownlie and Goodwin-Gill 2002: 402).

[5] Universal Declaration of Human Rights, 1948, Article 16 (Brownlie and Goodwin-Gill 2002: 21).

[6] Article 21, *Convention on Human Rights and Biomedicine*, Oviedo 4.4.1997: 'The Human Body and its parts shall not, as such, give rise to financial gain.'

[7] Australians may pay around $40,000 for fertilised eggs to implant and $170,000 for babies borne for them by American surrogates. At the same time, there are other Australian women who travel to the USA to sell their eggs for $20,000. In an unrelated development, protests were raised when Canadian medical students were offered free holidays in Australia in return for sperm donations.

[8] One US-based egg donation program says it has produced 2,500 children worldwide by the sale of eggs. Another program, based in Los Angeles, also supplies these services for gay men.

[9] Under proposals discussed in the United Kingdom in 2005, sperm donation could rise to £50+ expenses, which would offer the donor the chance of earning £2, 500. Egg donors could receive £1000+ expenses, in recognition of the risk and unpleasantness involved in egg retrieval. Expenses would include child-care and loss of earnings, so could be substantial. To determine whether this proposal is in fact a proposal to license the purchase of gametes and embryos, these figures can be compared with the payments made where purchase *is* allowed: Romanian women, for example, receive £150 ($300), i.e. a month's wages.

[10] These broader issues are discussed in my (2006) book. Much of the argument of this paper is also developed on ch. 6 of that book.

[11] Andrew Bainham describes Article 7 of the UN Convention of the Rights of the Child (1989) as a response to the Argentine experience: 'The child shall be registered immediately after birth and shall have the right from birth to a name, the right to acquire a nationality and, as far as possible, the right to know and be cared for by his or her parents.' (Bainham 1999: 37). See also Fortin 1998 and Le Blanc 1995.

[12] Weatherall (1994) reports authoritatively that 'in screening DNA from the PND programme for thalassaemia and other haemoglobin disorders which is run in Oxford for the whole of the UK we have found that this occurs in about 2% of all cases referred to us.'

References

Almond, Brenda. *The Fragmenting Family*. Oxford: Oxford University Press, 2006.

Archard, David. *Children, Rights and Childhood*. London: Routledge, 1993.

Bainham, Andrew. "Parentage, Parenthood and Parental Responsibility." In *What is a Parent? A Socio-legal Analysis*. Andrew Bainham, Shelley Day Schlater and Martin Richards (eds.). Oxford: Hart Publishing, 25–46, 1999.

Brownlie, Ian and Goodwin-Gill, Guy S. (eds.). *Basic Documents on Human Rights*. 4th ed. Oxford: Oxford University Press, 2002.

Dworkin, Ronald. *Life's Dominion: An Argument About Abortion and Euthanasia*. London: HarperCollins, 1993.

Fortin, Jane. *Children's Rights and the Developing Law*. London, Edinburgh and Dublin: Butterworths, 1998.

Goldstein, Joseph; Solni, Albert J.; Goldstein, Sonja and Freud, Anna. *Beyond the Best Interests of the Child: The Least Detrimental Alternative*. New York: The Free Press, 1996.

Harris, John. *Clones, Genes and Immortality: Ethics and the Genetic Revolution*. Oxford: Oxford University Press, 1998a.

Harris, John. "Rights and Reproductive Choice." In *The Future of Human Reproduction*. John Harris and Søren Holm (eds.). Oxford: Oxford University Press, 5–37, 1998b.

Harris, John. *The Value of Life*. London: Routledge & Kegan Paul, 1985.

Jackson, Emily. "Conception and the Irrelevance of the Welfare Principle." *Modern Law Review*, 65, 2, 176–203, 2002.

Le Blanc, Lawrence J. *The Convention on the Rights of the Child*. Lincoln and London: University of Nebraska Press, 1995.

Okin, Susan Moller. *Justice, Gender and the Family*. New York: Basic Books, 1989.

Robertson, John A. *Children of Choice*. Princeton: Princeton University Press, 1994.

Ryan, Maura A. "The Argument for Unlimited Procreative Liberty." *Hastings Center Report*, 6–12, 1990.

Skolnick, Arlene. "Solomon's Children: The New Biologism, Psychological Parenthood, Attachment Theory, and the Best Interests Standard." In *All our Families*. Mary Ann Mason, Arlene Skolnick and Stephen D. Sugarman (eds.). New York: Oxford University Press, 1998.

Velleman, J. David. "Family History." *Philosophical Papers*, 34, 3, 357–378, 2005.

Weatherall, David. "Human Genetic Manipulation." In *Principles of Health Care Ethics*. Raanan Gillon (ed.). London: Wiley, 971–984, 1994.

Assisted Reproduction and the Changing of the Human Body

Maurizio Mori

Let's imagine to be living in the first decade of the twentieth century, and more precisely after October 1st, 1908 when Henry Ford launched his famous "T model", a brand new automobile. In only one year, production and sales of the car doubled. It was a great success. But in the same year the sales of old coaches increased three times as much.[1] If one considers the relative numbers of coaches and cars already in existence, one should realize that in absolute terms coaches were much more requested and had a great advantage over cars. If one of us had to be a member of a commission (ethical or whatever you like) in charge of making a prospective analysis on the future of transportation, he or she would have established that cars would have gained a solid niche in the market, but that coaches would have continued to be in a greater demand. Hardly anyone in the first decade of the twentieth century would have foreseen that in a few years cars would have achieved total dominance and that coaches would disappear, becoming articles for museums.

Let's come now to the present, and think of assisted reproduction. What about the future of such a technique? Part of the answer depends on what we consider assisted reproduction to be. If you think that it is a therapy for infertility, then it may remain only therapeutic help exclusively for infertile people. I agree that assisted reproduction is a good therapy and ought to be used to help infertile people. But it is not only that: it is also a new opportunity for people to reproduce and to have control of human reproduction.

If this is the case, then the future of assisted reproduction will depend on what we think about increasing our capacity to control reproduction and reproductive opportunities.

Some people are against it, because they think that human reproduction should remain beyond human control, because children are a gift of God. In this perspective, they suggest that instead of speaking of "reproduction" it would be more appropriate to speak of "procreation" – as it actually is in French, Italian, Portuguese, and in some other languages. One should remember that "procreation" is (at least in origin) a theological term indicating "creating for": in procreation parents are simply

M. Mori
e-mail: maurizio.mori@unito.it

M. Düwell et al. (eds.), *The Contingent Nature of Life*,

cooperating with God in creating a new person, so that one should keep a veil of mystery over a man's origin. Even some secular thinkers concur on this viewpoint, maintaining that control of reproduction is dehumanizing or against human dignity.

However, the question remains: is control on reproduction morally good or bad? For which reasons do we really have to think that it is bad? To say that procreation or reproduction is a gift is no reason at all. It is verbiage, or playing with words, in order to say something which at the end is meaningless. It could be that such a way of speaking was meaningful when reproduction was a real mystery, as it was until a few decades ago, when humans knew very little about reproduction. In this sense, the thesis saying that we should let nature follow its natural course as a norm was a correct cultural response to that situation. But now historical circumstances have changed, and it is wrong to continue to present them as a gist of wisdom. Secular thinkers' views in this line are to be considered a sort of "cultural survival".

In fact I think we have many reasons to think that it is good to gain control of reproduction. People like to have control of such an important activity as their reproduction, because more control can enhance their lives. Moreover, it is good for offspring who can benefit from the opportunity to begin a life when they are accepted and desired. Birth control in the negative sense – "negative" because preventing life's transmission when undesired – produced an enormous benefit to humankind, in spite of protests on the part of religious traditionalism and conservatism.

If negative birth control was good, I do not see why positive birth control – "positive" because it aims at having new births instead of preventing them – could be bad. In fact it should be even more beneficial, because it permits prospective parents (I use the plural, but that is only a convention) to fulfill their life's plan and enjoy the birth of a new baby, and to the baby to be born.

Different modalities of conception and birth are irrelevant, because the essence is that a new person is there, and that he or she could not be born without such a technique. Carmen Shalev said that the first cloned child could have some psychological troubles. However, if we read the story of "extraordinary births" such as the first caesarean sections on living woman, we see that even in such cases people had problems.[2] More generally, anyone of us could have had or still has problems regarding one's birth, because each one could hope to have been born in a better situation, but as we are happy to be born, we have to accept the burdens connected with our human condition and social circumstances. For this reason I think it is good to achieve control of reproduction and try to have control.

If it is good to have control over reproduction, then my prediction is that in a few decades assisted reproduction will be used regularly in substitution of natural reproduction. This means that we are in a situation similar to the one I described concerning Henry Ford with his "T model". Even in early twentieth century most people would have not imagined that cars would completely replace coaches and other traditional means of transportation: at that time cars were not very efficient, not dependable, smelling and polluting, etc. In short, horses and coaches appeared more convenient and comfortable. However, car technology quickly improved and the situation radically changed.

My view is that something analogous will occur with assisted reproduction. Nowadays reproductive technologies are burdensome and even present more risks than natural reproduction, they have heavy side-effects, a low degree of efficiency and other difficulties, but they will be overcome and the practice will become more routine. It is not simply my wishful thinking that people should use assisted reproduction, but it is a factual statement referred to a future situation.[3]

There are at least three reasons for which assisted reproduction will replace natural reproduction. The *first* is that to have more opportunities to choose to have a child is best for the family. If parents can have more opportunities they can prepare better conditions, which is good for the child. Nowadays the situation is still defective.

The *second* reason is that assisted reproduction offers more opportunities on *when* to have a child. This is important because women will possibly extend their capacity to conceive and have a child later in life. Nowadays men can have children at 60 or 70 (and are praised for that), and technology can help more real equality in the field.

The *third* (and possibly the most important) reason for which assisted reproduction will replace natural reproduction depends on the possibility of pre-implantation genetic diagnosis (PGD) and has to concern the health of prospective children. This issue has direct connection with the topic of my presentation, the changing of the human body, i.e. changing of human nature or human bodies of future generations.

Why is this issue so important for the spread of assisted reproduction? One reason is that it is a deeply rooted idea that a society, as well as parents, should do the "best" for their children. It is important to make clear that "the best" in this context does not imply a duty to provide what is *optimum*. If such an *optimum* were required as a necessary condition, then no one would be born. The *optimum* is a limit-concept which never applies, because otherwise each of us could claim that some better situation was possible. The "best" that parents ought to do is to be understood, not in absolute terms, but simply as an average decent or adequate condition according to the given historical circumstances.

This duty implies that failing to guarantee such adequate conditions is to cause real "harm" to the child, who is deprived of some important component for his or her self-realization or welfare.

Until now the notion of "harm" to the child was limited to what I call "social harm", because the harm in question had to do with impoverished social conditions of the newborn, as for instance in cases where parents are unable to guarantee sufficient food and education.

Nowadays, however, parents also have a duty to prevent what I call a "structural or constitutive harm", i.e. a harm which depends on the very fact that the future child is physically constituted or structured in a certain way so that he or she will suffer more than can be expected in normal circumstances of life – for instance, if it has some sort of illness which can be detected and prevented. This is a new sort of parental duty and responsibility, because in the past the bodily or physical constitution of a newborn was totally dependent on "chancy nature", and parents had simply to accept the natural process. Now, we have the possibility to avoid some painful outcomes and for this reason we have a new responsibility concerning this issue.[4]

Now that prenatal diagnosis is a routine practice for most prospective mothers living in Western countries, parents have also a duty to prevent not only social harms but also constitutive harms. Since PGD is a great opportunity in this sense, and PGD presupposes IVF, I foresee that assisted reproduction will become a routine practice as well.

According to some authors this new possibility of controlling the bodily constitution of future generations is very problematic or simply wrong. They say that it is not at all a mere extension or enlargement of parental duties and responsibilities (as I presented), but it is a great divide in history, because parents have now a new form of power over the newborn. Until now parents could cause some sort of "social harm", but this "harm" is dependent on merely *external* conditions and could be repaired and reversed, while in the case of a "constitutive harm" what is done is *intrinsic* and therefore it cannot be repaired or reversed. Here is the deep difference between the two cases making prenatal diagnosis immoral as well as PGD. If I understand well, this is the core of Habermas' argument. In fact he says that "By means of the irreversible decision made by a given person about the 'natural' constitution of another person, we assist to the birth of a totally new interpersonal relationship. This new relation is a wound to our moral sensibility".[5]

One could dismiss what I call "the Habermas' objection" simply by observing that our current sensibility is not necessarily the top of morality, and that – therefore – it could not be wrong to "wound" it. I our common sense morality is defective and needs to be reformed, then a wound to our moral sensibility is beneficial and even due. I think that this is the case: for this reason, this part of Habermas' argument fails.

But Habermas' argument is invalid also for another reason. In order to see it, we should consider that it rests on the distinction between extrinsic and reversible conditions vs. intrinsic and irreversible conditions: my thesis is that such a distinction is untenable.[6] In fact, from the point of view of the child, both intrinsic and extrinsic factors are irreversible, once they have occurred. There is a sense in which historical constitution is as essential as our biological constitution. Both cannot be changed. If parents cause a "social harm" to a child, this harm is, from the point of view of the child, as irreversible as an alleged "constitutive harm" – dependent on genetics.

But this is not the end of the story. It may well be that in a sense "social harms" are more irreversible than "constitutive ones". For instance, if a child is abused in his or her family, then this harm will last for ever; but if a child is born with an illness, he or she can be cared for as much as we are able, and possibly a new therapy could be discovered in the meantime.

I know that Habermas' objection concerning irreversibility of genetic choice is aimed mainly against attempts to enhance our humanity, i.e., to the possibility to "design babies" as we want. This is the shocking part of the story. Here the distinction between "reversible" and "irreversible" conditions appears to be relevant.

But is it really so, or are we misled by our fear and concern? First of all, it is quite difficult to think about a situation so distant from our current capacities such as the one in which we will be able to "design babies". Now we can only detect some genetic illnesses, in order to prevent suffering, and we are totally unable to modify anything. The possibility of controlling a set of genes is far away from us. But let's imagine that in the future we will be able to "design a baby", i.e. we will be able to modify at will the genetic constitution of a new offspring. Is such a genetic choice in these future (alleged) conditions really "irreversible" and "more irreversible" than an educational choice?

I would like to draw your attention on the fact that, in a sense, already nowadays parents "design" their child through education. For this reason parents have a right to educate their children. So, for instance, religious parents teach their children to have a sense of God and send them to special schools, etc.

In some cases, education is successful and everybody is happy. But often, when the child has grown up, he or she rejects religion and becomes an agnostic or an atheist. Is in this case religious education reversible? Experience shows that in many cases the answer is "no!". There are millions of people suffering neurosis and other psychological diseases because of religious education they received. These persons are trying hard to get rid of it, but not always their efforts are successful. This shows that a parent's decision about religious education sometimes is not reversible.

However, one could retort, the situation would be even worse if parents could decide about genetic constitution of the child. But is it true? Let's assume that we know that a person's religiosity depends (at least in part) from a given gene or requires the cooperation of at least one gene. Granted that, suppose that our religious parents decide that their child will have the appropriate education as well as the special gene, so that he or she will be a religious person.

Our question is: is the parents' decision for their child's religiosity in this case more irreversible than in the former one in which their choice was limited to education?

I think that Habermas would answer a prompt "yes". In this case, the implicit presumption is that our thinking and feeling is totally dependent on proper genes, so that we are nothing but puppets. This is genetic determinism. But if it is so, then – like it or not – we are puppets nevertheless, even if now we do not know it. This means that if one is a religious person, he or she is so not because after reflection he or she decided to accept a faith, but simply because he or she has the relevant gene. The same thing if he or she is not religious. In this case it is true that parents' decisions about a child's genetic constitution is really irreversible, because it creates a religious person.

But, if it is so, what is wrong with this choice? Parents' decisions on genetic constitution is only instrumental to their eductional choice, so that they can successfully have a religious child. Such a choice is only good, because it prevents the many disasters of our times. Look what happens nowadays: we have many good religious parents trying to have religious children. They do all the best for that, and struggle hard, but in the end they fail, because their children become atheists. The tragedy is that this occur simply because these children do not have the proper gene for religiosity. Similar tragedies occur in the atheist's house: good atheists try to inculcate secular morality, but then their children, once grown up, become nuns or monks, simply because they have the gene for religiosity.[7]

If genetic determinism is true, then the parents' choice in one sense is irrelevant (because it is not even a real choice), and in another sense it is good, because it increases success in education.

Let's assume however that genetic determinism is false. Even if religiosity depends on the activity of a special gene, this does not mean that the gene is a sufficient condition for the outcome, but it provides only some sort of susceptibility. Religious attitude has a genetic basis, but it is the result of a decision and a virtuous habit.

If this is the case, how shall we answer our question? Is the parent's decision about the genetic constitution of their child more irreversible than their decision about his or her education?

The answer appears to be a clear: "no!". In fact, at least in one sense it is more reversible than the former. Assume that once grown up the religiously modified child chooses not to be religious. At this point, knowing that he or she has the gene for religion, he or she will procede to remove or disactivate it. At a time in which we shall have so much knowledge as to choose appropriate genes, it will also be possible to remove, activate and modify them. So parents' choice won't be as irreversible as it is assumed.

Conclusion. Assisted reproduction offers new opportunities of controlling timing and structural constitution of our children, a control which appears to be much needed for many people. For this reason, I foresee that assisted reproduction will take over as occurred with cars, and in the future natural reproduction will become obsolete – analogously to what occurred to coaches and horses.

Notes

[1] I had this piece of information during a visit at the Museum of automobile in Turin in 2003. A guide was explaining the history of the first cars and mentioned such a datum. I tried hard to find a more official reference but I wasn't able to get it. Even if it were not fully precise, it is reliable for our purposes and in any case it supports an idea which was very common until about the Second world war and that I have personally heard reported by many elders, i.e.: "Horses and coaches are so good that they will always be used: they always existed and they will continue to be with us as major means of transportation".

[2] On this issue there is an excellent book by N. Filippini (2000), *La nascite straordinarie*. Many stories are presented, and one is quite significant: at the end of XIX century in Cittadella (a town near Padua) a child was born through cesarean section on a living woman and both survived: the fact was so astonishing that the municipality granted a public pension to him in order to provide him with an adequate endowment to study.

[3] In this sense when I say "will replace" it is not a normative statement, but a statistical one: parents will find more convenient to resort to assisted reproduction in order to have children.

[4] Here we have to face the "doing – letting happen" issue, which is most controversial in bioethics. However, I hold that it is accepted that parents have a strong duty to provide a decent conditions of life for they children. Up to now these conditions applied only to the social and economical aspects of life, while now we can enlarge our view also to what pertains to "bodily consitution".

[5] Habermas 2002: 30 (my translation). The German text is: "Mit der irreversiblen Entscheidung, die eine Person über die 'natürlische' Ausstattung einer anderen Person trifft, ensteht eine bisher unbekannte interpersonale Beziehung. Diese Beziehung neuen Typs verletzt unser moralisches Empfindend".

[6] I take that this issue is the crucial one concerning the notion of "natural constitution" of a person, a notion which is ambiguous and evanescent. Other senses of "natural" are misleading unless more exactly defined, and I take that mine is a fair understanding of Habermas' point.
[7] I take that this outcome is a "tragedy" because it is contrary to the parents' original plans of life and it is a source of deep pain and suffering for the people involved.

References

Filippini, N. *La nascite straordinarie*. Milano: Franco Angeli, 2000.
Habermas, J. *Die Zukunft der menschlischen Natur: Auf dem Weg zu einer liberalen Eugenik?* Frankfurt am Mainz: Suhrkamp Verlag, 2002.

On the Limits of Liberal Bioethics

A 'Critical Ethics of Responsibility' Approach

Hille Haker

In their book *From Chance to Choice* the bioethicists Allen Buchanan, Dan Brock, Norman Daniels and Daniel Wikler offer a 'moral framework for choices about the use of genetic intervention technologies' (Buchanan et al. 2000: 14). They try to avoid the pitfall of a public policy model that subordinates individual autonomy to a societal, quasi-objective health model on the one hand, and the pitfall of a personal service model in which 'the choice to use genetic interventions is morally equivalent to the decision to buy goods for private consumption in an ordinary market' on the other hand (ibid.). In their view, the former model does not respect the autonomy of the individuals and inadequately prioritizes the prevention of health-related harm, whereas the latter model ignores the 'obligation to prevent harm as well as some of the most basic requirements of justice' (ibid.: 13). In the course of their book, they try to show that only a deontological, liberal moral framework that corresponds with the three principles of reproductive autonomy, harm-prevention, and justice, offers an adequate ethical answer to the new possibilities of genetic interventions.

Since this approach represents a thorough analysis of the questions of health, (parental) responsibility and the scope and limits of respect of individual autonomy in the context of genetic interventions, I will use it as a foil against which I argue for an alternative ethical approach, namely a *critical ethics of responsibility*. Before I can proceed with this approach, I will address the three main themes the authors of 'From chance to choice' deal with.

1 'From Chance to Choice'

1.1 The Subordination of Individual Autonomy or Reproductive Rights to State-Driven Policies

In a survey of historical studies on what they call 'old eugenics', the authors show that the state compulsion model to prevent specific couples from giving birth to

H. Haker
e-mail: H.Haker@em.uni-frankfurt.de

M. Düwell et al. (eds.), *The Contingent Nature of Life*,

children reveals all the flaws of a non-liberal public policy model – and this even in the case of those state-programs which did not go as far as Hitler in declaring a particular, racist health concept as imperative for the 'Third Reich'. While the 'vision' of a society gradually enhancing the health of its population might be correct in theory and only problematic in its specific features, so the authors argue, it is the denial of individual autonomy or reproductive rights of prospective parents and the ignorance of social justice's demands to create equal opportunities that make the old model useless for a moral approach to current and future genetic interventions. However, in their conclusion of the chapter of eugenics they state:

> Reprehensible as much of the eugenic program was, there is something unobjectionable and perhaps even morally required in the part of its motivation that sought to endow future generations with genes that might enable their lives to go better. We need not abandon this motivation if we can pursue it justly
>
> (ibid.: 60).

We can see in this quote an ambivalence that runs through the whole book: the authors clearly accord with the societal goal to enhance the health status of future generations, and they consider genetic interventions as possible means to achieve this goal. However, since history can teach us about the misuse of the eugenic paradigm, genetic interventions as a means of harm-prevention must be limited by two moral constraints: reproductive rights as rights of individuals, and the principles of justice.

1.2 Reproductive Rights and the Limit of State Intervention

Liberalism has been a strong moral and political theory to ensure the rights of individuals, especially against interventions by others or by the state. Situating the ethical reasoning in the US-context of the high cultural valuation of individual autonomy, the authors endorse a liberal bioethical approach of individual reproductive rights. Thus, since the goals of harm-prevention and/or enhancement of the overall health status of a population are affirmed by the authors, the state's role is to ensure that individuals have the means to achieve this goal within the limits of non-intervention. Contrary to other, more radical approaches within political liberalism – the concept of non-intervention and tolerance is not the authors' last word. In fact, the moral reasoning about parental responsibility and health-related justice complements the liberal principle of individual freedom.

Addressing parental responsibility, the authors dedicate one chapter of their book to the question of harm-prevention and enhancement of the health status via genetic intervention. They argue that parental liberties are limited by their obligation not to harm their offspring. As with born children, parental freedom is limited only in those cases, where the future children's harm can be determined, and demands the toleration of those parental decisions where no violations of rights can be proven. Critical about the ideological use of health concepts in the so-called old eugenics, the authors draw on a functional understanding of health and disease to determine harm, which they consider to be common in many medical and public health approaches but which in fact represents C. Boorse's concept that Norman Daniels applied to his approach to public health. According to this functional model, a harmful condition is the absence of 'general-purpose natural capacities' that enable a person to carry out 'nearly any plan of life' (ibid.: 168). These natural capacities are 'capabilities that are broadly valuable across a wide array of life plans and opportunities typically pursued in a society like our own' (ibid.: 174). Disease is understood as 'an adverse deviation from normal species function' (ibid.), calling for 'beneficial' genetic intervention or at least leaving enough room for what parents consider to be best for their offspring.

Surprisingly, this health/disease concept is not analyzed further;[1] although it expresses the liberal societies' value of leading a self-determined life, the focus on 'natural' capacities seems to offer more than that: Like, for example, Martha Nussbaum and Amartya Sen argue in their capabilities approach with regard to developmental economics and the quality of life in poor countries (Nussbaum 2000), Buchanan et al. in fact refer to a normative *anthropological* model of health that conceals its historical and cultural origin in the term 'normal *species* functioning'. Based on the normative determination of (a threshold of) life-quality that is needed to lead a 'good' life, parents are *morally* obliged to intervene in cases where 'deviations' could be treated medically, or where future children assumedly will not cross the threshold of a minimal quality of life. Political liberalism, on the contrary, seems to be doomed to toleration of whatever parents consider best for their (future) children, and hence to non-interference with individuals' reproductive choices. Seeing this tension between the 'good' for a future child and parental reproductive autonomy, the authors choose a middle-ground of 'encouraging' policies of harm-prevention and enhancement, to direct parents towards 'responsible choices'.[2]

As is well-known in the bioethical context, the term 'harm', which is defined as a violation of a right, is ambivalent in the context of prenatal and pre-implantation genetic diagnosis, if – as the authors acknowledge – the avoidance of 'harm' results in selection or abortion of a fetus. Taking up Hare's argument that future persons have no rights that can be harmed (Hare 1993), however, prospective parents can do no harm to embryos or fetuses in their *actual* state, but they can do harm to their *future* (born) children. Moral responsibility is thus addressing the expected future quality of life of a future child, and this predicted health-quality becomes the criterion for a responsible or irresponsible decision. In this understanding of the ethical analysis in line with a probabilistic medical prognosis of a future child's health status, parental responsibility is no longer defined by the partly emotional relation between a couple or, in the case of pregnancy, the woman's relation towards the embryo or fetus. Instead, the future child is considered as bearer of rights, a being who has no *actual* right but a *future* right to 'normal species functioning'. Birth changes the obligations radically: Whereas prenatal, pre-implantation or pre-conception interventions are left to the decision-making of prospective parents (however directed towards a particular choice, namely not to bring into existence children who will not cross the threshold of 'normal' species functioning), adverse intervention in cases of discrimination against disabled children or adults is indeed an obligation of state authorities. This shift in the analysis of parental responsibility entails many presuppositions that would need to be argued and articulated. As we will see later, among other things this unduly reduces the ethical reasoning of parental responsibility to the consideration of (future) rights and forecloses any responsibilities that might *not* correspond directly with the right of a right-holder.[3] The authors, however, argue that for reasons of justice, namely to provide or promote equal opportunities for every member of a given society, their concept of parental responsibility and state-intervention is to be adopted.

1.3 Justice as Commitment to Equal Opportunity

The tension between the principle of harm-prevention resulting in a normative concept of parental responsibility and the principle of reproductive freedom or non-intervention by the state becomes evident when the cases of 'persons with serious disabilities' are considered. The limitation of compulsory state-intervention, as we have seen already, does not leave the state altogether passive; while it calls for the tolerance of individual decisions, it may still establish policies directed at the societies' goal to improve the health status of its members:

> Indeed, a policy designed to encourage prospective parents to avoid the birth of persons with serious disabilities might be implemented while still pursuing strong antidiscriminatory policies to support people with disabilities, without any inconsistency.
>
> (Buchanan et al. 2000: 184)

In the discussion of genetic interventions, the disability movement has criticized that such a promotion of prenatal or pre-implantation genetic intervention is itself discriminative because it expresses the devaluation of disabled persons (Parens 2000; Graumann 2005). Buchanan et al. try to show at length that this claim is right if or insofar genetic interventions lead to new forms of moral exclusion of children or adults, and they praise the disability movement for raising the level of social sensitivity. However, the claim is considered wrong if and insofar it ignores the 'legitimate interest that people have in avoiding disabilities' (Buchanan et al. 2000: 270). Quite contrary to the disability movements' self-understanding, they argue, denying individuals the means to correct or prevent genetic defects that would put their equal opportunity at risk is in fact a violation of the principle of justice.

This reproach is rather serious and would, if correct, question a fundamental critique of genetic interventions that has been raised over and over again in the past decades. Therefore, a closer look is necessary. The authors understand 'justice' in line of Rawls' approach and Norman Daniels' application to health care (Daniels 1991). Against this broader theoretical background, justice is defined mainly as the commitment to equal opportunity. Genetic interventions, as we have already seen, do not violate the rights of any (actual) right-holders in the case of selection or abortion, and they do not violate the rights of future children, if their health-status is indeed improved by the techniques, or if the risk is considered 'minimal', which – as in the case of reproductive cloning – might not be the case today. Furthermore and more important for the discrimination argument, the claim that genetic interventions *indirectly* violate the *rights* of disabled persons, e.g. by reducing social support for them and decreasing efforts to improve their equal opportunities, is weak because it does not prove true in contemporary societies, because these have continuously improved the equal opportunities of disabled persons. Finally, the 'expressivist objection', claiming that genetic interventions express an attitude of repudiation of disabled persons, insinuates a disingenuous motivation of those who avoid having offspring with disabilities. This, however, is a mistake, as we have seen, because avoidance of disabled offspring is compatible with the moral recognition of disabled persons. Hence, this point is emphasized again, now from the perspective of social justice:

> We devalue disabilities because we value the opportunities and welfare of the people who have them. And it is because we value people, all people, that we care about limitations on their welfare and opportunities. . . .Thus there is nothing irrational, motivationally incoherent, or disingenuous in saying that we devalue the disabilities and wish to reduce their incidence while valuing existing persons with disabilities, and that we value them the same as those who do not have disabilities
>
> (Buchanan et al. 2000: 278).

Even though the offense that disabled persons may feel is understandable, 'a liberal society cannot count offense. . .as a sufficient ground for curtailing liberty' (ibid.: 281). Thus, while the justice-focus of liberal bioethics calls for the social and political obligation to improve the equal opportunities of persons with disabilities, it cannot deal with questions of recognition and esteem (considered to be 'expressivist' rather than strictly normative concepts) within its normative framework of rights and obligations, so that this dimension is excluded from the ethical reasoning of parental responsibility and social justice.

Nevertheless, the authors try to take the arguments of the disability movement into account, and distinguish between 'impairment', meaning the 'impairment of normal species functioning' (ibid.: 285), and relational, socially constructed 'disability', meaning the inability to 'perform some significant range of tasks or functions that individuals in someone's reference group . . . are ordinarily able to do, . . . where the inability is not due to simple and easily corrigible ignorance or to a lack of the tools or means ordinarily available for performing such tasks or functions' (ibid.: 286).

Further analysis would be needed, but important for me here is only the justice perspective, which the authors connect to the morality of inclusion, again trying to integrate the perspective brought forward by feminist political ethics and disability ethics. Within a so-called 'dominant cooperative framework' that constitutes the different structures of social interaction and that constitutes the cultural, social and political shape of a given society, the social status of its members is determined by way of their capability to participate in the social life – thus being included in the dominant cooperative scheme is a fundamental interest of individuals. From the moral point of view, the justice principle serves as the decisive force to establish equal opportunity and hereby the inclusion of all members. In line with the *Americans with Disability Act*, the authors emphasize that this 'commitment to equality of opportunity requires efforts to prevent disabling impairments' (ibid.: 292).

While it is certainly true that socially disabling practices need to be decreased, including those of practical or legal inequality, but also those of expressivist prejudice, the authors conceal what it practically involves to prevent the *impairments* resulting in the social disabling practices. In revealing their reluctance to follow the view of the social construction of disability by emphasizing the biological impairment, not societies' lack of adaptation to the needs of persons with disabilities, as cause for the lack of chances, the authors whitewash what – at least for the time being – prevention of harm and enhancement of health on the genetic level in fact implies: the prevention of handicapped children to be born and (if possible) the promotion of procreating the birth of children with specific 'enhanced' characteristics. The central argument, however, namely that 'life' in particular circumstances becomes *objectively* intolerable because participation in social practices is impossible by reason of a given health status – is more assumed than argued for. Neither can the authors claim that the assumed health status of future children can be determined precisely nor do they elaborate on what participation in social practices means in detail. The dependence of a right to life on normal species functioning, and the definition of justice as equal opportunity to participate in social practices are too narrow concepts for the determination of the question of parental responsibility and future policies of genetic interventions.

2 Respect vs Recognition?

2.1 The Unsolved Problems of Liberal Bioethics

My response to *From Chance to Choice* will proceed rather indirectly than directly. However, let me start with three points of disagreement:

1. *The functional concept of health and disease* conceals its origin in the social construction of 'normal species functioning'; as such it serves as a normative basis for the determination of harm that is valid on the basis of persons' rights but changes dramatically if the harm principle is used to determine thresholds of qualities of life. There are implicit rather than explicit normative assumptions in this concept, hidden, for example, in the term 'normality'. To conceive of normality as criterion for having a 'decent' quality of life or being able to live a 'good' life, does not only run the risk of a naturalistic fallacy but moreover makes a fundamental mistake of ethical reasoning: it confuses the normative claim to guarantee human beings the conditions for a (minimal or average) standard of living that is necessarily linked to a concept of 'normal functioning', i.e. including access to food, housing, education, etc. with the *valuation* of a human being who (for whatever reasons) does not have access to or does not have the means to reach this standard of normality. This devaluation was exactly the turn the old eugenics movement took, long before compulsory measures to prevent the birth of particular human beings for reasons of a social health ideology were even discussed. Language played and plays an important role in this shift,

euphemizing the implementation of the quality-of-life-threshold criterion in ethical decision-making as 'responsibility' or 'altruism'. Furthermore, the authors fail to see that through the shift from the social concept of prevention to the individualistic concept of prevention the objective of prevention does not change a bit. Furthermore, via this redefinition of parental responsibility a particular group of parents is becoming vulnerable to social misrecognition and pressurized to 'choose' the apparently *socially desired* and *morally required* selection and abortion. This is clearly in tension with the concept of reproductive autonomy.

2. *The ethical concept of parental responsibility* is not only dependent on the concept of health and harm but also on the moral status of embryos and fetuses. The authors do not discuss the moral status from a relational perspective or from the perspective of pregnant women. Instead, they reduce the pre-birth development as mere (biological) condition to become a person of (moral) rights. This, however, reduces not only the debate on the moral status but also suppresses the concerns of feminist ethics, namely that women are pushed into the background of the ethical consideration.
3. The authors present a reductionist concept of the interaction and interdependence of the private sphere of individuals, social interactions, and state actions. But between the individual and the state, social imaginaries play a decisive role that need to be considered in ethical analyses (Taylor 2004). *From Chance to Choice* offers almost no analysis of the *social constructions and imageries* that constitute both social norms and the hermeneutical frameworks of interpretation. It does not consider evaluative differences and tensions between economic, scientific and cultural understandings of, for example, life, health, quality of life, disease or suffering. Leaving the social dynamics arising from the different interpretations unanalyzed, the authors suppress the discussion of what has been called 'struggles for recognition' as part of the social life.[4] It is, however, this interaction that many critics of genetic interventions try to bring forward to the bioethical discourse (Haker 2002).

As we have seen, the liberal deontological ethical framework strives to endorse the Kantian notion of respect for the moral subject's freedom, taking her autonomy and equality as starting point for reciprocal relations. To respect the other means not to interfere with her 'striving for a good life' (negative obligations), or to promote his or her well-being where this is needed (positive obligations). The latter obligations are, of course, highly controversial, but within the Rawlsian framework, the authors of *From Chance to Choice* favor the welfare-state view of society's obligations to promote equal opportunities in order to compensate given inequalities.

However, since the 1980s, communitarian ethics criticized liberalism not only for its 'atomistic' concept of the self but also for its generalizing view on the needs of individuals. The individuality and concreteness of the other, they claim, easily slips away, if respect for the other is defined more or less in line with the egalitarian perspective, focusing on shared interests and rights. In Rawls' *Theory of Justice*, for example, it is evident that he presupposes that every member of a society shares certain (basic, moral) interests with others; and Nussbaum's *Human*

Capability Approach endorses a common anthropological concept that emphasized 'sameness' rather than 'difference'.[5] On the one hand, one could say that the reference to a shared humanity or common interests at least provides us with a normative criterion to decide who is the addressee of respect, what respect calls for and, in the borderline cases, where actions of 'patronizing altruism' are necessary for the well-being of the addressee. On the other hand, however, this approach easily ignores that 'main' interpretations of basic social concepts express only the dominant views within societies. Normativity disguises not formally but very well in its content, so to say, its interdependence with social norms creating standards of normality. Foucault and others have shown convincingly that 'health', 'disease' and 'impairment' are by no means anthropological or biological 'givens' but historically situated and evaluative concepts, based on struggles of interpretation – and social exclusion. The litmus test for an ethics of the 21st century is precisely its ability to reflect upon its own situatedness within given societies and this interdependence of normativity and social norms, and its ability to be self-critical to its own potential of exclusions and violence.

Hence, if liberal ethics of genetic interventions tries to learn from history, it is not enough to look at the history of eugenics; it is also necessary to look at liberalism's own history, which at times contributed to or connived with those who endorsed the exclusion of human beings from the moral community – human beings whom moral subjects today would 'grant' this status without hesitation, as slaves and women, or persons with disabilities. The struggle for recognition these groups have faced in the modern political history, was also in part a struggle for the inclusion within the moral community of moral agents. These experiences should remind us that an egalitarian ethics is still dependent on the interpretation of who counts as equal. The ethics of moral inclusion Buchanan et al. correctly endorse must hence be *critically* analyzed against the background of social, political and ethical blindness – in the case of prenatal selection on the basis of a predicted lack of 'normal species functioning', the question whether a fetus is to be treated equally to all other human beings and hence to be granted at least his or her right to life, is not trivial.[6] Within the liberal framework, I will show that the 'respect for the equality of others' must be complemented by 'recognition of the concrete situation of the other', in order to escape the trap of generalization.

I will now proceed with my own approach of a *critical ethics of responsibility*. This approach takes up the Kantian deontological perspective of respect and dignity, spelled out in the social sphere as justice, i.e. the inclusion by means of fair distribution and compensation for inequality on the basis of moral equality. But I will go beyond this specific framework in two important ways: Firstly and very much in line with liberalism but reversing it from within, I will turn to the social-ethical concept of a recognition ethics that is sensitive to who is excluded from a particular social framework. Here, only the *methodological negativism* of critical theory can provide ethics with the tools of permanently reassessing the social and moral norms. Secondly, I am searching for a more appropriate basis of the moral self-other-relation than both the concept of respect and recognition allow us to address; this basis I find in the Levinasian concept of re-sponsibility, because it allows me to transcend the rights-based concept and establish a responsibility based on the fundamental claim to take the other (whoever this might be) serious *as* another.

2.2 The Ethics of Recognition

In recent years, the ethics of recognition has been brought forward as a valuable complement of either liberalism or communitarianism (Honneth 1992; Taylor 1994; Markell 2003; Christman 2005; Ricoeur 2005). Recognition theories start with the acknowledgment that the moral self is constituted as a social self, emerging from a long-lasting forming of identity that begins with almost complete dependence on the emotional devotion and recognition of (biological or social) parents and only gradually leads to the transcendence of these asymmetrical relations. With view to our context of genetic interventions, I should note that this dependence and indeed subjection to the power of others is a fact independent of the 'full' moral status and starts long before birth, the origin of which is possibly more in the dark than is often assumed in the debate on the moral status of the embryo. In view of the Western aging societies, we could also add that the social self-model and the relational autonomy of the recognition ethics is perhaps all the more needed to deal with the decline of capabilities in the later phases of life, which may leave the self again dependent on the care, devotion and recognition of others.

Whereas recognition theory does indeed regard the social self as constituted by personal and social relations, it still aims at autonomy as self-determination within the social interactions. According to Axel Honneth's concept, for example, the task of ethics is to analyze the structures of injustice and misrecognition, to call for the decrease of the potential and actual violence of personal relations, social structures and institutions, and finally to call for the compensation of injustices.

Honneth's approach enables us to analyze the vulnerability of the self and her need to be recognized by others; but it also provides a framework how over and above determining the violation of rights it is possible to examine the social dynamics of exclusion via different forms of misrecognition and denial of esteem. If the concept of universal respect calls for the acknowledgment of *equality* of others with respect to her needs and rights, the ethics of recognition calls for the acknowledgment of others' *individuality* and *particularity* by taking their subjective and affective expression of social experiences serious. For example, the political struggle of the disability movement can be interpreted (a) as a struggle for legal equality, (b) as moral critique of their exclusion from social life – and, in the extreme, from humanity, and (c) as plea to be made and have the chance to be visible in their distinct individuality.[7] Their critique is not just the expression of 'offense' but also the outcry against the social norms assuming that specific life-conditions are not considered as possible conditions *of* the human species but rather labeled as 'deviation' *from* it. It is the normativity of a constructed normality that creates the misrecognition; and on this level 'exclusion' is not merely an empirical fact that can be countered with statistics of improved life-conditions, precisely because the excluding force is the *norms* that reiterate the 'otherness' of certain life-conditions as 'deviant otherness'.

Recognition ethics is not to be separated too far from the ethics of respect, however. Rather, it is the other side of it. Its goal is clearly the same as the one for example stated in *From Chance to Choice*: to establish a morality of ever-greater inclusion as demand of an egalitarian understanding of justice as participation. It differs in its 'methodological negativism' (Deranty 2004) or *critical* hermeneutics, taking much more serious than traditional liberalism the experiences of injustice, shame and stigmatization. As Deranty says:

> The truth of the social is not to be found in the consciousness of those who dominate, but in the experience of the dominated
>
> (Deranty 2004).

The difficulty, however, with the approach of recognition ethics is the understanding and status of the other. While recognition ethics provides us with the tools to acknowledge the need for recognition in order to develop a self-identity, it is exactly this teleological framework of identity that risks to reiterate Hegel's mistake of regarding the other in the first place as necessary dimension of *self*-consciousness. Furthermore, the emphasis on reciprocal relations as goal of any social interaction, while taking serious the concreteness of disrespect or misrecognition, assimilates recognition ethics to its liberal counterpart, and maintains the sovereignty of the self which remains the underlying ideal of personal identity as well as of moral identity.

I agree with Judith Butler and poststructuralist theories that the social self-model must be interpreted even more radically than recognition theory does (Butler 2004a; 2005). In this radicalized version of the social constitution of the self, the self not only faces a life-long dependence in different degrees and aspects, but entails a constitutive opacity. The self remains vulnerable not only to all kinds of life contingencies but also to different kinds of power, violence and misrecognition. Power and violence are part of the self-other relation, and they should not too quickly be set aside by turning to the normative model of respect as response to the potential violence. Quite contrary, violence needs to be addressed as necessary element of interaction, and criticized where and when it goes beyond the necessary power over the other, where necessary power becomes arbitrary and unjustified.[8] As Butler holds, the overpowering violence in the self-other relation can only be transcended if the self exposes herself to the other in her vulnerability (and vice versa!), in being interrupted in a particular addressing, and in responding to the other by way of a speech other than judgment.

Nevertheless, in maintaining the important insight of the theory of recognition, namely the necessity to integrate the critical, experiential hermeneutics of injustice, I will now reinterpret the moral subject's vulnerability and non-sovereignty from the perspective of a critical ethics of responsibility

3 A Critical Ethics of Responsibility in the Age of Genetic Diagnosis

If ethics must be designed as well as an ethics of respect as a critical theory of social norms, it will make use of the methodological negativism and *listen* to the concrete experiences of injustice and misrecognition for an ongoing reassessment of social and ethical interpretations.[9] Those who suffer from injustice and misrecognition are better experts in their own cause than ethicists. But ethicists can – and sometimes must – speak on their behalf, and hold their narratives against ethical, political and social frameworks. Neither side may be right or have the 'truth' – but to ignore either side might turn out to be the worst solution.[10] To concretize this turn of perspective, I will reconsider the concept of parental responsibility in the age of genetic interventions, which in fact creates a particular group of parents who *themselves* become vulnerable to social stigmatization and misrecognition.

3.1 The Origin of Morality: The Responsible Response to the Other

As the term indicates semantically, responsibility is the response of the moral self to a situation where she is faced with being addressed by another, leaving open 'who' this other might be. The most important systematic approach to responsibility in recent philosophy has been brought forward by Emmanuel Levinas. He has shown most convincingly that the moral self is first and foremost constituted as responsible self in the encounter with the other, being addressed in a specific way, namely in a plea to act in a way that is a responsible response to the other (Levinas 1998). It is sometimes overlooked that Levinas is not so much interested in a phenomenology of the self or a psychological theory of self-development, but rather tries to grasp the coming-into-existence of the *moral* self. Levinas' concept does not know, so to say, the non-responsible self, and he is not interested in any developmental perspective. In encountering the other, the ('adult') self is in a specific way 'ruptured' in his or her identity – an identity that, as Levinas holds, without the moral perspective is bound to self-centeredness, the (mere) perseverence of one's own life, the 'conatus essendi', to put it in Spinoza's term. This identity is 'ruptured' if the encounter takes place; *if* the other enters the horizon of the self, the egocentric self is conversed into a new kind of identity.[11] For our present concern, I only want to stress that Levinas' interpretation changes ethics, the theory of morality, radically: he argues for the departure from the framework of contractarian-like *reciprocity of the self-other-relation*, the departure from *autonomy* understood as sovereignty of the moral agent, and the departure from the concept of the *symmetry of rights and obligations* as reason for morality.[12]

In contrast to liberalism as well as communitarianism but quite in line with recognition ethics, the ethics of responsibility focuses much more on the encounter itself – what is morally right in a given situation can only be determined by 'listening' to the voice of the other (to be addressed before one addresses the other), by acknowledging the other's presence and plea. This emphasis on the situation of the encounter is not at all a romantic appraisal of harmony between the self and the other; neither is it the abandonment of the other's right to be respected or the turn to situation ethics only. Quite contrary, it puts the burden of moral judgment on the moral agent like it is the case in Kant's concept of autonomy, but the agent is 'craved' by the mere presence of the other not to disrespect her by reducing her to a means of one's own purposes and self-fulfillment.[13] *In short, the encounter with the other is the occasion and reason for morality to become a dimension of the self's horizon of his or her personal identity, and all other spheres of morality are derivations of this origin, the* Ursprung *of responsibility*.[14] To respond to the other as other first and foremost implies to *endure* her otherness, the difference and the gap between me and her; to endure the lack of certainty of what she might demand of me but also to be open to how the encounter might change my own self-understanding, my own self-perception and identity; to question my moral judgments; to interact, to listen, to keep still.[15] Due to our cultures of sovereignty, independence and action, this respect for the other calling for non-sovereignty, dependence and passivity has been more and more alienated from the liberal concept of autonomy – and it is one of the reasons for the critical ethics of responsibility to emphasize this side of the 'otherness' over against the 'sameness' of equality.

In this concept, respect is not primarily constituted by the *rights* of the other; neither are the self and the other primarily occupied in a struggle of reciprocal recognition; acting on behalf of the other, acknowledging the other's otherness, has not much in common with the *patronizing* acting in the 'best interest of the other', which Buchanan et al. perhaps unintentionally argue for. Rather, acting on behalf of the other must be seen in light of a necessary gap between the self and the other, a gap that may in deed unsettle the self in his/her own identity, exposes her to her own vulnerability and impotence as much as to the other's, a gap that maintains the question of how to respond responsibly instead of answering it with reference to a general normative concept. Crucial for the understanding of this approach to responsibility is the fact that impotence, uncertainty about one's right response is part of *any* moral interaction – and not only of those interactions and relations where the other cannot articulate his or her interest.

To be sure – the mere reference to the 'origin' of responsibility in the experiential encounter does *not* solve the normative problem of a morally right action – or at any rate not without further mediations. Different categories of encounters need to be distinguished, however: If the other can articulate his or her interests, he or she can be listened to and her/his claims towards the self can be held against his or her own interests and judgments of what is right. This is also the underlying sense of the liberal political concept of tolerance that confirms the negative freedom of any agent and that leaves room for the interpretation of 'positive' claims to be supported in securing, sustaining or expanding the scope of action (Gewirth 1978, 1996). The more difficult categories of cases are, as we know however, asymmetrical relations: relations in which the 'other' cannot (for whatever reason) articulate his or her interests and is dependent on the interpreted advocacy of the addressee.[16] Here, one might assume, the 'patronizing altruism' is a necessary moral effect of the relation between the self and the other, one that cannot be escaped. All depends, in my view, whether the criteria for the 'acting on behalf of another' are based upon a *general concept of a 'good life'* or quality of life – an approach that is contradictory to liberal ethics, or based upon a *given situation*, a given personal relationship, and based upon the judgments of the persons involved and affected by the different decisions. I hold that only the latter stance is both consistent with the ethics of respect, recognition and the ethics of responsibility – and thereby escapes the reproach of paternalism.

In calling my approach a *critical* ethics of responsibility, I am trying to acknowledge that the moral self, in striving to respect the other, needs to permanently reassess her judgments, either in light of a new situation, or because of our tendency to assume that our own experiences, needs and desires are the same as the other's. We certainly need these assumptions to orient our actions – however, since there will always be a gap that separates me from the other, a hermeneutics of doubt must be a part of my moral identity, and this even more so in cases where I must act on behalf of another.

In order to go beyond the level of personal relationships, I would need to argue now on the level of social ethics. What impact has my approach for the analysis of social imaginaries? How does it change the concept of justice? How is the liberal effort to balance tolerance of reproductive choices and state interventions on behalf of members of a given society affected? To answer these questions, much more than this article would be needed.[17] Instead, I will now concretize shortly why specific attention to the concept of responsibility is needed.

3.2 Parental Responsibility in the Age of Genetic Diagnosis

Parental responsibility is perhaps the most striking paradigm of an encounter with the other in an asymmetrical relation (Jonas 1984; Haker 2002). Furthermore, it entails the acting on behalf of the other's well-being and the un-conditioned respect as inherent normative element of responsibility. Responsibility in parenthood means to acknowledge the otherness of one's child, his or her strangeness and individuality. It does not mean, however, that parents must give up their own desires of self-fulfillment in favor of the well-being of their child or children. However, different to what *From Chance to Choice* suggests, the situation of responsibility before and after birth differs radically not only from the public perspective but also for parents: after birth, a child is not only in a different developmental stage, as it is sometimes argued, but rather a child becomes a 'social' person, a *citizen* of a state who takes (and must take) the responsibility to protect and support it in his or her well-being. The presupposition, of course, is that the state is in a position to be able to take this responsibility. It can even *replace* the biological parents' responsibility. In situations when parents come to the conclusion (or state representatives force them to accept) they cannot care for a child, other persons may become the primary caregivers or social parents of a child.

But what about the phase before birth, during pregnancy? The replacement of responsibilities is obviously not possible. Generally, we presuppose that during pregnancy women want to and can take the responsibility to care for a 'child' (whether we call it embryo, fetus or child does not matter in this respect!), and envision that they can continue to do so after birth. But as we also know well, this is not always the case. In fact, responsibility during pregnancy is probably the best-surveilled category of inter-human relationship – and perhaps a type of relationship that could not be more subjected to social norms and rules of conduct. Nevertheless, the decision to 'refrain from' the socially well-defined responsibility, even if it involves the termination of pregnancy, can only be made by the pregnant woman herself, exactly because *nobody else* can take her place of responsibility. More and more, the flaws of either of the polarizing slogans that defined the political debate around 'abortion' for the last decades, can be seen – assuming on the one hand that in view of human life, there cannot be a moral dilemma but only one answer to the obligations in question, no matter what circumstances the pregnant woman may face, or assuming, on the other hand, that the decision-making resembles a mere 'choice', concealing, sometimes ignoring the moral conflict countless women face. Both positions may silence women, both may disrespect them in their concrete situations. And yet, liberal ethics has argued that with the *right* to decide (or choose) left to the women, liberalism in fact promotes the respect of reproductive autonomy. Contrary to this and in line with my previous argumentation, I hold that respect or negative freedom is far from being innocent – if it flattens the moral interpretation by way of a (necessary, but not sufficient) concept of tolerance, it may easily contribute to social taboos, in fact may make moral experiences unheard of, or invisible in the public realm. In view of a future child's health status, women are left with probabilistic statistics rather than specific information, and the uniqueness of their situation, their fetus and its future prospects is easily subsumed under the generalizing categories of health and disease or handicaps.

In yet another stage, i.e. in the in vitro, pre-implantation phase of embryonic life, the situation of responsibility is different again: As in the time after birth, it is not the woman alone who is responsible for the child, but everyone directly involved in the procedure of assisted reproduction. It is in this phase that the embryo – or rather what he/she is projected to be in the future – is extremely exposed and in fact vulnerable to the 'normativity of normality', expressed in uncontrolled anxiety (disease or health risks) as well as in desired preferences (sex, race, health, other characteristics). If one speaks of social responsibility, it starts here: rather than promoting a specific concept of enhancement (of health) or prevention (of future children with predicted disabilities), ethics should take its task serious to (a) critically analyze the underlying social concepts of 'normality' in line of their potential to disrespect, and (b) advocate for those who might not meet the standards of these social concepts, and (c) elaborate procedures to enable those who are directly concerned by a given situation to make responsible decisions – including those groups of prospective parents who belong to the group of facing realistic risks. The concept of informed consent clearly has helped patients and clients to become visible as agents in the medical context (and not only subjects who are acted upon). Today, however, these medical-ethical achievements are in danger to become absorbed by a (medical, social, and sometimes political) movement of hysteria with respect to health, enhancement and perfection, so that we need to think and go further, and take a second look whether these movements empower prospective parents or rather intensify their and their potential offspring's vulnerability.

Prospective parents are perhaps much more vulnerable to social interpretations than is often assumed. Since we must assume that they desire to do the 'best' for their child, they are looking for interpretations of 'good parenthood'. If they are advised or encouraged to take responsibility for the genetic make-up of their offspring, and if bioethics argues that responsibility means not to have children with particular characteristics that might 'impair their capabilities', then the moral judgment violates the respect for the other as another, and instead defines her as not 'eligible' to respect because of specific characteristics. The normative standard of justice as providing every member of society equal opportunities to participate in social practices turns against those parents whose offspring is considered not to be able to participate in any social practice. Furthermore, in theoretically (or rather, socially) defining thresholds of life-quality, the respect for the other *as* other is transformed into qualifications *we* consider necessary in order to be included in the community of rights.

4 Conclusion

Prospective parents never set the agenda of their interaction with each other and/or their future child alone. Cultural, social and medical frameworks mediate prospective parental imagination. Nevertheless, an ethics of responsibility calls for the respect of reproductive decision-making as much as liberal ethics does; the difference is that 'autonomy' in the approach of a critical ethics of responsibility is not understood in line of the sovereign self's action; it is also not only understood in line of the Kantian concept of moral autonomy, but rather in line with the recognition of and responsibility towards the other, creating a burden of decision rather than a 'choice' for the prospective parents, or, in the case of pregnancy, for the pregnant woman. To argue that these decisions should be made dependent on the 'impairing of equal opportunity' caused by a particular health status, reduces the parents' decision to but one option. The concept of political tolerance only conceals that in this concept, morality does not give the parents a choice. In calling the choice to give birth to a child with a particular health condition, such as Down syndrome, irresponsible, bioethics may echo the self-understanding of a society that is threatened by persons who do not seem to fit into the mainstream understanding of 'normal species functioning'. In creating the threshold of quality of life at the price of life itself, bioethics does not serve the morality of inclusion; it rather contributes to a culture of fear that might well stigmatize those who do not meet the standards of this threshold – and prospective parents deciding on their behalf. Furthermore and perhaps more important even, instead of maintaining the ethics of inclusion and broadening it to the time before birth, many societies have already constructed prospective parenthood and pregnancy as a time of fear, resulting in evermore and ever-better surveillance of the embryo and fetus, in order to comply with the cultural norms, and in order not to be excluded from the social and moral community. My concern is that this culture, backed by utilitarian and liberal bioethics alike will, however unintended it may be, create the grounds for an extended heteronomy, extended injustice and misrecognition rather than creating a culture of autonomy, justice and recognition. Therefore, only a critical ethics of responsibility can maintain the autonomy and freedom of the self as well as the respect for oneself and the other – without knowing beforehand, what this responsibility may mean in a concrete situation of moral judgment.

Notes

[1] Boorse, Christopher: Health as a Theoretical Concept: *Philosophy of Science* 44 (1977), 542–62, for an alternative approaches cf. Lanzerath, Dirk:*Krankheit und ärztliches Handeln. Zur Funktion des Krankheitsbegriffs in der medizinischen Ethik* (München: Alber, 2000); Bobbert, Monika: Die Problematik des Krankheitsbegriffs und der Entwurf eines moralisch-normativen Krankheitsbegriffs im Anschluss an die Moralphilosophie von Alan Gewirth: *Ethica* 8, no. 4 (2000), 405–40.

[2] In her Tanner Lectures, M. Nussbaum explicitly deals with this problem, especially in relation to persons with disabilities Nussbaum, Martha C.: *Frontiers of Justice. Disability, Nationality, Species Membership* (Cambridge: Cambridge University Press, 2006). Her approach, however, does not escape this fundamental tension within the liberal framework.

[3] The use of the term 'responsibility' is chosen deliberately: Rather than referring to an obligation-approach that needs to argue in what extend it goes beyond the response to rights of others, the term responsibility is broader and not immediately determined by obligation's counter-term, namely the concept of rights. The indeterminacy might be considered a weakness, but I will argue later, that in fact, it is more adequate a concept to analyze the moral relation of the agent and the other.

[4] Cf. Honneth, Axel: *The struggle for recognition: the moral grammar of social conflicts*, 1st MIT Press ed., Studies in contemporary German social thought (Cambridge, Mass.: MIT Press, 1996), who introduced the Hegelian concept anew, without applying it to the context of bioethics. In recent years, Honneth has emphasized the role of misrecognition in form of social exclusion and practices of making persons socially 'invisable' Honneth, Axel: *Unsichtbarkeit: Stationen einer Theorie der Intersubjektivität*, Suhrkamp Taschenbuch Wissenschaft; 1616 (Frankfurt: Suhrkamp, 2003).

[5] Sameness is not the same as equality, however; whereas the latter emphasizes a shared normative status, sameness alludes to attributes of personhood or identity concepts people may or ma not share. Part of the problem is the unsolved distinction between the normative and the evaluative sphere within the concept of morality. Cf. for this discussion Haker, Hille: *Moralische Identität. Literarische Lebensgeschichten als Medium ethischer Reflexion. Mit einer Interpretation der 'Jahrestage' von Uwe Johnson* (Tübingen: Francke, 1999).

[6] This theoretical stance does not exclude the context-sensitive application of the principle, for example in light of the tenet 'ultra posse nemo obligatur', which may turn out to be central in the case of pregnancy.

[7] Here, even the generalization of 'the handicapped' is as problematic as any other generalization like 'women', 'blacks', 'Jews', or 'Germans'. In bioethics, persons with handicaps are rarely considered in their particular individuality; often, they are reduced to their particular health status (e.g. the 'blind', 'Down Syndrome', 'Breast cancer').

[8] In her important book, Butler at times is not so clear on this point; nevertheless, I believe that this is exactly what she is arguing for. Cf. Butler, Judith: *Precarious life: the powers of mourning and violence* (London; New York: Verso, 2004), Butler, Judith: *Giving an account of oneself* (New York: Fordham

University Press, 2005) and my further elaboration on this departure from liberal ethics in Haker, Hille: The fragility of the moral self: *Harvard Theological Review* 97, no. 4 (2004), 359–82.

[9] 'Listening to' can be spelled out in different ways: beyond the concrete communication, listening can be practiced in several forms; with respect to listening to the particularities, quantitative empirical studies certainly have the disadvantage of generalizing what is to be taken beyond generalization. Narratives (either biographical or literary) may be a better medium for this 'purpose' to transcend one's own horizon. Haker: *Moralische Identität. Literarische Lebensgeschichten als Medium ethischer Reflexion. Mit einer Interpretation der 'Jahrestage' von Uwe Johnson*, Haker, Hille: Narrative Bioethics, in: *Biomedicine and the Future of the Human Condition*, ed. Christoph Rehmann-Sutter, Mieth, Dietmar (Doordrecht: Kluwer, 2005).

[10] In this respect it is more than revealing that *From Chance to Choice* does not once quote the actual experiences of individuals or couples confronted with the new technologies. This is also echoed in most studies on twentieth century eugenics; here, too, the victims are invisible, silenced. Instead of engaging with the clients of genetic interventions, the authors of *From Chance to Choice* create science fiction-like stories and scenarios with little significance to counter their own perspective. This is a secure manner not to be 'disrupted' in one's own thinking by the 'other'.

[11] In a theological understanding, this conversion is interpreted as metanoia. The religious concept can well serve as an (experiential, formal) foil for the ethical understanding of the moral conversion in the encounter of the self and the other: It cannot be argued for; it can only be experienced. This arbitrariness is a weakness that the moral experience shares with the religious experience. It is not, however, overcome in an ethics that argues for the *reasonability* or taking the moral perspective. As much as I share the arguments for the necessity of a rational foundation of ethics, it cannot miss out the fact that the other must enter the 'horizon' of the self not only normatively (which is obvious) but factually. Therefore, I consider the turn to the experience of the encounter of the self and the other as the only promising way to better understand why in many cases the moral point of view is or is *not* taken.

[12] For a much more elaborated argumentation of this point cf. Haker: The fragility of the moral self.

[13] Obviously, this starting point should be compared with (and distinguished from) existential ethics, especially from Sartre's ethics that remains within the thematic horizon of Hegelian self-consciousness.

[14] With Walter Benjamin, the German term 'Ursprung' not only refers to an initial point of something in time, but also alludes to the 'idea' of something. Thus, in the experience of the other 'interfering' with the self by way of addressing him or her morally, not only the origin but also the very idea of morality is revealed.

[15] As Paul Ricoeur puts it, this passivity, with its linguistic connotation to 'passion' is part of the self's agency. Cf. Ricoeur, Paul: *Oneself as another* (Chicago: University of Chicago Press, 1992).

[16] This does not mean that every asymmetrical relation is dependent on advocatory actions. However, they go beyond the scope of the contractual construction of morality and entail a power-relation that needs to be addressed.

[17] I have, however, made some suggestions in my previous work on prenatal diagnosis and pre-implantaton diagnosis that might serve as indications how one could proceed on the level of social ethics Haker, Hille: *Ethik der genetischen Frühdiagnostik. Sozialethische Reflexionen zur Verantwortung am menschlichen Lebensbeginn* (Paderborn: mentis, 2002). For a strong argumentation on 'positive' social obligations within a liberal framework cf. Gewirth, Alan: *The community of rights* (Chicago: University of Chicago Press, 1996).

References

Bobbert, Monika. Die Problematik des Krankheitsbegriffs und der Entwurf eines moralisch-normativen Krankheitsbegriffs im Anschluss an die Moralphilosophie von Alan Gewirth. *Ethica* 8, 4, 405–40, 2000.

Boorse, Christopher. Health as a Theoretical Concept. *Philosophy of Science* 44, 542–62, 1977.

Buchanan, Allen; Brock, Dan; Daniels, Norman; Wikler, Daniel: *From chance to choice. genetics and justice.* Cambridge: Cambridge University Press, 2000.

Butler, Judith. *Precarious Life: The Powers of Mourning and Violence*. London: New York: Verso, 2004a.
——. *Undoing Gender*. New York: Routledge, 2004b.
——. *Giving an Account of Oneself*. New York: Fordham University Press, 2005.
Christman, John, Anderson, Joel, ed. *Autonomy and the Challenges to Liberalism: New Essays*. Cambridge: Cambridge University Press, 2005.
Daniels, Norman. *Just Health Care*. Cambridge: Cambridge University Press, 1991.
Deranty, Jean-Philippe. Injustice, Violence and Social Struggle: The Critical Potential of Axel Honneth's Theory of Recognition. *Critical Horizons* 1, 2, 2004.
Gewirth, Alan. *Reason and Morality*. Chicago; London: University of Chicago Press, 1978.
——. *The Community of Rights*. Chicago: University of Chicago Press, 1996.
Graumann, Sigrid, Grüber, Katrin, ed. *Anerkennung, Ethik und Behinderung. Beiträge aus dem Institut Mensch, Ethik und Wissenschaft*. Münster: Lit, 2005.
Haker, Hille. *Moralische Identität. Literarische Lebensgeschichten als Medium ethischer Reflexion. Mit einer Interpretation der "Jahrestage" von Uwe Johnson*. Tübingen: Francke, 1999.
——. *Ethik Der genetischen Frühdiagnostik. Sozialethische Reflexionen Zur Verantwortung am menschlichen Lebensbeginn*. Paderborn: mentis, 2002.
——. The Fragility of the Moral Self. *Harvard Theological Review* 97, 4, 359–82, 2004.
——. Narrative Bioethics. In *Biomedicine and the Future of the Human Condition*, edited by Christoph Rehmann-Sutter, Mieth, Dietmar. Doordrecht: Kluwer, 2005.
Hare, Richard M. Possible People. In *Essays on Bioethics*. Oxford: Oxford University Press, 67–83, 1993.
Honneth, Axel. *The Struggle for Recognition: The Moral Grammar of Social Conflicts*. 1st MIT Press ed, Studies in Contemporary German Social Thought. Cambridge, Mass.: MIT Press, 1996.
——. *Unsichtbarkeit: Stationen einer Theorie der Intersubjektivität*, Suhrkamp Taschenbuch Wissenschaft; 1616. Frankfurt: Suhrkamp, 2003.
Jonas, Hans. *The Imperative of Responsibility: In Search of an Ethics for the Technological Age*. Chicago: University of Chicago Press, 1984.
Lanzerath, Dirk. *Krankheit und ärztliches Handeln. Zur Funktion des Krankheitsbegriffs in der medizinischen Ethik*. München: Alber, 2000.
Levinas, Emmanuel. *Otherwise Than Being, or, Beyond Essence*. Pittsburgh, Pa.: Duquesne University Press, 1998.
Markell, Patchen. *Bound by Recognition*. Princeton: Princeton University Press, 2003.
Nussbaum, Martha C. *Women and Human Development: The Capabilities Approach*. Cambridge: Cambridge University Press, 2000.
——. *Frontiers of Justice: Disability, Nationality, Species Membership*. Cambridge: Cambridge University Press, 2006.
Parens, Erik, Asch, Adrienne, ed. *Prenatal Testing and Disability Rights*. Hastings Center Studies in Ethics. Washington: Georgetown University Press, 2000.
Ricoeur, Paul. *Oneself as Another*. Chicago: University of Chicago Press, 1992.
——. *The Course of Recognition*, Institute for Human Sciences Vienna Lecture Series. Cambridge: Cambridge University, 2005.
Taylor, Charles. *Modern Social Imaginaries*. Durham: Duke University Press, 2004.
Taylor, Charles, Gutman, Amy et al., ed. *Multiculturalism: Examining the Politics of Recognition*. Princeton: Princeton University Press, 1994.

The Human Embryo as Clinical Tool

Private Tragedy or Public Good?

Sheila A.M. McLean

1 Introduction

At the outset of this chapter it is important to make it plain that what follows is a contribution from a lawyer rather than a philosopher. This is because the approach adopted will essentially draw on legal and jurisprudential considerations rather than on purely or even substantially philosophical ones. It is also important to define the context within which this discussion is situated; the focus will be on the use of human embryos for research purposes, broadly within the area of assisted reproduction. That is, the discussion focuses on embryos which are 'spare' or 'surplus' and therefore will not be implanted irrespective of what else is done with or to them.[1] However, given these disclaimers, it is still necessary to identify and briefly discuss some ideas as to the status which is or should be accorded to the human embryo.

2 The Status of the Embryo

There are a number of ways in which the human embryo can be viewed. The recent report from the House of Commons Select Committee on Science and Technology[2] in the UK identified three main possibilities. First, the embryo can be seen as a person (or a person in waiting), and therefore deserving of the protection of the full panoply of human rights. This perspective appears to have informed the recent Italian law on assisted reproduction which permits In Vitro Fertilisation (IVF), but which requires that clinicians restrict themselves to the creation of only three embryos, all of which must be implanted. Second, the embryo can be seen as just a collection of cells. On this view, absolutely no rights or interests attach to it, and theoretically it could be used for any purpose and dealt with in any way. Finally, there is what is called the gradualist approach, which recognises – as did the report of the Warnock Committee[3] – that the embryo of the human species is worthy of some respect, and that the level and nature of that respect increase as the embryo

S.A.M. McLean
e-mail: s.mclean@law.gla.ac.uk

M. Düwell et al. (eds.), *The Contingent Nature of Life*,

develops. Normally, in law, respect is evidenced by the attribution of enforceable rights, and the Warnock Report, for example, recommended that, "the embryo of the human species should be afforded some protection in law" (63, para. 11.17). However, it also concluded that this "does not entail that this protection may not be waived in certain specific circumstances . . ." (63, para. 11.18) and in any event protection need not be the equivalent of rights. Rather, it may simply imply that there are limitations on the use(s) to which the protected entity may be put. This gradualist approach commended itself to the Select Committee and arguably reflects what many, if not most, people feel is the appropriate approach to the human embryo.

3 Regulating the Use(s) of the Human Embryo

It is generally assumed that adopting any approach – including the gradualist one – which sees the embryo as being more than a mere collection of value-neutral cells means that it is permissible, perhaps even required, that legal restrictions are placed on the uses to which it can be put. For example, it was argued by the Warnock Committee, whose report formed the basis of current United Kingdom law – The Human Fertilisation and Embryology Act 1990 – that:

> .... research conducted on human *in vitro* embryos and the handling of such embryos should be permitted only under licence.... We see these controls as essential to safeguard the public interest and to allay widespread anxiety (54, para. 11.18).

Thus, UK law places limitations on the use of embryos not because it attributes rights or interests to the embryo, but rather out of respect for society. Presumably, therefore, these restrictions reflect our interests; not theirs. However, the gradualist position does entail the demonstration of some respect for the embryo of the human species. The question, therefore, is, what does owing respect to the very early embryo mean? Bearing in mind that the embryos on which I am focusing are destined for research only – that is, there is no intention to implant them – does this have any impact on what 'respect' might actually entail? Might it, for example, be argued that respect is specific to the status of the embryo on a narrower basis? That is, if the embryo specifically and unequivocally will never have the opportunity to become a person, because it has been deliberately designated solely for research, does it make any sense to treat it with the same kind of respect which would be owed to one which will be implanted and may develop into a human person? Before seeking to answer this question, it is useful to consider the general position of the embryo in law.

4 The Legal Position

In many jurisdictions there is no direct protection offered to the human embryo: it has no legal status, nor indeed does the more developed, even viable, foetus of the human species. The law permits, for example, termination of pregnancy within defined legal conditions;[4] indeed, in some countries there are no restrictions on the termination of pregnancy up to about 12 weeks,[5] and even where restrictions apply, termination up to term may be permitted within certain parameters.[6] Of course, some countries, such as the Republic of Ireland and, more recently Italy, do regard the human embryo as possessing certain rights (for example, by prohibiting pregnancy termination or requiring the implantation of every embryo that is created) these are arguably not in line with mainstream jurisprudence.

here the embryo has no legal status, it is accorded neither rights nor interests. Any 'rights' attributed can only be inferred or acquired after birth. Thus, it is possible for a child *once born* to claim compensation for harm caused at the embryonic or foetal stage, but this is dependent on live birth. Indeed, in English Law, the child must survive for more than 48 hours before any compensation can be awarded.[7] Courts have been clear that no rights are attributable before the child is born, both in claims for pre-natal harm and also where the 'interests' or 'rights' of the embryo or foetus appear to conflict with the rights of the pregnant woman. It must be conceded, however, that the law in the United Kingdom and other jurisdictions, such as the United States, has until recently been occasionally obscure or inconsistent on this latter point, but it is widely assumed by most academic commentators[8] that in the United Kingdom the case of *Re MB*[9] finally brought any confusion to an end. In this case, the judge was clear that:

> The fetus up to the moment of birth does not have any separate interests capable of being taken into account when a court has to consider an application for a declaration in respect of a caesarian section operation. The court does not have the jurisdiction to declare that such medical intervention is lawful to protect the interests of the unborn child even at the point of birth (227).

A secondary legal device which can co-exist with human rights is the attribution of interests, and this will be returned to in what follows below.

5 A Gradualist Approach to the Human Embryo: Rights or Interests

We have seen that in most jurisprudence the embryo of the human species is accorded no rights. It is probably also true to say that there are no rights accorded to it from those ethical positions which do not regard the embryo as the equivalent of an actual person, which is the position adopted in this discussion and, commonly, in law. However – unlike the law – the gradualist position seems to imply that there is *something* about the embryo which means that we cannot adopt the rather sweeping position adopted by the law. As the House of Commons Select Committee on Science and Technology said:

> Adopting a gradualist approach, we believe, recognises the special status of the embryo of the human species, while at the same time respects the legitimate interests of intending parents and the wider society. It does not, therefore, exclude other considerations such as seeking to provide treatment for the infertile or discovering the causes of infertility or the genesis of serious illness. However, it does require that embryos should not be used without carefully evaluating the reasons and rationales for their use in a specific manner or for a specific purpose (17, para. 29).

Thus, for both the Warnock Committee and the Select Committee, some concern is due to the management and use of embryo from the moment it exists, but this is not sufficient to prevent other competing considerations from permitting its use in certain ways and for certain purposes, nor does it require or mandate the attribution of rights. Rather, the gradualist approach – since it does not adopt the view that the embryo is just a collection of cells – intends or implies that some *interests* may exist. The question is, what kind of interests, or – perhaps more appropriately – whose interests are they? Some help in providing a legal answer to this question may be drawn from the example of the patient in a permanent vegetative state (PVS).

6 A Relevant Example?

We do not doubt that born people have both interests and rights. When they become unable to experience them, such as in the case of a patient in PVS, they do not disappear. However, by and large when a person is no longer able to exercise his or her rights, attention shifts from rights to interests; specifically the interests of the person him or her self. For example, in the leading case on Persistent or Permanent Vegetative State (PVS) in the United Kingdom, the House of Lords felt impelled to ask not what were the patient's rights, but rather what might be in his 'interests' or 'best interests'. In *Airedale NHS Trust v Bland*,[10] although Their Lordships used a variety of devices to reach their conclusion that it was permissible legally to discontinue assisted nutrition and hydration, one which played a significant role, at least for some of them, was the concept of what might serve Anthony Bland's interests or, as the legal test is most commonly phrased, his 'best' interests. However, this approach was not regarded with favour by all of Their Lordships, with Lord Mustill for one unhappy about applying this criterion. As he said:

> Stress was laid in argument on the damage to his personal dignity by the continuation of the present medical regime, and on the progressive erosion of the family's happy recollections by month after month of distressing and hopeless care .... But it seems to me to be stretching the concept of personal rights beyond breaking point to say that Anthony Bland has an interest in ending these sources of others' distress. Unlike the conscious patient, he does not know what is happening to his body, and cannot be affronted by it; he does not know if his family's continuing sorrow. By ending his life the doctors will not relieve him of a burden become intolerable, for others carry the burden and he has none. What other considerations could make it better for him to die now rather than later? None that we can measure, for of death we know nothing. The distressing truth which must not be shirked is that the proposed conduct is not in the best interests of Anthony Bland, for he has no best interests of any kind (141).

Of course, Lord Mustill was but one of the judges in this case, and some of his colleagues seemed more comfortable with the use of 'interests' and 'best interests' in reaching their conclusion. Therefore, despite Lord Mustill's concern, there seems to have been a general level of comfort with the idea that it was either in, or not against, the interests of Anthony Bland that he should be allowed to die. To an extent, these interests might have been inferred from the assumption that Anthony would have preferred not to continue living in the condition he was in; in other words, because of the fact that Anthony had had an existence, we could impute to him a plausible interest in not subsisting in a permanent vegetative state. On the other hand, our *own* likely perception of existence in that condition is just as likely to inform our assessment of his best interests. In other words, our interests dictate both that he cannot be treated in an abusive manner and equally that he should not continue to live.

Alternative decision-making approaches are used in other jurisdictions. For example, in some states in the United States, the preferred test is not dependent on identifying interests, but is rather founded in the concept of substituted judgement. Using this method of concluding what should be done requires the decision-maker to try to identify what the person would have wanted had they been able to express it. While UK courts have shown no patience with this test, it does have some obvious benefits, not least that it seems likely to focus more clearly on the rights of the individual under consideration, assuming, of course, that the proxy decision-maker is not entirely wrong. However, while it may be that substituted judgement is relevant where a person has already had an existence, its relevance to an unimplanted embryo seems somewhat tenuous.

To continue with the PVS analogy, to confirm that the person in PVS is not someone to whom the concept of rights is of primary importance, we can derive a similar conclusion from the case of *NHS Trust A vs M, NHS Trust B vs H*;[11] a case which was – unlike the *Bland* case – heard after the incorporation of the European Convention on Human Rights into United Kingdom law.[12] In this case, two women were deemed to be in PVS and the question for the court was whether or not removal of nasogastric nutrition and hydration was in breach of their right to life under Article 2 of the Convention and/or Article 3 which prohibits inhuman and degrading treatment.[13] It was first held that the Article 2 right was not breached by removing assisted nutrition and hydration. More importantly for our purposes, it was also held that their Article 3 right was not infringed, because:

> . . . the proposed withdrawal of treatment . . . . has been thoroughly and anxiously considered by a number of experts in the field of PVS and is in accordance with the practice of a responsible body of medical opinion. The withdrawal is for a benign purpose in accordance with the best interests of the patients not to continue life-saving treatment . . . . I am, moreover, satisfied that art 3 requires the victim to be aware of the inhuman and degrading treatment which he or she is experiencing or at least to be in a state of physical or mental suffering (100).

Two relevant factors emerge from this judgement. First, that there is no absolute right to life; or at least no right to have continued life facilitated when to do so would involve ‘futile’ or ‘burdensome’ treatment. Why is this relevant to the embryo? Substantially because, if it cannot be said that there is an obligation to preserve or maintain a life which already exists, there is – legally at least – no basis from which to argue that the potential life of an embryo should receive any more special consideration. If we do not require existing life to be continued when it is insensate, why should we attach more importance to the unimplanted embryo? For the purposes of this discussion, there can be absolutely no reason on this basis to require that a right to life should be attributed to the human embryo; in particular, this must be so for the embryo which is not intended for implantation and which is the subject of this chapter. Second, if awareness of inhuman or degrading treatment is necessary before article 3 is breached, manifestly this cannot apply to the embryo which – irrespective of one's perspective on its standing – has no awareness.

However, a more recent judgement refused to accept the Article 3 argument,[14] and although this judgement was overturned on appeal,[15] this section of the judgement was not considered or contradicted by the Court of Appeal. In this case, Oliver Leslie Burke challenged the lawfulness of the General Medical Council's (GMC) guidelines on withholding and withdrawing medical treatment. Suffering from a degenerative condition, he was afraid that at the end of his life doctors would take the decision to withdraw assisted nutrition and hydration when he was in no position to object, but over his current wishes. At the first hearing of the case, Mr Justice Munby argued (amongst other things) that Dame Elizabeth Butler Sloss had been wrong to suggest that Article 3 of the Convention was inapplicable when the person concerned could not experience the inhuman or degrading treatment in question. For Munby:

> ... however unconscious or unaware of ill-treatment a particular incompetent adult or a baby may be, treatment which has the effect on those who witness it of degrading the individual may come within art 3. Otherwise ... the Convention's emphasis on the protection of the vulnerable may be circumvented (177).

Although the Court of Appeal warned against 'cherry picking' from parts of Munby's judgement, it would have consequences for consideration of the embryo if we believed it to be a rights-bearer. For example, if rights are attributable even to the insensate, then the embryo would presumably qualify. However, since in most jurisdictions the embryo is not legally regarded as a bearer of rights, this conclusion would be flawed. This, of course, does not affect the question of whether or not we have an interest in either constraining or facilitating the use of the human embryo. Thus, where the analogy between the patient in PVS and the status of the human embryo is particularly appropriate is that both reflect one common characteristic; namely, that even if we struggle to identify the interests held, for example, by a person now in PVS, and even if we concede that the embryo has no such interests, that does not mean that *we* have no interests in them.

Unlike Anthony Bland or Leslie Burke, the embryo has never had (and will never have) prior rights to be respected, nor will it – for the same reason – have any prior interests. However, irrespective of this, the respect which everyone seems to agree is due to it requires that we treat it carefully, but in so doing, it is *our* interests that are being protected; not those of the embryo (and arguably also not those of the person who is in PVS).

However expressed, in the case of someone in PVS it seems that the interests which subsist are arguably no longer those of the person him or herself when we do not know what they are or would have been. Rather, they become *our* interests – primarily, it could be said, in ensuring that the person (who one day might be yourself) is treated with respect. This assumption of interests could depend either on what we have previously said about what we would want to be done to us or by the inferences drawn by the law as to what we would have wanted.

Unlike the person in PVS, the embryo is generally held in law to have no pre-existing or current rights or interests. Like the person in PVS, however, most would agree that *we* have an interest in treating the embryo with some respect (however that is defined) even if the embryo itself has not, never has had and never will have any such interest. But, given that any respect or concern owed is about *our* interests and not those of the embryo, what it is permissible to do will presumably vary according to those interests: the embryo's status is static even if our views as to what it is acceptable to do with or to it are not. So, if it were deemed to be in our interests to undertake specific kinds of research – even those which are currently impermissible – then arguably there is no basis, beyond any prohibitions outlined in the law, for preventing them from being carried out.

Again, on analogy with the PVS patient, there may be limitations placed on what we can or should do, but these would also result more directly from *our* interests than those of the embryo – not to treat carelessly or with disrespect is an interest for us because it also affects us. However, as we can never re-experience being an embryo, our interest in the embryo is of a different order from our interest in the person in PVS. It relates more generally to concern for something that is potentially a human being, but since it is not, never has been and never will be an actual human being, the kind of concern owed is different from that accorded to born, but comatose, people, and perhaps also to embryos destined for implantation. However, as has been hinted at already, even if we are comfortable with the idea that the embryo is worthy of some concern, the question remains whether or not this is because it may become a person, or simply because it exists.

The embryo which is designed to be used for research never in reality had the potential to be a human being. Its designation as a research embryo means that it will never be implanted. Arguably, therefore, the 'spare' or 'surplus' embryo (once it acquires that status) need not be subject to the same kind of legal regime that we would wish to apply in the case of a person who has lived a life, but is now insensate and in PVS, nor even to the embryo destined for implantation. Therefore, the scrutiny of the uses and treatment of the research embryo can differ from that applied in the case of the patient in PVS and indeed from that which would inform our consideration of the embryo which is designated for implantation (whether or not this actually occurs).

Thus, we can say that – on analogy with the person in PVS – the embryo which is destined for research use has no interests beyond the interests that we have in treating it in an ethically sound manner. We may, on the other hand, concede that the embryo which is designed to be implanted should be attributed with interests which are, if not equivalent to, then on a parallel with, those with which we endow those who are irreversibly comatose and at the end of their life. This inevitably means that there are limitations and restrictions imposed on how we can or should treat them. However, it has also been proposed that these limitations derive not from anything inherent in them, but are rather inherent in the interests that we have in behaving humanely and in a respectful manner.

However, this still leaves the question of the embryo for which there is no possibility of implantation; no possibility of becoming a person. If it is the potential to become, or the fact of having been, a person that generates respect, then might we not simply treat the research embryo in any way we choose? If the debate was about characteristics inherent in the embryo per se, then it could be said that this conclusion would be entirely logical. However, because I have argued that the real issue is about *our* interests rather than the interests of the embryo, then it is legitimate – perhaps even obligatory – to take account of what common morality holds to be acceptable. Of course, views on what is or is not permissible will vary, but one way of considering this is to ask not just what is done to or with the embryo, but also what benefit may or will arise from what is done. This could be conceptualised in terms of benefit for people already in existence or even for future embryos. Limitations on the use of the human embryo then can be permissible, but their legitimacy rests on our attitudes and expectations, not on the fact that the research tool is an embryo. This leads to one final consideration.

7 Is Current Regulation Proportionate?

In the UK, the oversight of the treatment of people in PVS rests with the courts. At the moment on the other hand, legislation limits the uses to which embryos can be put, and requires their destruction no later than 14 days if they have been used for research. Given what has gone before, it might be thought rather strange that this differentiation exists, or rather that it seems to accord more stringent protection to the embryo which is not destined for implantation than it does to the person who has been born, has acquired human rights and whose interests are more likely to be discoverable and, perhaps, discovered.

The question then is whether or not this level of legal intervention, with its associated restrictions, is appropriate or necessary to demonstrate sufficient concern for the human embryo. Essentially, the use of the embryo is subject to tighter and more rigid regulation than is the way in which we can treat a living but comatose person. Arguably, this is disproportionate, especially if the human embryo would be used for the potential benefit of living people. On this analysis, it is possible to show concern for the human embryo, yet also see it as a clinical tool, its use governed not by restrictive, invasive legislation but rather by the ethical oversight generated by professional ethics, peer review and ethics committee consideration, and – importantly – the views of society. Indeed, given that the research embryo will otherwise never be implanted it could be said to show sufficient concern for it to permit its use to obtain benefit for others rather than condemning it to what would otherwise be useless destruction.

This, of course, begs the question as to whether or not there should be limits on what we can do with the human embryo and, if so, how they should be identified or drawn? Should they rest on public distaste – the so-called 'yuck factor' – on scientific merit or some other consideration? A classic situation in which answering this question is important is the case of human embryonic stem cell research. Hailed by some as the 'holy grail' of medicine in the twenty-first century, if stem cell therapies are to develop it is widely conceded (by scientists at least) that it will be necessary to create and use human embryos. Leaving aside objections based on faith, might there be other objections to this scientific development?

Sadly, recent events which have shocked the scientific community might suggest that there are.[16] The recent revelations that cutting edge research in South Korea has in fact been a major scientific fraud may cast doubt on the actual benefits to be derived from stem cell research. Most importantly, if it is not in fact possible to generate stem cells for specific patients, the aspirations of science may be overstated. Indeed, one eminent UK infertility doctor recently cautioned against making too many assumptions about the likely benefits of stem cell research.[17] If we are to evaluate the ethics of research uses of the human embryo by reference to the likely benefits to others now or in the future, it may be that this set-back will have consequences for our justification to proceed. On the other hand, if there is a realistic expectation of benefit there would seem to be no fatal objection to continued research and development in this area.

8 Creating Embryos for Research

Finally, it must be asked whether or not there are any spin-off consequences from what has gone before in respect of the creation of embryos specifically for research. The Council of Europe's Convention on Biomedicine[18] seems to differentiate between the two kinds of embryo by specifically prohibiting the deliberate creation of embryos for research (Article 18), while not directly outlawing research on 'spare' embryos. This might seem to imply that the two kinds of embryos can be differentiated based on the *reason* for their creation, but on what grounds? It could, of course, be argued that the intention to create solely for research purposes is disrespectful, whereas this is not so when the embryo is already created but surplus to requirements.

On this argument, the same problems do not arise in the case of embryos created for a 'good' purpose – implantation – but which do not for one reason or another end up being implanted. Therefore, the embryo of the human species, once created for implantation but now surplus or spare, seems, in terms of the Council of Europe at least, to be seen as available for use as a clinical tool, and its research use would not be inappropriate so long as the proposed research is otherwise ethically and scientifically sound. Of course, it could be said that if it is permissible to use the human embryo for legitimate, ethical research then arguably it is should be permissible specifically to create them for that purpose. It is not within the scope of this chapter to take this particular discussion further, beyond noting that there may be a paradox in the Council of Europe's position (and that of a number of countries which do not permit the deliberate creation of embryos for research purposes). In those countries where this is permissible, it would doubtless be argued that the important issue is the quality (scientific and ethical) of the research rather than how and for what reason the embryo was created. Our interests in the way in which embryos are treated would then be satisfied.

9 Conclusion

Gradualism obviously imports no absolutes. Inevitably, therefore, it is potentially permissive as to the way(s) in which the embryo can be used. This is, of course, dependent on agreed ethical principles which may mandate some limitations on these uses. However, the restrictions – if any – which can be legitimately imposed are best defined by characteristics and interests which *we* hold; not interests which can be specifically attributed to the embryo *in se*. Thus, there may be some limitations which we would wish to impose; for example, that the embryo should not be treated in a callous or cavalier manner. Equally, any legitimate research use need not offend our ethical senses.

Second, it can be argued – in my view appropriately – that the current UK legal situation is not proportionate to the interests to be served. Although purporting to recognise the gradualist approach to the human embryo, in fact it prioritises the alleged interests of the embryo or the interests we have in the human embryo over the interests we have in a person who has been – and still is – alive, albeit in PVS.

Finally, the fact that we may regard some uses of the human embryo as being unacceptable allows the moderation and prevention of any outlandish or maverick uses of them; because we are required to ask ourselves what is the appropriate thing to do with human embryos, there will virtually inevitably be a consensus against some uses. This moderating capacity, when combined with ethical review, peer review and the standards required by professional ethics, acts as a brake on manifestly unacceptable (ab)use. The research embryo can therefore be instrumental in achieving a social good.

Notes

[1] There will also be some consideration where appropriate of the embryo deliberately created for research use.

[2] Human Reproductive Technologies and the Law, Fifth Report of Session 2004–2005, NC 7–1.

[3] Report of the Committee of Inquiry Into Human Fertilisation and Embryology, Cmnd. 9314/1984.

[4] Abortion Act 1967 (as amended).

[5] E.g. in the United States. See, for example, *Roe vs Wade* 93 S. Ct. 705 (1973).

[6] For example, it is permissible in the United Kingdom to terminate a pregnancy up to term if the foetus is likely to be severely handicapped (s. 1 (1) d).

[7] Congenital Disabilities (Civil Liability) Act 1976.

[8] But see McLean and Ramsey 2002: 239–258.

[9] 8 Med. LR 217 (1997).

[10] 12 BMLR 64 (1993).

[11] 58 BMLR 87 (2000).

[12] Human Rights Act 1998.

[13] Article 8 rights were also considered, but for reasons of space this will not be discussed here in any depth.
[14] *R (on the application of Burke) vs General Medical Council* (2004) 79 BMLR 126.
[15] *R (on the application of Burke) vs General Medical Council* (2005) EWCA Civ. 1003.
[16] See, for example: "S Korea research was fake". http://news.bbc.co.uk/1/hi/world/asia-pacific/4554422.stm. Accessed on 28/12/05.
[17] See: "Caution over science claims urged". http://news.bbc.co.uk/1/hi/health/4214800.stm. Accessed on 29/12/05.
[18] Council of Europe, Convention for the Protection of Human Rights and Dignity of the Human Being with Regard to the Application of Biology and Medicine. Convention on Human Rights and Biomedicine, Oviedo, 4.IV.1997.

References

McLean, S.A.M. and Ramsey, J. "Human Rights, Reproductive Freedom, Medicine and the Law." *Medical Law International*, 5, 4, 239–258, 2002.

The Naked Emperor

Bioethics Today and Tomorrow

Michiel Korthals

1 Introduction

Much is expected of bioethics. Its activities, publications, and professional organizational networks are greatly influenced by this. Simultaneously, the problems confronting the dominant style in bioethics are multiplying, and the gaps between expectations, opportunities and achievements seem insurmountable. There are large ethical conflicts in modern, globalizing societies: consumers are confronted with different medical, agricultural and food systems, with privatization of health systems, with diversified economic, social and cultural implications and with differences in treating animals, plants, and humans – all of these difficult to reconcile in an ethical acceptable way. Answers will surely be forthcoming, for better or for worse; in the decades ahead of us we will see a proliferation of ethical committees, consultants and even ethical businesses. But the trouble is that the dominant style in bioethics is suffering from severe theoretical and conceptual shortcomings. Topics such as the relationship between ethics and technology, between ethics and (plural) dynamics of norms and values, and between ethics and diversity of norms, values and ethical intuitions receive scant attention, and are left out of popular treatises on bioethics altogether.

This essay starts with a brief indication of the substantive ethical issues that the emergence and expansion of the life sciences forces us to face in the (very) near future. Next I discuss some of the main deficits of our current ethical theories in tackling those issues. Thirdly, I analyse some of the problems that confront bioethics as reflection on these issues, in particular the question of its philosophical status and its quality standards.

2 Four Ethical Issues Challenging the Life Sciences

In the course of the past thirty years the life sciences have come to face, more or less successively, at least four major ethical issues which are so pervasive that the

M. Korthals
e-mail: Michiel.Korthals@wur.nl

M. Düwell et al. (eds.), *The Contingent Nature of Life*,

traditionally 'neutral' position cannot be maintained. Societal concerns regarding their implicit or explicit ethical agendas cannot be ignored.

2.1 Public versus Private

In his famous 1942 essay on the ethics of science, Robert Merton described the so-called CUDOS ethos that stands for the values of *communism, universalism, disinterestedness, and organized skepticism*. At the time 'communism' was a dirty word in the USA, but Merton wanted to stress the idea that in science there is no private ownership of facts, ideas, insights or methods (Merton 1968). I don't know enough of the fate of the other CUDOS values, but the values of communism and disinterestedness are, at least in biology, no longer dominant. In 1980, members of the two large political parties in the US took steps to ensure that the scientific community would become more directly involved in the creation of economic wealth and the US Congress passed the *Bayh-Dole Act*, which together with subsequent legislation, allowed and encouraged universities and public research institutes to file patents on the results of federally funded research and licence them to private companies (Thursby and Thursby 2003). The aim was to facilitate the transfer of knowledge and technology from academe to industry. The move proved singularly successful, resulting in thoroughgoing privatization of the life sciences (Etzkowitz and Leydesdorff 2002). In approximately twenty years the proportion of public research compared to that of private companies shifted from 9/10 to 1/10 (Thackray 1998). By 2005, 9/10 of biomedical research was privately financed. This huge shift caused a change in strategies of openness and secrecy, connected with the rise of patenting, bio banking, and new aspects around conflicts of interests (Krimsky 2004).

Secrecy is one of the main characteristics of modern life sciences. It means that data are in principle not shared and sometimes even research priorities are not made public (Graff et al. 2003; Science 2000). *Patenting* is the main vehicle of implementing this new way of severely restricting public access to information. The *ethics of data banking* (access, management, aftercare of data) becomes an urgent thing to regulate (Shorett et al. 2003). Moreover, *conflicts of interests and of commitment* are increasingly frequent occurrences (Krimsky 2004). The chance is rather high that in case of conflicts of interests the outcome of research is distorted, as research has found out (Nestle 2002). Research funding and the possession of shares and other stakes in research can distort research results. The urge to do fundable research can prevent the full exercise of teaching tasks and responsibilities (Bulger 2002). Even when funding is not at stake, but other vehicles of distributing status and research opportunities are involved, such as membership of journal boards, or participation in review and evaluation committees, the danger of conflicts of interests appears and is very often silenced due to lack of reasonable standards of research ethics.

Last but not least, political battles about controversial innovations emerge when research priorities are questioned, as in the case of genetic modification, genomics and nanotechnology, because often the private interests behind those research decisions do not correspond with public interests (Nestle 2002).

2.2 Health and Food Production

A second major issue confronting the life sciences is the emergence of different systems of health and food production, which makes the stakes high in the life sciences, earning them the epithet 'post-normal' (Ravetz 2002, 2004).

In the health sector the silent and sometimes open competition between two systems, regular and alternative medicine, is becoming more prominent than ever, due to distrust of patients with the regular system, the role of insurance companies, the overly optimistic promises made by the regular system, the globalization of information and other trends that, although originating elsewhere, have an impact on the health and food sectors.

In the food sector the differences between a system using genetic modification (GM) and a non-GM system is taking shape in Europe. On other continents, and partially in Europe as well, this is complicated by the already existing differences between intensive, high-tech agriculture and extensive, artisanal agriculture (Lyson 2002). The different systems partially corresponding with different policy perspectives like risk-driven and precautionary-driven policies. The issue of the coexistence of these systems is becoming more urgent, because there are many interferences and often unfriendly border crossings between the two. Information and labelling is frequently demanded, and it is interesting to see if and how trust in these systems rises or falls depending on the transparency and integrity of the information given (Nestle 2002). The life sciences until recently functioned mainly by tackling the scientific problems in the regular health system and the intensive food sector; however, due to increasing consumer concerns with these systems, they are challenged to contribute to the alternative systems as well. It is questionable in how far the life sciences can contribute to all systems in an equal and fair way. However that may be, the emergence of these different systems makes it impossible for the life sciences to remain neutral.

2.3 Consumer Involvement

A third major ethical issue concerns the role of participation, communication and consultation. In the 1960s the involvement of users of the services and products of the life sciences rested on (and was limited to) the principle of 'informed choice'. Later, it seemed increasingly necessary to launch broad information and consultation offensives (Shrader-Frechette 1994). According to the European General Food Law, Articles 6, 7, 8 and 9, policymakers have the obligation to inform the public on innovations and to ask for informed choice. However, it is far from clear what form this involvement will get. The functions and forms of consultations depend on democratic traditions and questions to be resolved. Sometimes consultations are (one-sidedly) meant to postpone unwelcome decisions, or to placate certain groups that are too powerful to ignore, but not powerful enough to determine the outcome of a policy process (Food Ethics Council 2004). Sometimes the function is only to inform the public on innovations and decisions already made. Known as the 'deficit model', this approach has severe drawbacks, e.g., it reduces trust (Hansen et al. 2003). However, often consultations are indeed organized because scientists and policy makers are aware of their lack of knowledge about the concrete social and other implications of life science innovations and technologies. This cognitive motive coincides with sometimes indeed present genuinely ethical motives to acquire informed consent from the people involved. In the analysis of the various forms the deficit model is generally contrasted with the communication and consultation model (see Rowe and Frewer 2000, 2004).

However, we can question the relevance of consultations, not only in terms of their agenda and the exclusive participation of some groups, but also with respect to the way the original, problematic innovation is made concrete in problems that the consultation is supposedly meant to deal with (Curtin 2003). The British arrangement of the public debate on genetic modification (GM Nation, see gmnation.org.uk) singled out an issue for scientific fact finding that in the end was not about GM food and its ethical acceptability, but about different regimes of using pesticides. So, the whole debate of GM Nation was in fact framed towards something else.

In general, the ethical challenges posed by the demand for consultations cover the problems of who to include and who to exclude, the agenda setting of consultations, the framing of the problems to be the subject matter of consultations and the policy effects of consultations (Korthals 2004).

Especially relevant for the future of bioethics is the issue of the institutionalization of bioethics, not only through ethical committees, but also by way of the several types of consultation and ethical regulation (see below).

2.4 Gray Zones of Health and Food

A fourth major ethical issue is the emerging overlap between food and drugs, nutrition and medicine, food consumption and drugs use. Health care and food production systems based on pharmaco-genomics and nutria-genomics give rise to new gray areas and change the interfaces between scientists, producers (food and medical professionals and their intermediaries), government regulators and consumers/patients (Castle et al. 2004). In each of these groups and between these groups changes take place (like the shifting relationship towards one's relatives in preparing and consuming a meal in the group of consumers/patients; the role of informed consent in the relationship between consumers/patients and food and medical professionals). Numerous conflicts and controversies on economic, financial, cultural and other issues arise, inspired by various political and ideological worldviews (Sharp et al. 2004).

The preventive turn to food and health is increasingly mapping these gray zones through personalized or profiled nutrition. It implies that individuals are urged to select their lifestyle and nutrition habits in correspondence with their vulnerabilities and susceptibilities as discerned through continued genetic testing. The ethical twist of this individualizing trend is the fact that it has far-reaching collective implications. Information from genetic testing is mostly not only valid for the individual applicant but also for his or her relatives and poses questions of ethics as for example concerning the right to be or not to be informed.

These four major problems challenge bioethics. The private–public nexus calls for new reflection on the role of public health and demand new types of research ethics tackling conflicts of interests; the different systems in food and health urge new analysis of the interaction between experts and lay persons, and demand reflection on what in these systems is the right balance between protection of consumers and patients and encouraging them to take their own responsibility. The involvement of consumers and patients, ranging from giving them more information to engaging them in communication to inviting them to complete participation, requires thorough ethical re-evaluation of the role of deliberations resulting in ethical assessments. The new gray zones between food and health put the issue of responsibility on the agenda for governments, health officials, food industries and consumers/patients.

3 Major Shortcomings in Ethical Principles and Standards

Is the currently dominant ethical approach in the life sciences conceptually able to address these issues? I doubt it, as long as bioethics is focussed on individual autonomy and informed consent (or choice). *Autonomy and informed consent* are the core ethical concepts of applied ethics and bioethics since the World Medical Association Declaration of Helsinki of 1964 (Beecher 1966). At that time, the Holocaust uppermost in many minds, the main issue was to erect an ethical barrier against state-led infringements of individual rights. The declaration, in which the ethical principles for medical research involving human subjects were stated, expressed the general agreement that concern for the interests of the individual subject must always prevail over the interests of science and society. The main ethical principles at work here, individual autonomy and informed consent, served to protect the individual. Ethicists, ethical committees and numerous ethically concerned people have done a great job in elaborating these principles in many concrete cases.

However, these ethical guidelines have not only yielded much progress in the field of ethics, but also have drawbacks when used in contexts in which the highest priority is not the protection of the individual patient versus the state, as in the case of bio banking, research agenda setting, and preventive food and medicine strategies (O'Neill 2002). These three issues, more recently coming to the fore because of new developments in the life sciences, require guidelines other than those aimed at protection of the individual patient. The deficit model of bioethics conceptualizes patients and consumers as passive with respect to research agenda setting, to health and food intervention schedules, and to preventive food and medicine projects where the role of health in good life is at stake and not only the curative role of medicine.

Needed here are rules of commitment, involvement and consultation rather than merely protective rules.

Certainly, there are ethicists who deviate from this dominant style, criticize it severely even. In particular Onora O'Neill (2002) has impressively argued that individual autonomy and informed consent are not the most fruitful perspectives to tackle issues of trust, of consultation and the ethical commitment that they are supposed to stimulate. The development of the life sciences in the direction of personalized medicine and nutrition has huge collective implications, as when common or public goods like genomes, knowledge, and the privatization of research and services are turned into commodities. These developments require guidelines, structures and organizational perspectives that respect forms of solidarity and ethically acceptable collective arrangements. Moreover, it becomes more and more necessary to conceptualize the different relationships between different types of good life, and to find out what types are compatible with universally respected moral values such as justice and fairness.

Another deficit of the perspective of informed consent and individual autonomy is that it fails to consider the dynamics of norms and values developed in response to new technologies in the life sciences (Keulartz et al. 2002). It turns out that the norms and values originally triggered by new life science technologies are subject to unexpected transformations. To give some examples: the pill (or anti-baby pill as it is often called) was initially seen as a device to regulate menstruation; reproductive technologies were considered to be technologies for improving happiness in heterosexual marriage but changed the relationship between homosexual partners as well, and as a consequence of that, the meaning of marriage in general. A related drawback is the model's neglect of the pluralism of values and norms that are gaining considerable weight in ethical debates on global networks and worldwide relationships in the medicine and food sectors. It is simply not the case that reasonable people all over the world agree on the role of health and nutrition in life, or on reproductive technologies.

Finally, the deficit perspective fails to reflect on its relationship with technologies. For the most part its adherents only stimulate introduction of restrictions by drawing red lines for technologies and they initiate efforts to limit technologies, rather than (carefully) opening Pandora's box to look for technological alternatives and to discuss technological priorities (Keulartz et al. 2002).

4 Reflective Problems of Bioethics

I turn form the major issues to problems that confront bioethical reflection as such, in particular the issue of its social and philosophical status and its quality standards. The questions here are of the following order: What kind of expertise can ethicists claim to have? What can progress in ethics mean? What can count as successful ethical advice? What ethical insights are to be considered seriously in all events? Can bio ethicists live up to the high expectations? Can we find better ways to guarantee that societal expectations of ethicists are met?

4.1 From Expert to Committee Member to Communicator

In 1972, Peter Singer wrote an influential paper in which he argued that ethicists should present themselves as experts (Singer 2000). He stated that in their role as authors ethicists should influence the public by writing on ethical issues, and he acted accordingly with enormous success; he contributed in particular to the now popular theme of animal welfare with his book *Animal Liberation* (1975). Due to his public presentations his reputation allowed him to tackle less popular themes without losing his independence. He comments in positive and negative terms on ethical problems and gives a green, a red and an orange light to ethically acceptable, negative and dubious solutions respectively – the expert policing the highway along which the life sciences should travel. Another well-known example is Francis Fukuyama, who in his 2002 study explicitly justifies the red lines between ethically acceptable and ethically unacceptable biotechnology: 'there are certain things that should be banned outright . . .'(p. 207).

In the 1980s however, mainstream applied ethics moved in a different direction. Its representatives became members of ethical committees in hospitals, medical research institutions and medical organizations. It should be noted that the majority of these committee members were not ethicists in the sense of academics with a degree in that discipline; rather, they were medical scientists interested in ethical issues. It is not surprising therefore that ethical problems in their field tended to be tackled by reducing them to concerns that could be handled in meetings of physician and patient. Moreover, very general principles and standards were formulated, rather tenuously connected with the life problems that emerged in the consulting sessions. I have in mind here the often rather casual references to the well-known book of Beauchamp and Childress that formulated four ethical principles (justice, respect for autonomy, beneficence and non-malfeasance) in its various editions (1979, 1983, 1989, 1993 and 2001).

With the rise of controversial technologies in the 1980s, such as reproductive technologies, and the innovations in the food sector, like genetically modified crops and cloned animals, the situation again changed, and ethicists became involved in organizing public debates with various forms of citizen participation. In this new role they act neither as experts nor as committee members. They now function as facilitators and moderators of debates and various forms of consultation. The new reproductive technologies and genomics have collective implications that go far beyond the relationship between physician and patient. The most common matrix for involvement for ethicists is no longer the relationship between physician and patient, but the much more complex relationship between science and society. Although many ethicists are still functioning as experts and committee members, ethical facilitators and moderators connected to public deliberations and consultations are numerous indeed. In particular the programs of social and ethical aspects of the latest in genomics and nanotechnology give a boost to these kinds of activities. Core issues of these programs are: What shall be the agenda of these consultations? Who shall be included? What kind of information and scientific expertise is necessary to make up the mind of the participants? What kind of long-term policies can earn the trust of participants (Burgess 2004)? A principled approach like that of Beauchamp and Childress is ill suited to assist in organizing public processes, procedures, and orientations when it comes to wholly new developments and ways to cope with them. Moreover, the expert seeks a verdict, the facilitator initiates a discussion. To be sure, the expert role and the principled approach have their merits, and can be of great help in elucidating ethical aspects; but they are in my view not really helpful as overall orientations for deliberations and debate.

To meet the expectations of the public more adequately, ethicists should make it clear that the still widespread idea of ethicists as neutral meta-ethical experts who on basis of this lofty status can provide answers is simply not valid. The moderator role, however, is still not really institutionalized in programs of education and should be more encouraged.

4.2 Three Types of Standards in 'Doing Ethics'

Depending on the social and historical context, different standards have been advocated and observed. In the period that ethicist were seen as experts, the main standards were those of meta-ethics, in the sense of standards of objectivity and analytical acuity. Again, Peter Singer is a very good example of someone who adheres to these standards. His writings are well known and he lectures to every available audience, be it Greenpeace, the World Economic Forum Annual Meeting for the rich and wealthy in Davos, or an animal protection movement's protest meeting.

Ethicists in their role of committee members operate according to standards that relate to social cooperation and accommodation, intent on forming influential networks so that reports will find their way to the right office desk. One such standard concerns avoidance of conflicting interests, which certainly reflects the position of the ethicist, because being a member of an ethics committee puts one constantly in jeopardy of losing one's integrity. It is difficult to reject a research proposal of a colleague scientist.

It is not hard to be critical of committee ethicists, and many are. But I would like to draw attention to a feature of this practice that is importantly new compared to the role of the ethicist as an expert. As a committee member the ethicist has to take into account the intricacies of a scientific or technological project, and to find out where, ethically, it hurts most. It is not sufficient to act from the outside like a policeman. Indeed, the committee member's hands are likely to be dirty, but this is to his credit, for he has not shunned involvement in the details of scientific projects.

For facilitators and moderators the standards are different again, because inclusiveness and reconciliation are so important. It means that not analysis but experimentation with new and promising vocabularies and objects, metaphors and languages that can transcend existing frameworks and controversies is explored. Agendas are often obscured by vocabularies that silence other, possibly fruitful ways of approaching certain technologies. For example, the debate on biotechnology is often dominated by the vocabulary of risk and safety, and not on questions about what kind of society modern biotechnology is promoting. Ethicists intent on harmonizing vocabularies have to use their imagination and simultaneously listen very carefully to what is being said, whispered or otherwise overheard. Scenarios, ethical experimenting, and interdisciplinary cooperation are to be stimulated. Bioethical progress should be conceptualized in terms of social and experiential experimentation and ways of evaluating them.

4.3 Relationship with Other Branches of Ethics and the Social Sciences

When we look at the various branches of applied ethics, like environmental ethics, animal ethics and global ethics, it seems that there is an uneven development, partly driven by outside demand and funding. What repercussions does this unevenness have, and is there a better way to determine research priorities in bioethics?

In the new period of bioethics as facilitator and moderator, we will need to address the relationship with the social sciences, in particular now that applied ethics is 'going empirical' – which all too often only means doing bogus interviews and low-quality empirical research. How to improve the empirical input of bioethics? If ethics is indeed empirical, it can be tested, which poses the challenging question of evidence-based ethics (Halpern 2005).

4.4 The Social Role of Bioethics

Finally, as a consequence of the broad societal interest in bioethics, we should discuss the social role of bioethics, e.g. its relationship with governments and with private actors. Can we do without regulations or codes of conduct, to prevent conflicts of interests and to ensure integrity and transparency in affiliations? Do we have to consider these issues in ethics programs of communication, education, schooling, and training?

For some time now ethics and philosophy have been taken a an essential, but ancillary discipline. By and large they have been drawing their problems from society and neighbouring disciplines, using methods that are continuous and overlapping with theirs. However, philosophy and in particular ethics is not merely a handmaiden, it is a full partner, posing the core questions of justification, legitimization and quality. Ethics can and should play a central role in grounding and shaping the most basic commitments of human practice.

5 New Concepts and Approaches

In correspondence with the challenges, the new situations, and the high expectations of society, the future of bioethics depends on how far it is able to develop new concepts, tools, organizational structures and communication channels with both science and society.

The concept of responsibility for complex problems in complex societies is still in its infancy (Bovens 1998). Responsibility is very often connected with being punished for wrong deeds (*retrospective responsibility*), but accountability does not only cover these kinds of things, but also more simple attitudes and organizational virtues like integrity, openness and transparency. We need a broader concept of *prospective responsibility*, which will include consideration for future consequences, a way of assessing benefits and losses that moves beyond calculation to deliberation. How far in time and space can human responsibility be stretched? To produce knowledge is to create new responsibilities, as in bio informatics and in particular the holistic, systemic progression of knowledge in x-omics (genomics and all the other branches). Much depends also on the particular technologies used: these too need to be taken seriously in distributing and allocating responsibilities.

We need tools for taking into account the fallibility, variability and ambiguity of knowledge. Too often our tools are directed only towards the happy outcome of scientific results, but the systematic nurture of positive expectations in science has as its counterpart a system of silencing the failures. The variability and ambiguity of knowledge permeates the production of knowledge at all stages. From the start, ethically responsible analysis of alternative pathways to knowledge and technology should be made more systematic and structural. In particular genomics produces knowledge that is uncertain, ambiguous and complex: the huge mass of data is difficult to decipher, the resulting conclusions on causal connections between genes, proteins, life styles and certain propensities for health risks are very provisional. Nevertheless, people manage to live with these dubious data, and all kinds of decisions are based on them, without court of appeal.

Thirdly, bioethics is in need of new structures, like regional and global organizations and platforms. UNESCO (United Nations Educational, Scientific and Cultural Organization) tried to initiate these forms, until now however not much attention has been paid by national and international bioethics organizations. EurSafe, the European Association for Agricultural and Food Ethics, should take the initiative to establish a bioethics platform in countries and continents that until now have been unable to start these, like African and Latin-American countries. These new types of communication should be accompanied by a *consensus gentium* document on best practices in bioethics (be it consultancy, teaching, or research). The questions raised in Section 4 above concerning the standards in ethical involvement need global consideration and evaluation.

Fourthly, we need other types of communication channels with both science and society other than the present-day committees and policy organizations. Cultural and art performances and organizations could be taken as an example of how bioethics can reach out to sectors of science and society until now not reached. Art and other cultural activities can people make aware of opportunities and barriers by framing scientific and technological innovations (or promises of innovations!) in a new light. George Orwell's novel 1984 probably adds nothing, cognitively speaking, that pessimistic philosophers of technology did not write before; still, the way the novel frames the fate of the main characters explores the depths of human experiential potencies and the message is therefore much more understandable. This respected form of art is since the 1950s accompanied by entirely new art forms such as working with video or information technology to tackle the issues of science and society.

6 Conclusion

The picture of bioethics sketched above is a gloomy one: bioethics stands at the brink of a new era, but is short of valid tools and concepts to tackle the challenges. Emperor Bioethics is still naked although the social expectations are large. Mainstream bioethics continues to concentrate too much on individual autonomy and informed consent, notions that we need to transcend. We need to move away from the individualistic fixation and to analyze issues of good life in a global and plural world, issues of coexistence of radically different value systems and technological systems and issues of consultation for radically disagreeing partners.

Moreover, the establishment of accountability networks that regulate responsibilities of stakeholders in the drug and food sectors is an urgent requirement for increasing trust in these sectors.

Finally, the upcoming global role of bioethics should be accompanied by codes of best practice for bio ethicists. The new role of ethicists as communicators and facilitators of public debate should certainly count as one of these best practices.

References

Beecher, H. "Ethics and clinical research." *New England Journal of Medicine*, 274, 1354–1360, 1966.

Beauchamp, T. and Childress, J. *Principles of Biomedical Ethics*, 5th ed. Oxford: Oxford University Press, 2001.

Bovens, M. *The Quest for Responsibility, Accountability and Citizenship in Complex Organizations*. Cambridge: Cambridge University Press, 1998.

Bulger, R. (ed.). *The Ethical Dimensions of the Biological Sciences*. Cambridge: Cambridge University Press, 2002.

Burgess, M. "Public consultation in ethics: an experiment in representative ethics." *Journal of Bioethical Inquiry*, 1, 1, 4–13, 2004.

Castle, D., Cline, C., Daar, A.S., Tsamis, C. and Singer, P.A. "Nutrients and norms: ethical issues in nutritional genomics." In *Nutrigenomics: Concepts and Technologies*. J. Kaput and R.L. Rodriguez (eds.). New Jersey: John Wiley & Sons, 2004.

Curtin, D. "Private interest representation or civil society deliberation? A contemporary dilemma for European Union governance." *Social & Legal Studies*, 12, 1, 55–75, 2003.

Etzkowitz, H. and Leydesdorff, L. (eds.). *Universities and the Global Knowledge Economy: A Triple Helix of University-Industry-Government Relations*. (Science, Technology and International Political Economy). London: Continuum, 2002.
Food Ethics Council. *Just Knowledge*. Brighton: Food Ethics Council, 2004.
Fukuyama, F. *Our Posthuman Future: Consequences of the Biotechnology Revolution*. Farrar: Straus and Giroux, 2002
Graff, G.D., et al. "The public-private structure of intellectual property ownership in agricultural biotechnology." *Nature Biotechnology*, 21, 9, 989–995, 2003.
Halpern, S.D. "Towards evidence based bioethics." *British Medical Journal*, 331, 901–903, 2005.
Hansen, J., et al. "Beyond the knowledge deficit: recent research into lay and expert attitudes to food risks." *Appetite*, 41, 111–121, 2003.
Keulartz, J., Korthals, M., Schermer, M. and Swierstra, T. (eds.). *Pragmatist Ethics for a Technological Culture*. Dordrecht: Kluwer, 2002.
Korthals, M. *Before Dinner: Philosophy and Ethics of Food*. Dordrecht: Springer, 2004.
Krimsky, S. *Science in the Private Interest*. Lanham: Rowman, 2004.
Lyson, Th. "Advanced agricultural biotechnologies and sustainable agriculture." *Trends in Biotechnology*, 20, 5, 193–196, 2002.
Merton, R. *Social Theory and Social Structure*. New York: The Free Press, 1968.
Nestle, M. *Food Politics: How Food Industry Influences Nutrition and Health*. Berkeley: University of California Press, 2002.
O'Neill, O. *Autonomy and Trust in Bioethics*. Oxford: Oxford University Press, 2002.
Ravetz, J. "Food safety, quality, and ethics: a post-normal perspective." *Journal of Agricultural and Environmental Ethics*, 15, 3, 255–265, 2002.
Ravetz, J. "The post-normal science of precaution." *Future*, 36, 347–357, 2004.
Rowe, G. and Frewer, L.J. "Evaluating public-participation exercises: a research agenda." *Science, Technology, & Human Values*, 29, 512–556, Autumn 2004.
Rowe, G. and Frewer, L.J. "Public participation methods: a framework for evaluation." *Science, Technology, & Human Values*, 25, 3–29, Winter 2000.
Science. "Biomedical ethics on the Front Burner." *Science*, 290, 5500, 2225, 2000.
Sharp, R.R., Yudell, M.A. and Wilson, S.H. "Shaping science policy in the age of genomics." *Nature Reviews Genetics*, 5, 311–315, 2004.
Shorett, P., Rabinow, P. and Billings, P. "The changing norms of the life sciences." *Nature Biotechnology*, 21, 123–125, 2003.
Shrader-Frechette, K. *Ethics of Scientific Research*. Lanham: Rowman, 1994.
Singer, P. *Writings on an Ethical Life*. New York: Harper, 2000.
Singer, P. *Animal Liberation*. New York: New York Review/Random House, 1975.
Thackray, A. (ed.). *Private Science: Biotechnology and the Rise of the Molecular Sciences*. Philadelphia: University of Philadelphia Press, 1998.
Thursby, J.G. and Thursby, M.C. "University licensing and the Bayh-Dole Act." *Science*, 301, 1052, 2003.

Part IV
Abilities and Disabilities

Disability: Suffering, Social Oppression, or Complex Predicament?

Tom Shakespeare

Traditional accounts of disability have focused on deficits of body or mind. Disabled people have been defined in terms of their impairments. In different historical and cultural contexts, there have been competing understandings of disability, drawing on discourses of religion, law, or medicine (Stiker 2000). In the post-war period, accounts of disability from social policy, medical sociology and bioethics also failed to understand disability as a human rights issue. By seeing disability as a problem of individuals with flawed bodies, the collective injustice and discrimination suffered by disabled people as a minority group remained invisible (Barnes 1998). While it would be wrong to suggest that researchers have entirely failed to pay attention to environments, contexts and social relations, it is certainly the case that traditional academic research has not addressed the political imperatives which emerge from the structural disadvantage that many disabled people face in most societies across the globe.

In this chapter, I explore how the disability rights movement have challenged the traditional medicalisation of disability. I then analyse the limitations of the radical position. Finally, I argue for a more complex account of disability, which has space for both biological and socio-political dimensions of disabled people's experience (Shakespeare 2006).

1 Making Disability Political

The emergence of civil rights, citizenship or equal opportunities approaches to understanding disability has owed most to the actions of disabled people themselves, rather than to those researchers and professionals who have traditionally worked in the field of disability. A wave of direct action and self-organisation rippled around the world from the early 1970s onwards. The Independent Living pioneers of Berkeley, California were amongst the initiators of this movement, along with those US activists who occupied Secretary of State Califano's office to achieve implementation of Section 504 of the 1973 Rehabilitation Act, prohibiting

---

T. Shakespeare
e-mail: T.W.Shakespeare@newcastle.ac.uk

M. Düwell et al. (eds.), *The Contingent Nature of Life*,

discrimination against disabled people in federal programmes (Dreidger 1989). In Britain, it was Paul Hunt and the activists of the Union of Physically Impaired Against Segregation (UPIAS) who developed the new account of disability, which later became known as the social model of disability (Campbell and Oliver 1996). In many other countries across the developed and developing world, disabled people rallied and campaigned to have greater control over their lives, better services in the community, and a new approach to understanding disability (Charlton 1998).

Common to all these movements of disabled liberation is the changed conception of the nature of disability. Rather than inhering solely in the individual, or their medical problems of body or mind, disability is reconceived as a political issue, in the same way as other issues such as class, race, gender and sexuality. Attention shifts from care, cure and rehabilitation to barrier removal, anti-discrimination statutes and independent living. Rather than being ashamed or invisible, disabled people can develop an identity politics approach, in which they can express solidarity, pride and resistance. Key strategies in this shift include new forms of research, commonly labelled “disability studies” and explicitly modelled on feminism, post-colonial and gay and lesbian studies; also new political tactics such as direct action, again using precedents from other struggles; and finally, disability arts, expressing the changed cultural values and vision of disabled people. Each of these strategies both works instrumentally, but also stands symbolically to define disability as a matter of politics and difference.

There are, of course, differences in the way that the civil rights ideology has been conceived. For example, in the US movement, emphasis was placed on a minority group approach, seeing disabled people in the same way as African Americans had been seen, as a group excluded from the mainstream and the market, and hence denied prosperity and respect (Hahn 1985). In the Nordic countries, with a strong welfare state tradition, disability was viewed in relational terms (Tøssebro 2004). In the UK, Marxist-inspired disability activists developed the social model, arguing that disability itself should be redefined. Rather than people with disabilities suffering exclusion, British social model theorists argued that disability should be defined as exclusion: disability was a relationship between people with impairments, and a society which had failed to include or support them. As the UPIAS activists argued:

> In our view, it is society which disables physically impaired people. Disability is something imposed on top of our impairments, by the way we are unnecessarily isolated and excluded from full participation in society
>
> (UPIAS 1976: 3).

Despite these different emphasis and terminologies, there was agreement on the focus of disability research and disability policy: the new approach mandated environmental and social change not medical or social care.

The redefinition of disability in the social model parallels the feminist movement’s redefinition of women’s experience in the early 1970s. Anne Oakley (1972) and others had distinguished between *sex* – the biological difference between male and female – and *gender*, the socio-cultural distinction between men and women, or masculine and feminine. The former was biological and universal, the latter was social, and specific to particular times and places. Thus it could be claimed that sex corresponds to impairment, and gender corresponds to disability. The disability movement followed a well-established path of de-naturalising forms of social oppression, demonstrating that what was thought throughout history to be natural was actually a product of specific social relations and ways of thinking.

2 Losing the Body

> It is not individual limitations, of whatever kind, which are the cause of the problem, but society's failure to provide appropriate services and adequately ensure the needs of disabled people are fully taken into account in its social organisation
>
> (Oliver 1996: 32).

The social model defines disability in terms of oppression and barriers, and breaks the link between disability and impairment. This has led to the common criticism that social model approaches have neglected the role of impairment. For example, in her book *Pride Against Prejudice* (1991) Jenny Morris discussed features of disability which had been neglected by the dominant, UPIAS-inspired ideology of the British disability movement: culture, gender, personal identity. Most importantly, she acknowledged that impairment itself created pain and difficulties which were not solely attributable to disabling factors in society:

> While environmental barriers and social attitudes are a crucial part of our experience of disability – and do indeed disable us – to suggest that this is all there is to it is to deny the personal experiences of physical and intellectual restrictions, of illness, of the fear of dying
>
> (Morris 1991: 10).

Following this lead, in 1992 Liz Crow published a paper in *Coalition*, the Journal of the Greater Manchester Coalition of Disabled People (subsequently published as Crow 1996) in which she criticised the social model for failing to encompass the personal experience of pain and limitation which is often a part of impairment. While she expressed commitment to the social model itself, she called for it to be developed in order to find a place for the experience of impairment:

> Instead of tackling the contradictions and complexities of our experiences head on, we have chosen in our campaigns to present impairment as irrelevant, neutral and, sometimes, positive, but never, ever as the quandary it really is
>
> (1996: 208).

Crow did not suggest that impairment was an explanation for disadvantage, but that it was an important aspect of disabled people's lives. In the 1993 Open University course book, *Disabling Barriers Enabling Environments*, Sally French also wrote an important and careful article about the persistence of impairment problems:

> I believe that some of the most profound problems experienced by people with certain impairments are difficult, if not impossible, to solve by social manipulation
>
> (French 1993: 17).

As a person with visual impairment, she gave the example of being unable to recognise people, and failure to read non-verbal cues in interaction, explaining how these aspects of being a visually impaired person caused problems interacting with neighbours and with her students. According to French, no amount of barrier removal or social change could entirely remedy or remove the problem of visual impairment. French also explored the reasons for resistance to these alternative perspectives:

> It is no doubt the case that activists who have worked tirelessly within the disability movement for many years have found it necessary to present disability in a straightforward, uncomplicated manner in order to convince a very sceptical world that disability can be reduced or eliminated by changing society, rather than by attempting to change disabled people themselves...
>
> (French 1993: 24).

Most recently, Carol Thomas (1999), from within the materialist social model tradition, has developed an approach to disability which makes space for the exploration of personal experience, of the psycho-emotional dimensions of disability, and for the impact of what she calls "impairment effects". She uses the latter concept "to acknowledge that impairments do have direct and restricting impacts on people's social lives" (Thomas 1999: 42).

All of these writers have argued from within a social model perspective, calling for reform or development of the model, rather than its abandonment. Nevertheless, many of these critical voices have encountered strong opposition from within the British disability movement and disability studies. Indeed, there have an implicit tendency to deny the reality of impairment: for example, Dan Goodley (2001) seems to cast doubt on the existence of learning difficulties:

> social structures, practices and relationships continue to naturalise the subjectivities of people with 'learning difficulties', conceptualising them in terms of some a priori notion of 'mentally impaired'
>
> (Goodley 2001: 211).

Advocates frequently use scare-quotes and phrases such as "labelled with learning difficulties". The BCODP briefing document on the social model refers to "... barriers encountered by people who are viewed by others as having some form of impairment" (BCODP, n.d.). Michael Oliver talks of "people who are viewed by others as having some form of impairment" (Oliver 2004). Mental health campaigners use the terminology "survivors of the mental health system" (Beresford and Wallcraft, 1997). While attention to labelling and discourse is important, there is a danger of ignoring the problematic reality of biological limitation. Linguistic distancing serves as a subtle form of denial.

There are a number of reasons why it might be important for disability studies to engage with impairment:

1. Disability studies should pay attention to the views and perspectives of disabled people, rather than accepting medical claims about the nature and meaning of impairment. Many respondents say that impairment is a central and structuring part of their experience.

2. Disability studies should be concerned with medical responses to impairment. Is treatment effective? Are there side effects? Is research funded effectively? Does the NHS prioritise disabled people's impairment needs?
3. Disability studies should be concerned with the prevention of impairment. If there is an interest in the quality of life of disabled people, then this includes minimising the impact of impairment and impairment complications.
4. Disabling barriers both cause and exacerbate impairment. For example, poverty and social exclusion make impairment worse and create additional impairments, particularly risk of mental illness.

Many social modellists deny the relevance of impairment to disability studies. For example, Rachel Hurst (2000) makes a familiar comparison of disability to gender, claiming that just as it would be inappropriate to analyse details of women's biology in political debates, so there is no need to analyse individual characteristics of disabled people:

> to concentrate on the personal characteristics of the disabled individual and the functional limitations arising from impairment is itself, disablism
>
> (Hurst 2000: 1084).

It is common to hear such analogies being made between the experiences of disabled people and those of women, minority ethnic communities and lesbians and gays (e.g. Gordon and Rosenblum 2001). For example, Carol Thomas sees the concept "disablism" as on a par with concepts such as sexism, racism and homophobia. The term "disablism" has also been deployed by the disability charity, Scope, defined as "discriminatory, oppressive or abusive behaviour arising from the belief that disabled people are inferior to others", an approach which borrows from the concept of racism. But how far can the analogy between different movements and oppressions be sustained?

As social movements, women's liberation, gay rights, disability rights and anti-racism are similar in many ways. Each involves identity politics, each challenges the biologisation of difference, each has involved an alliance of academia and activism. There are parallels between the theorisation of disability, and the theorisation of race, gender and sexuality, as the many citations of other oppressions within disability studies literature demonstrate. Yet the oppression which disabled people face is different and in many ways more complex, than sexism, racism and homophobia. Women and men may be physiologically and psychologically different, but it is no longer possible to argue that women are made less capable by their biology:

> Gender, like caste, is a matter of social ascription which bears no necessary relation to the individual's own attributes and inherent abilities
>
> (Oakley 1972: 204).

Similarly, only racists would see the biological differences between ethnic communities as the explanation for their social differences. Nor is it clear why being lesbian or gay would put any individual at a disadvantage, in the absence of prejudice and discrimination. But even in the absence of social barriers or oppression, it would still be problematic to have an impairment, because many impairments are limiting or difficult, not neutral.

Comparatively few restrictions experienced by people with impairment are "wholly social in origin". If someone discriminated against disabled people purely because they had an impairment, and imposed exclusions which were solely on this basis, and nothing to do with their abilities, then this would be a wholly social restriction. Examples clearly exist of this form of discrimination: nightclubs which exclude disabled people because they cater only to attractive young people; the notorious "ugly laws" in early twentieth century Chicago and elsewhere which prohibited disfigured people from public spaces. Here, disability discrimination parallels racism, sexism and other social exclusions exactly. But in most cases, disabled people are experiencing both the intrinsic limitation of impairment, and the externally imposed social discrimination.

For example, disabled people have problems in employment for a variety of social factors: inaccessible workplaces; prejudiced employers; lack of training or education opportunities; lack of confidence. But as well as these external barriers, many disabled people are limited to some extent in the type of work they can do, or the amount of work they can do. Some disabled people are unable to do any work at all for some or all of their lives, due to the impact of their impairment. Disabled sociologist Paul Abberley noted that:

> even in a society which did make profound, genuine attempts, well supported by a financial provision, to integrate profoundly impaired people into the world of work, some would be excluded
>
> (Abberley 2001: 131).

Early disability activists were aware that employment was not a possibility for all. For example, Paul Hunt (1966) challenged the focus on work, arguing that disabled people outside the labour market contributed to society in different ways, not least by challenging utilitarian values.

Impairment often has explanatory relevance in ways that the colour of someone's skin, or their sex, or their sexual orientation usually does not. If social model approaches fail to allow for the role of impairment, they will fail to understand the complexities of disabled people's social situation. Moreover, disability rights academics and activists risk creating stories about disability which many disabled people will not recognise as describing their own experience. As disabled feminists have argued, impairment is an important part of the disability experience. Impairment affects individuals in different ways. Some people are comparatively unaffected by impairment, or else the main consequences of impairment arise from other people's attitudes. For others, impairment limits the experiences and opportunities they can experience. In some cases, impairment causes progressive degeneration and premature death. These features of impairment cause distress to many disabled people, and any adequate account of disability has to give space to the difficulties which many impairments cause. As Simon Williams has argued,

> ...endorsement of disability solely as social oppression is really only an option, and an erroneous one at that, for those spared the ravages of chronic illness
>
> (Williams 1999: 812).

Of course, impairment may also lead to opportunities: for example, to experience the world in a different way, or to develop one sense or aptitude because others are unavailable. Moreover everyone, even the supposedly able-bodied, experiences limitations: it's not just the wheelchair user who is unlikely to climb Everest. It is not necessary to claim that all impairments are negative, or that impairment is only and always negative. But for many, impairment is not neutral, because it involves intrinsic disadvantage. Disabling barriers make impairment more difficult, but even in the absence of barriers, impairment can be problematic.

3 The Complexity of Disability

Disability is always an interaction between individual and structural factors. Rather than getting fixated on defining disability either as a deficit or a structural disadvantage, a holistic understanding is required. The experience of a disabled person results from the relationship between factors intrinsic to the individual, and extrinsic factors arising from the wider context in which she finds herself. Among the intrinsic factors are issues such as: the nature and severity of her impairment, her own attitudes to it, her personal qualities and abilities, and her personality. Among the contextual factors are: the attitudes and reactions of others, the extent to which the environment is enabling or disabling, and wider cultural, social and economic issues relevant to disability in that society.

This approach to disability seems plausible, and relates closely to the arguments of a range of recent writers (Williams 1999; Danermark and Gellerstedt 2004; Gabel and Peters 2004) as well as to the Nordic relational approach. Anders Gustavsson talks about this relative interactionist perspective as an alternative to both biological and contextual essentialism,

> a theoretical perspective that rejects assumptions about any primordial analytical level and rather takes a programmatic position in favor of studying disability on several different analytical levels
>
> (Gustavsson 2004: 62).

A recent paper by Van den Ven and a Dutch team (Van den Ven et al. 2005) came to a similar conclusion based on qualitative research. They concluded:

> Both the individual with a disability and others in society have a shared responsibility with respect to the integration of people with disabilities into society. Each must play their part for integration to occur: it takes two to tango. An individual with a disability should be willing to function in society and adopt an attitude towards others in society in such a way that they can join in with activities and people in society. On the other hand society should take actions to make functioning in society possible for people with disabilities. In other words, society should be inclusive with respect to people with disabilities by passing laws on anti-discrimination, ensuring accessibility of buildings and arranging appropriate care facilities for people with disabilities
>
> (Van den Ven et al. 2005: 324).

These authors, like myself, balance medical and social aspects. They refer to three issues which influence integration: individual factors, which include personality and skills as well as impairment; societal factors, referring to accessibility, attitudes etc., and factors within the system of support, by which they mean social support, professional care and assistive devices. The interrelation of these three sets of factors determine or produce disability.

The interactional approach to understanding disability as a complex and multi-factorial phenomenon necessitates coming to terms with impairment. Critiquing the social model, I argued that impairment was important to many disabled people, and had to be adequately theorised in any social theory of disability. Until now, only two alternatives have been offered: what Michael Oliver calls "medical tragedy theory", and on the other hand, the denial or neglect of impairment within social model theory. Impairment is not the end of the world, tragic and pathological. But neither is it irrelevant, or just another difference. Many disabled people are unable to view impairment as neutral, as Michael Oliver and Bob Sapey concede in the second edition of *Social Work with Disabled People*:

> Some disabled people do experience the onset of impairments as a personal tragedy which, while not invalidating the argument that they are being excluded from a range of activities by a disabling environment, does mean it would be inappropriate to deny that impairment can be experienced in this way
>
> (Oliver and Sapey 1998: 26).

Instead of the polarised and one-dimensional accounts in both traditional research and disability studies, a nuanced attitude is needed, involving a fundamental ambivalence. Disability studies needs to capture the fact that impairment may not be neutral, but neither is it always all-defining and terrible.

One way of capturing the complexity of impairment is to view it as a *predicament*.

The Concise Oxford Dictionary defines predicament as "an unpleasant, trying or dangerous situation". Although still negative, this does not have the inescapable emphasis of "tragedy". The notion of "trying" perhaps captures the difficulties which many impairments present. They make life harder, although they can be overcome. The added burdens of social oppression and social exclusion, which turn impairment into disadvantage, need to be removed: this seems to me very much the spirit of the original UPIAS approach to disability. Everything possible needs to be provided to make coping with impairment easier. But even with the removal of barriers and the provision of support, impairment will remain problematic for many disabled people.

The predicament of impairment – the intrinsic difficulties of engaging with the world, the pains and sufferings and limitations of the body – mean that impairment is not neutral. It may bring insights and experiences which are positive, and for some these may even outweigh the disadvantages. But that does not mean that we should not try and minimise the number of people who are impaired, or the extent to which they are impaired.

It is not only impairment which is a predicament. For example, Zygmunt Bauman argues that:

> The postmodern mind is reconciled to the idea that the messiness of the human predicament is here to stay, This is, in the broadest of outlines, what can be called postmodern wisdom
>
> (Bauman 1993: 245).

Other aspects of embodiment – for example, the pains of menstruation or childbirth for women – could also be understood through the predicament concept, as could the inevitability and tragedy of death. To call something a predicament is to understand it as a difficulty, and as a challenge, and as something which we might want to minimise but which we cannot ultimately avoid. As Sebastiano Timpanaro suggests,

> physical ill . . . cannot be ascribed solely to bad social arrangements: it has its zone of autonomous and invincible reality
>
> (Timpanaro 1975: 20).

But this is not to fall into the trap of regarding impairment as a tragedy or an identity-defining flaw.

Some people will object to what appears to be a negative approach to impairment. They should note that I am not denigrating disabled people, nor claiming that impairment makes disabled people second class citizens or less worthy of support and respect. Disabled people are often inferior to non-disabled people in terms of health, function or ability, but they are not lesser in terms of moral worth, political equality or human rights. The suffering and happiness of disabled people matters just as much as that of non-disabled people.

If impairment truly was neutral – or beneficial – then we could have no objection to someone who deliberately impaired a child. If impairment was just another difference, then, as John Harris (1993, 2001) points out, there would be nothing wrong with painlessly altering a baby so they could no longer see, or could no longer hear, or had to use a wheelchair. Even if no suffering or pain was caused in the process, we would surely consider this irresponsible and immoral. Something would have been lost. The implication of this must be that impairment prevention should have an important role in social responses to disability. This does not undermine the worth or citizenship of existing disabled people. It suggests that because impairment causes predicaments and is limiting in various ways, we should take steps to prevent or mitigate it, where possible.

Furthermore, the connection to other embodiment predicaments underlines a commonplace observation which was made central in the work of Irving Zola (1989), and which has great significance for a post-social model approach to disability (see also Bickenbach et al. 1999). Impairment is a universal phenomenon, in the sense that every human being has limitations and vulnerabilities (Sutherland 1981), and ultimately is mortal. Across the life-span, everyone experiences impairment and limitation. Impairment is more likely to be acquired than congenital: ageing is particularly associated with increased levels of impairment. The ubiquity of impairment is underscored by the Human Genome Project which has shown that everyone has approximately hundreds of mutations in their genome, many of which may predispose the individual to illness or impairment. As Armand Leroi has written, "We are all mutants. But some of us are more mutant than others." (Leroi 2005: 19). In this sense, genetic diagnosis is toxic knowledge, which has the power to turn healthy people into pre-impaired people.

To claim that "everyone is impaired" should not lead to any trivialising of impairment or the experience of disabled people. As I have stated previously, impairments differ in their impact. It is important to respect real differences – particularly the extent to which people are affected by suffering and restricting. At the extreme, as Alasdair Macintyre argues, are very severely impaired people:

> such that they can never be more than passive members of the community, not recognizing, not speaking or not speaking intelligibly, suffering, but not acting
>
> (1999: 127).

It would be wrong to neglect the particular needs which arise from these different differences.

Not everyone is impaired all the time. Taking a life course view of impairment highlights the ways that impairment is manifested over time: disabled children grow up to be non-disabled adults, non-disabled people become impaired through accident or in old age. Impairments can be variable and episodic: sometimes people recover, and sometimes impairments worsen. The nature and meaning of impairment is not given in any one moment. Not all people with impairment have the same needs, or are disadvantaged to the same extent. Moreover, different people experience different levels of social disadvantage or social exclusion, because society is geared to accommodate people with certain impairments, but not others. Everyone may be impaired, but not everyone is oppressed.

The benefits of regarding every human being as living with the predicament of impairment are that it forces us to pay attention to what we have in common; it counsels us to accept the inextricable limitation of life, rather than to deny or fight against it; it suggests the need to re-evaluate disabled people; it focuses attention on the social aspects of disability. For example, if everyone is impaired, why are certain impairments remedied or accepted, and others not? Why does impairment result in exclusion in some cases and not others? These processes and choices are largely social and structural and can be changed. In policy terms, a universal approach would use the range of human variation as the basis for universal design, and aim for justice in the distribution of resources and opportunities. Disabled people would not be expected to identify themselves as separate and incompetent, in order to qualify for provision.

Human beings are not all the same, and do not all have the same capabilities and limitations. Need is variable, and disabled people are among those who need more from others and from their society. Alasdair Macintyre begins to explore the political implications of this reality:

> a form of political society in which it is taken for granted that disability and dependence on others are something that all of us experience at certain times in our lives and this to unpredictable degrees, and that consequently our interest in how the needs of the disabled are adequately voiced and met is not a special interest, the interest of one particular group rather than of others, but rather the interest of the whole political society, an interest that is integral to their conception of their common good
>
> (1999: 130).

In an era where enthusiasts for biotechnology, nanotechnology, stem cell research and pharmacology seem to be reasserting a perfectionist project, and when the boundaries between therapy and enhancement are beginning to be blurred, asserting the ubiquity of disability is a radical and important step. Disabled people show that good quality of life does not depend on being free from limitation or suffering (Albrecht and Devlieger 1999). Understanding what it is to be disabled is a vital part in accepting what it is to be human.

References

Abberley, P. "Work, disability and European social theory." In *Disability Studies Today*. C. Barnes, M. Oliver and L. Barton (eds.). Cambridge: Polity, 2001.

Albrecht, G.L. and Devlieger P.J. "The disability paradox: high quality of life against all odds." *Social Science and Medicine*, 48, 977–988, 1999.

Barnes, C. "The social model of disability: a sociological phenomenon ignored by sociologists." In *The Disability Reader*. T. Shakespeare (ed.). London: Cassell, 1998.

Bauman, Z. *Postmodern Ethics*. Oxford: Blackwell, 1993.

Beresford, P. and Wallcraft, J. "Psychiatric system survivors and emancipatory research: issues, overlaps and differences." In *Doing Disability Research*. C. Barnes and G. Mercer (eds.). Leeds: The Disability Press, 1997.

Bickenbach, J.E., Chatterji, S., Badley, E.M. and Ustun, T.B. "Models of disablement, universalism and the international classification of impairments, disabilities and handicaps." *Social Science and Medicine*, 48, 1173–1187, 1999.

British Council of Disabled People (n.d.). "The social model of disability and emancipatory disability research" – briefing document, downloaded from www.bcodp.org.uk/about/research.shtml. Last visit: 23 March 2004.

Campbell, J. and Oliver, M. *Disability Politics: Understanding Our Past, Changing Our Future*. London: Routledge, 1996.

Charlton, J. *Nothing About Us Without Us: Disability, Oppression and Empowerment*. Berkeley: University of California Press, 1998.

Crow, L. "Including all our lives." In *Encounters with Strangers: Feminism and Disability*. J. Morris (ed.). London: Women's Press, 1996.

Danermark, B. and Gellerstedt, L.C. "Social justice: redistribution and recognition – a non-reductionist perspective on disability." *Disability and Society*, 19, 4, 339–353, 2004.

Dreidger, D. *The Last Civil Rights Movement*. London: Hurst, 1989.

French, S. "Disability, impairment or something in between." In *Disabling Barriers, Enabling Environments*. J. Swain et al. (eds.). London: Sage, 17–25, 1993.

Gabel, S. and Peters, S. "Presage of a paradigm shift? Beyond the social model of disability toward resistance theories of disability." *Disability and Society*, 19, 6, 585–600, 2004.

Goodley, D. "'Learning difficulties', the social model of disability and impairment: challenging epistemologies." *Disability and Society*, 16, 2, 207–231, 2001.

Gordon, B.O. and Rosenblum, K.E. "Bringing disability into the sociological frame: a comparison of disability with race, sex and sexual orientation statuses." *Disability and Society*, 16, 1, 5–19, 2001.

Gustavsson, A. "The role of theory in disability research: springboard or strait-jacket." *Scandinavian Journal of Disability Research*, 6, 1, 55–70, 2004.

Hahn, H. "Towards a politics of disability: definitions, disciplines and policies." *Social Science Journal*, 22, 4, 87–105, 1985.

Harris, J. "Is gene therapy a form of eugenics?" *Bioethics*, 7, 178–87, 1993.

Harris, J. "One principle and three fallacies of disability studies." *Journal of Medical Ethics*, 27, 6, 383–388, 2001.

Hunt, P. (ed.). *Stigma*. London: Geoffrey Chapman Publishing, 1966.

Hurst, R. "To revise or not to revise." *Disability and Society*, 15, 7, 1083–1087, 2000.

Leroi, A.M. *Mutants: On the Form, Varieties and Errors of the Human Body*. London: Harper Perennial, 2005.

Macintyre, A. *Dependent Rational Animals: Why Human Beings Need the Virtues*. London: Duckworth, 1999.

Morris, J. *Pride Against Prejudice*. London: Women's Press, 1991.

Oakley, A. *Sex, Gender and Society*. London: Gower, 1972.

Oliver, M. *Understanding Disability: From Theory to Practice*. Basingstoke: Macmillan, 1996.

Oliver, M. "The social model in action: if I had a hammer." In *Implementing the Social Model of Disability: Theory and Research*. C. Barnes and G. Mercer (eds.). Leeds: The Disability Press, 18–47, 2004.

Oliver, M. and Sapey, B. *Social Work with Disabled People*. 2nd ed. Basingstoke: Palgrave Macmillan, 1998.

Shakespeare, T. *Disability Rights and Wrongs*. London: Routledge, 2006.

Stiker, H.-J. *A History of Disability*. University of Michigan Press, 2000.

Sutherland, A. *Disabled We Stand*. London: Souvenir Press, 1981.

Thomas, C. *Female Forms: Experiencing and Understanding Disability*. Buckingham: Open University Press, 1999.

Timpanaro, S. *On Materialism*. London: New Left Books, 1975.

Tøssebro, J. "Understanding disability." *Scandinavian Journal of Disability Research*, 6, 1, 3–7, 2004.

UPIAS. *Fundamental Principles of Disability*. London: UPIAS, 1976.

Van den Ven, L., Post, M., de Witte, L. and Van den Heuvel, W. "It takes two to tango: the integration of people with disabilities into society." *Disability and Society*, 20, 3, 311–329, 2005.

Williams, S.J. "Is anybody there? Critical realism, chronic illness and the disability debate." *Sociology of Health and Illness*, 21, 6, 797–819, 1999.

Zola, I.K. "Towards the necessary universalizing of a disability policy." *The Milbank Quarterly*, 67, Suppl. 2, Pt. 2, 401–428, 1989.

Disability and Moral Philosophy: Why Difference Should Count

Sigrid Graumann

1 Introduction

Disability plays an important role in applied ethics, particularly in Medical Ethics and Bioethics. However, the perspectives of disabled people differ radically in several important respects from those of professional ethicists. Furthermore, the relationship between ethicists and disabled activists is rather tense for two reasons. The first has to so with the level of applied ethics, the second with moral theory building.

The perception of disability in *applied ethics*, particular in medical ethics and bioethics, is primarily reduced to functional defects and suffering according to the so-called medical model of disability (Hirschberg 2004). The debate about prenatal selection (Wolbring 2001) and euthanasia – as exemplified by the Peter Singer debate (Dederich 2003) – are only two examples. In these debates representatives of the disability movement have accused (non-disabled) ethicists of taking a discriminatory view of disability. Involved ethicists, on the other hand, argue that the feelings of disabled people are understandable but that there is no violation of their individual rights and hence no discrimination (Birnbacher 1999; Hoerster 1997). Nevertheless, it is doubtful whether the problem has been settled by this reply. Perhaps the topics "selection" and "quality of life" can not be satisfactory discussed in terms of the violation of individual rights but rather in terms of the "cumulative consequences" of biomedical innovations (Lübbe 2003).

As many feminist theorists have pointed out, most *moral theories* are formulated from the perspective of more or less wealthy and honoured, independent and autonomous, healthy, usually male and white citizens. This perspective generally neglects the specific needs and interests of members of subordinated groups and, in particular, of vulnerable and dependent individuals (Wendell 1996; Nussbaum 2002), and those caring for them (Kittay 1999). That means that many ethical theories are remarkably biased and thus probably indeed involved in stigmatising and discriminating against members of subordinated groups, particularly if they are

S. Graumann
e-mail: graumann@imew.de

M. Düwell et al. (eds.), *The Contingent Nature of Life*,

vulnerable in a unique way as many disabled people are. Consequently, I think we have to switch to a sort of "metadiscourse" allowing us to analyse the role played by ethics in stigmatising and discriminating against disabled individuals.

Before I come back to the analysis of the perception of the phenomenon of disability in moral philosophy I would like to make two presumptions:

Firstly, the relevance of the perception of disability in ethics is not restricted to a fringe group issue. It is (or it should be) rather a general question of importance for political and social philosophy as well as for medical ethics and bioethics for both anthropological and sociological reasons: Only certain periods of life are marked by independence and more or less unrestricted autonomous agency. Other periods of life are characterized by dependency and vulnerability (Noddings 1986; Kittay 1999; MacIntyre 2001). Because of the neoliberal political reorganisation of institutions and social structures taking place in our time, the question of how we deal with vulnerable and dependent members of society becomes a rather crucial one accompanied by a particular need for normative orientation. Furthermore, the dividing line between disabled and non-disabled individuals seems to be fluid (Link 2004; Waldschmidt 2003). If we want to decide – for example, with the aim of regulating access to preimplantation and prenatal selection – in which cases we can speak of "real" disabilities which are worthy of being prevented and in which cases we can speak of variations within the normal range which should generally be tolerated, we are already faced with several problems (Lübbe 2003).

Secondly, it is not as easy to start an investigation into the perception of disability in ethics as to initiate debates on ethical issues in other, thematically related fields. In the latter case the ethicists involved refer to one or the other ethical school, pointing out its strong and weak points, exploring a line of argumentation which seems to be defensible for good reasons and then applying it to the concrete question at hand. The reason for this is that it is not quite clear whether ethics, as a theory-building enterprise, is "innocent" in this context. From the point of view of representatives of the disability movement and of scholars working in the area of disability studies, ethics as such is accused of being involved in disregarding and discriminating against disabled members of society. Authors who adopt the perspective of disabled people thus tend to discuss the relation between ethics and disability exclusively in terms of power relations, usually referring to Michel Foucault and Emmanual Lévinas (Rösner 2002). On the one hand, this approach provides a framework for analysing forms of injustice which have hitherto remained invisible. On the other hand, it is unsatisfactory because, in most cases, the author's own normative standpoint remains in the dark. It is scarcely possible, however, to determine whether claims asserted in societal and political conflicts are justified without referring to a normative standpoint.

Now, if it is true that many ethical theories and debates in applied ethics are remarkably biased and thus involved in stigmatising and discriminating against people with disabilities, I think, moral philosophy should reflect on the views of people with disabilities.

2 What Can Moral Philosophy Learn from the Disability Movement?

Representatives of the disability movement usually formulate their experiences in terms of oppression and devaluation ("Abwertung") and their demands in terms of empowerment and recognition ("Anerkennung") (Arnade 2003). The experience of oppression and devaluation includes, for example, paternalistic incarceration in sanatoriums, homes and other institutions especially created for disabled people such as schools and "protected" working places; unemployment and poverty or the risk of these because of disability, physical and other barriers which restrict mobility and thus social, political and cultural participation; humiliating treatment by medical and educational professionals who continually convey the message that the disabled individual is defective and/or malfunctioning; and the widespread use of language and of cultural symbols which discriminate against people with disabilities.

Demands for empowerment and recognition are usually connected to demands for autonomy, which are often expressed by the phrase "independent living". The demands voiced are for deinstitutionalisation, for full social, political and cultural participation, for the removal of barriers (e.g. language, buildings), for the provision of assistance in professional and private life, and for the guarantee for an income on a level appropriate for meeting the individual's specific needs; all of this, if necessary, with public assistance. Other demands for recognition are focused on the necessity to change values and norms regarding the social perception of disability.

Obviously, not all of these experiences of devaluation and not all of these demands for recognition can be easily formulated in terms of equal individual rights and duties. The focus of most contemporary moral theories, however, is restricted to individual interests, rights and duties. At the same time, the broader social and cultural context is tuned out (Sherwin 1992). Sociological analysis of social movements describe the phenomenon that subordinated groups usually require not only equal rights and social justice but also equal valuation as persons with specific attributes and particular needs. Critical analyses of this kind claim that current social movements pursue two different strategies which can be described as politics of equal rights on the one hand and politics of difference on the other (Rommelspacher 2002).

In the case of the disability movement, politics of equal rights demand full social, cultural and political participation for disabled people and respect for their autonomy. In concrete terms, this means liberation from the patronizing, marginalisation and oppression caused by the medicalization of disability and institutionalisation of disabled people on the one hand and the entitlement to social support which makes independent living and participation first possible on the other. Subsequently, politics of equal rights demand the recognition of negative rights as well as of positive rights (Held 1993: 207). This is of great importance in our context because many ethical theories clearly give priority to negative rights (Rawls 2003), going so far in some cases as to totally neglect positive rights (Stemmer 2000).

Instead, the politics of difference claim equal respect and value for individually different life plans and life situations. They reject the simplifying identification of disability with suffering, misfortune and severe burden. And they question the aesthetic ideals of beauty based on an uninjured body, the medical norms of functioning and perfectibility, and the cultural norms of capacity, autonomy and independence which are in contradiction to the life situations and life plans of many people with disabilities. This means that the politics of difference rejects not only disregard and psychological mistreatment of disabled people but also indifference to vulnerable persons and lack of concern about their specific needs. It follows that such politics of difference demand not only recognition of difference but also recognition of neediness. (Neediness in this context means the unavoidable dependence on others to satisfy basic needs.) Unfortunately, most normative ethical theories display little interest in the recognition of difference and the recognition of neediness (Benhabib 1995).

In contrast, the theories of recognition developed by Charles Taylor, Axel Honneth and Nancy Fraser include phenomena of oppression based on negative stereotypes linked to differences separating certain social groups from the majority and from the mainstream norms they represent. They understand themselves explicitly as critical theories of society which are able to expose not only obvious but also masked forms of exclusion and oppression. Owing to their self-understanding, these theories are normative in the sense that they have the potential to show how institutions and social structures should be changed to abolish exclusion and oppression. These theories have been discussed mainly in the context of multiculturalism and feminism. In the following I want to explore the question of whether they can also help us to more adequately understand the phenomena of deprivation and devaluation experienced by disabled people and the claims and demands related to this experience.

3 Towards a Theoretical Framework to Understand Deprivation and Devaluation Experienced by Disabled People

3.1 The Politics of Recognition (Charles Taylor)

The article "Multiculturalism and the Politics of Recognition" by Charles Taylor was the starting point for the political debate on the politics of multiculturalism which took place in many Western countries. In this article Taylor defends the demands for recognition of a cultural identity which require the assertion of specific rights by cultural minorities against a strict policy of equal treatment. Taking a Hegelian perspective, he refers to the relationship between recognition and personal identity. His central thesis is that mis- or non-recognition shows not only a lack of respect for a person or a group of people. In addition, a person or a group of people can suffer real damage to their identity because of misrecognition by others (Taylor 1992: 25).

Taylor argues that cultural minorities which are afflicted by mis- or non-recognition adopt a deprecatory view of themselves. As a result, he says, they are "incapable of taking advantage of new opportunities" and are "condemned to suffer the pain of low self-esteem" (ibid.: 26). With this thesis he pleads for a change in politics to make the survival of particular cultures in our multicultural Western societies possible; this new policy is necessary, he says, for members of racial, cultural and other minorities to survive without suffering damage to their identities. His ideal is a non-discriminating and non-homogenising society which tolerates cultural differences and is open to diverse perceptions of the common good.

Taylor's normative point of reference is a teleological concept of self-realisation for the individual person as well as for cultural groups. Taylor takes the Hegelian view that interaction with others constitutes the starting point for the genesis of the authentic self. Along with the fundamental dependence on social relationships goes the possibility of either success or failure in developing personal identity.

In modern times politics are dominated by the leading idea of equal dignity requiring universal and different blind principles. But these principles, as Taylor points out, are only apparently neutral because they are marked in reality by particularity due to the hegemonial culture, i.e. they are nothing else then "a particularism masquerading as the universal" (ibid.: 44). Taylor rejects such politics of equal dignity which insist on neutral, different blind principles and uniform application of rules and rights. Instead he refers to the necessity of caring about esteem, which he associates with the concept of honour based on strong common goals. According to Taylor, it is only possible to gain esteem by displaying an orientation towards collective goals. In a multicultural society, however, not all citizens share the same collective goals. This is, as Taylor points out, one of the main problems facing liberal societies. There is obviously a tension in the idea of equal dignity which requires neutrality with respect to common goods but at the same time confers esteem only on members of the hegemonial culture who share the covert common norms. Subsequently, Taylor pleads for a conscious adoption of collective goals.

For Taylor the demand for recognition of equal value formulated by cultural minorities is reasonable because it is the prerequisite for developing an intact personal identity. One could expect that Taylor would, on these grounds, conclude that there is a general duty to respect other cultures equally. With his normative orientation towards particular collective common goals, however, he has to totally reject the idea of universal reasoning of normative principles. Consequently, the demand for recognition of equal value can not be universally binding for him. What we do owe to any other culture, nevertheless, is the presumption that it has equal value as a "starting hypothesis with which we ought to approach the study of any other culture" (ibid.: 68). The peremptory demand for favourable judgments, however, could be in itself homogenising.

There can be no doubt that Taylor's great achievement with this work was to set off a debate about forms of injustice against cultural minorities and other subordinated groups since such injustice had previously been deeply etched into the norm systems of the hegemonial Western cultures (Habermas 1997: 150) and their practical philosophical theories. It is equally to Taylor's credit that he pointed out that the formal equal treatment of all citizens in liberal societies is not enough to exclude oppression and discrimination if at the same time ways of gaining social esteem are denied. It follows that, in certain cases, a policy of recognition of difference will require particular rights, e.g. to prevent or counteract disadvantages. This is, incidentally, of great importance for disability politics.

With his clear condemnation of mis- and non-recognition of cultural minorities, Taylor stakes out a rather strong normative position on the one hand while adopting a relativistic attitude with his teleological normative concept oriented towards a particular perception of common goods on the other. Consequently, this means that demands for recognition of equal value remain contingent. I don't think that this is satisfactory for the affected individuals. By describing the obstacles to developing an undamaged identity which have to be overcome by individuals subject to mis- and non-recognition, however, Taylor has virtually laid the groundwork for a stronger normative position regarding universal claims.

3.2 The Struggle for Recognition (Axel Honneth)

With his work emphasising "the struggle for recognition", Axel Honneth wants to present a normative theory of social relationships, as he explicitly states. However, it is not quite clear what exact purpose a "normative" theory of recognition should serve. It seems to me, however, that he wants to show in what respects phenomena of mis- and non-recognition within social relationships are wrong and subsequently how social institution and structures should be changed. With Taylor he shares the Hegelian concept that certain forms of recognition are necessary for the development of undamaged identities and thus constitute subjectivity. Contrary to Taylor, however, his central aim is to provide a normative foundation for a theory of recognition instead of Hegel's speculative reasoning. In his own words, his aim is to provide a naturalistic foundation referring to psychoanalytical knowledge which shows that the development of personal identity is fundamentally dependent on recognition within social relationships. Accordingly, he espouses a teleological concept of self-realisation similar to Taylor's but with a much stronger normative claim.

Honneth's starting point, again in contrast to Taylor, is the experience of mis- and non-recognition of every individual and not of particular social groups. According to Honneth, the experience of frustration of demands for recognition leads to confrontations between individuals by means of which they gain reciprocal recognition; this is well known as the struggle for recognition. In this way, the individuals constitute themselves as subjects on the one hand and trigger the progress of social morality on the other. Honneth's line of argumentation proceeds in three steps: Firstly, he reconstructs the early concept of recognition developed in Hegel's writings in Jena; secondly, he sets out to actualise the early Hegelian theory of recognition by referring to the social psychology of George Herbert Mead. Thirdly, he proposes a "naturalistic" foundation for this theory of recognition.

Along with Hegel and Mead, he distinguishes three forms of recognition: love, equal respect and esteem. He understands the development of personal identity as an unfolding of recognition in these three steps. Firstly, the recognition of neediness, by the mother, in the loving mother-child relationship enables the child to gain *self-confidence*. Secondly, through the recognition of equal rights in his or her social life, the individual obtains *self-respect*. Thirdly, with the recognition of individual achievements in the cultural realm, the individual develops *self-esteem*. All of these three forms of self-relation are, according to Honneth, constitutive aspects of personal identity. The integrity of an individual's psyche can be endangered by failures at any step in the development of personal identity. Honneth intends to prove this Hegelian thesis, which has been rather speculative up to this point, by referring to the psychoanalysts Donald Winnicott and Jessica Benjamin. Their works can be interpreted to show that the comprehensive recognition of the child's neediness from birth on by his or her mother is absolutely essential for the development of self-confidence. Any disturbance in this form of recognition may lead to severe emotional disorders.

Even after the individual has successfully developed subjectivity, his or her personal identity continues to be precarious. Personal identity is always endangered by the experience of mis- and non-recognition. Self-confidence can be damaged by violation and cruelty, self-respect can be disturbed by deprivation, and self-esteem by humiliation and psychological mistreatment. All of these experiences of mis- and non-recognition are associated with strong feelings of shame and rage. If such experiences occur systematically and if a vehicle for articulating them is available in the form of a social movement, the motivation for political resistance can emerge. According to Honneth, this is the source of social progress.

Honneth provides us with an understanding of forms of recognition in social relationships which is analytically rather sharp and complex. His model of the three different types of recognition in diverse forms of social relationships paves the way for a more complex and differentiated understanding of the claims asserted by subordinated groups. There are two reasons, however, why he fails to reach his own goal, which ultimately seems to be, to show that the demands for recognition in social conflicts are justified and thus trigger moral progress.

Firstly, of course, he can perhaps show that certain forms of recognition are necessary for an undamaged identity; however, he can not distinguish between right and wrong demands for recognition. As Fraser points out in a rejoinder to Honneth, according to this concept, claims for recognition formulated in a chauvinist, racist or sexist way would also have to be tolerated. Obviously, this can not be accepted (Fraser 2003b: 56–61).

Secondly, his normative foundation is in itself not convincing. It is based on the recognition of neediness in the mother-child relationship and thus on a relationship which is inevitably characterized by emotionality and asymmetry. According to Honneth, claims which are based on emotionality and asymmetry in social relationships can not be asserted in a reciprocal manner. Thus, claims for recognition of neediness in asymmetric relationships are by definition in contradiction to the reasoning of common duties and therefore can not be universally binding (Honneth 1998: 174). Consequently, there is no justification for concluding on these grounds that claims for recognition asserted in symmetric relationships should be binding.

Nevertheless, I think that social relationships have a normative character in the understanding of Taylor that certain kinds of mis- or non-recognition are experienced as violation by the affected individuals. This is quite evident, even without a psychoanalytical foundation. Furthermore, the experience of damage due to mis- or non-recognition can count as a strong indication that at least some claims for recognition could be justified. A normative foundation of particular demands for recognition, however, requires more.

Generally, Honneth tends to reduce the legitimacy of demands for recognition to those of equal rights. Recognition of neediness is for him only the prerequisite for the recognition of rights. Recognition of difference refers, as in Taylor's concept, primarily to solidarity within one's own group and not to society as such. Consequently, recognition of difference has for Honneth only a mediating function for moral progress regarding a better recognition of rights in the general public realm. I think this can not be satisfying for people with disabilities and members of other subordinated groups. As I've already pointed out, members of marginalised groups are demanding equal valuation not only within their groups but within society as a whole. Furthermore, people who are dependent on outside assistance to meet their basic needs, as is the case for many people with severe disabilities, would not readily agree with the opinion that demands for recognition of neediness in asymmetric relationships are not binding at all.

3.3 Redistribution and Recognition (Nancy Fraser)

In contrast to Taylor and Honneth, Nancy Fraser does not understand recognition as a general philosophical category in her writings on "redistribution and justice". According to Fraser, demands for redistribution and demands for recognition are two aspects of justice within a critical theory of society.[1]

She starts with the observation that social movements usually demand redistribution and esteem independently of each other (Fraser 2003b: 228). Related to these two types of demands are two forms of experience of injustice. This is, firstly, the experience of socio-economic injustice due to an unjust distribution of resources, and secondly, the experience of devaluation and disregard because of traits or attributes marking certain status groups. Status groups are, for example, women, lesbians and gays, ethnic and national minorities, and disabled people. The example typically given for socio-economic injustice is the exploitation of the working *class*. The adequate answer to exploitation is a politics of redistribution. The typical example of injustice attributable to membership in a certain *status group* is the social devaluation and disregard of lesbians and gays. The adequate answer to devaluation and disregard is a politics of recognition (of difference). According to Fraser, the fight for recognition of status groups has proved to be the paradigm for social movement in our times.

Fraser explicitly rejects the Hegelian perception of recognition as an ideal reciprocal relationship by means of which two individuals constitute themselves as subjects. This notion of recognition refers to the Hegelian "Sittlichkeit" and thus to self-realisation, which makes it incompatible with questions of justice to be discussed in the Kantian concept of "Moral" (Fraser 2003a: 19, 43–44). Fraser avoids the problematic immanent reasoning of claims for recognition and refers instead to an external deontological principle which she names "parity of participation". Parity of participation means that all adults, at least, have a universal right to equal social, political and cultural participation. According to Fraser, both demands, for redistribution as well as for esteem, are justified if they meet the principle of "parity of participation". They are not justified, however, if they contradict the parity of participation of others, as is the case, for example, with groups asserting chauvinist, sexist or racist claims for recognition. Generally speaking the principle of "parity of participation" requires two preconditions: an "objective" and an "inter-subjective" one. The objective precondition is that "forms and levels of economic dependency and inequality which render the parity of participation difficult are excluded". The inter-subjective precondition is that "all institutionalised patterns of estimation are excluded which withhold the status of fully entitled partners in social interaction for some people" (Fraser 2003a: 55).

However, for the purpose of an adequate analysis of claims and demands for recognition in different social relationships, Fraser's concept is rather thin compared with Honneth's distinction of three forms of recognition, i.e. of neediness, of rights and of difference. Instead Fraser distinguishes only between demands for redistribution and demands for recognition (in the sense of requiring esteem). She does not discuss violations of fundamental rights apart from questions related to the distribution of resources, and totally tunes out demands for recognition of neediness.

The advantage of Fraser's concept is situated, however, on the normative level. The reference to a deontological normative principle instead of a theory-immanent teleological concept of self-realisation is more convincing. The principle of parity of participation corresponds strikingly to the political demands of the disability movement. I have doubts, however, whether the principle of parity of participation is comprehensive enough to cover all questions of justice which are important in this context, in particular if we think about indifference to existential needs of dependent and vulnerable individuals or about the withholding of rights apart from the distribution of resources. However, what we can learn from Fraser is that the questions which claims for recognition are justified has to be answered within the framework of deontological moral theory.

4 Conclusion

With the distinction between recognition of neediness, recognition of rights and recognition of difference in social relationships we have an analytical framework which seems to be adequate for evaluating the political claims and demands of people with disabilities. The evidence for the assumption that mis- or non-recognition might cause damage to the personal integrity of afflicted individual on all three levels of recognition is a rather strong indication that certain related claims for recognition are justified. According to Fraser, certain of these different claims can be justified within a broader concept of justice. The question of which normative principle may be useful for this purpose, however, has to be discussed further in the light of normative moral theory.

Claims for *recognition of neediness* refer to relationships between "care-givers" and "cared-fors", which are (usually) marked by emotionality and by an asymmetry between taking and giving (Tronto 1993). Nevertheless, Honneth's opinion that universally binding duties in asymmetric relationships are not reasonable at all is not convincing. Obviously, Honneth assumes that emotionally asymmetric personal relationships, particularly the mother-child relationship, are predetermined by nature. Consequently, he seems to take the satisfaction of the basic needs of the cared-for child for granted. Such an essential understanding of the mother's role, however, has been deeply questioned by feminist theorists (Noddings 1986: 79). Moreover, if motherhood is more adequately understood as socially constructed and historically contingent, and if the division of care-giving labour between women and men is being critically scrutinized (Noddings 1986: 97), Honneth's position is not easily defensible. Furthermore, motherhood is, of course, a paradigmatic asymmetric relationship but nevertheless one with a unique character. There are other asymmetric relationships in which recognition of neediness can not be taken for granted at all, like other family relationships involving taking care of elderly people as well as professional care-giving relationships. Consequently, claims for recognition of neediness, like other claims for recognition, should be subject to critical reflection. The satisfaction of the basic needs of dependent individuals by caring for them is obviously necessary for the life, as well as the physical and emotional integrity, of the cared-for. I think this fact is a rather strong indication that certain universal claims and duties within asymmetric relationships can be defended with good reasons in the light of normative moral theory. The adequate way to do this might be to refer to positive rights and to the reciprocal interdependence of caring and being cared for in the course of a lifetime.

The question of whether claims for *recognition of equal rights* in the public sphere are justified is much less problematic since the reasoning of rights and duties is genuinely a subject of normative moral theory. From the perspective of disabled people, the perception of equality seems to be particular important in this context. For example, a concept of rights and duties which requires strict equal treatment referring to difference-blind principles obviously might not be adequate. The same might be true for concepts which strongly favour negative rights in contrast to positive rights.

To examine the question of whether claims for *recognition of difference* are justified, finally, we have to deal with a lot of potential misunderstandings since related forms of mis- and non-recognition are deeply inscribed in the hegemonial norm systems. Obvious forms of mis- and non-recognition, such as arbitrary insults and various kinds of psychological mistreatment, might be generally understood as illegitimate. Instead, more subtle and covert forms of systematic mis- and non-recognition of members of certain status groups are often tuned out. All three theorists of recognition – Taylor, Honneth and Fraser – provide us with tools for understanding such forms of injustice due to difference which have hitherto remained invisible. Nevertheless, Honneth's and Taylor's notion of recognition of difference, which tends to be restricted to solidarity within status groups, is certainly not sufficient since claims for recognition of social movements are usually directed against society as a whole. Fraser instead refers, more appropriately, to institutional patterns which are required for the status of fully entitled partners in social interaction and thus for social inclusion. As follows, such institutional patterns should be evaluated in the light of normative moral theories as well.

Note

[1] She uses the term recognition exclusively in the sense of requiring or obtaining *esteem* and not in the context of rights which might be a source of misunderstanding.

References

Arnade, Sigrid. "Zwischen Anerkennung und Abwertung: Behinderte Frauen und Männer im biopolitischen Zeitalter." In *Politik und Zeitgeschichte*, B 08, 2003. http://www.bpb.de/publikationen/72MPN5,0,0,Zwischen_Anerkennung_und_Abwertung.html (Last visit 20 August 2004).

Benhabib, Seyla. *Selbst im Kontext*. Frankfurt a.M.: Suhrkamp, 1995.

Birnbacher, Dieter. *Selektion am Lebensbeginn – ethische Aspekte*. Vortrag auf dem Kongress für Philosophie. Konstanz, 1999.

Dederich, Marcus (ed.) *Bioethik und Behinderung*. Klinkhardt: Bad Heilbrunn, 2003.

Fraser, Nancy. "Soziale Gerechtigkeit im Zeitalter der Identitätspolitik. Umverteilung, Anerkennung und Beteiligung." In Fraser Nancy and Honneth Axel, 2003a, 13–128. (Fraser, Nancy. "Social Justice in the Age of Identity Politics." In Fraser Nancy and Honneth Axel 2003.)

Fraser, Nancy. "Anerkennung bis zur Unkenntlichkeit verzerrt: Eine Erwiderung auf Honneth." In Fraser Nancy and Honneth Axel, 2003b, 225–270. (Fraser, Nancy. "Distorted Beyond all Recognition: A Rejoinder to Axel Honneth." In Fraser Nancy and Honneth Axel 2003.)

Habermas, Jürgen. "Anerkennungskämpfe im demokratischen Rechtsstaat." In Charles Taylor (ed.) *Multikulturalismus und die Politik der Anerkennung*. Frankfurt a.M.: Fischer, 147–196, 1997.

Held, Virginia. *Feminist Morality: Transforming Culture, Society and Politics*. Chicago, London: University of Chicago Press, 1993.

Hirschberg, Marianne. *Die Klassifikation von Behinderung der WHO*. IMEW expertise 1, Berlin, 2004.

Hoerster, Norbert. "Lebenswert, Behinderung und das Recht auf Leben." In Riccardo Bonfranchi (ed.) *Zwischen allen Stühlen: Die Kontroverse zu Ethik und Behinderung*. Erlangen: Harald Fischer Verlag, 43–56, 1997.

Honneth, Axel. *Kampf um Anerkennung: Zur moralischen Grammatik sozialer Konflikte*. Frankfurt a.M.: Suhrkamp, 1998. (Original work published 1992). ("The Struggle for Recognition: On the Moral Grammer of Social Conflicts." Cambridge: Polity Press, 1995).

Honneth, Axel. "Die Tugenden der menschlichen Kreatur. Eine philosophische Würdigung der Fürsorge: Über Alasdair MacIntyres "Die Anerkennung der Abhängigkeit." *Die Zeit* 30/2001. http://zeus.zeit.de/text/archiv/2001/30/200130_st-macintyre.xml (Last visit 6 August 2004).

Kittay, Eva Feder. *Love's Labor: Essays on Women, Equality, and Dependency*. New York: Routledge, 1999.

Link, Jürgen. "'Irgendwo stößt die flexibelste Integration schließlich an eine Grenze' – Behinderung zwischen Normativität und Normalität." In Sigrid Graumann, Katrin Grüber, Jeanne Nicklas-Faust, Susanna Schmidt and Michael Wagner-Kern (eds.) *Ethik und Behinderung: Ein Perspektivenwechsel*. Frankfurt a.M.: Campus, 130–139, 2004.

Lübbe, Weyma. "Das Problem der Behindertenselektion bei der pränatalen Diagnostik und der Präimplantationsdiagnostik." *Ethik in der Medizin* 15, 3, 203–220, 2003.

MacIntyre, Alasdair. *Die Anerkennung der Abhängigkeit: Über menschliche Tugenden*. Hamburg: Europäische Verlagsanstalt/Rotbuch Verlag, 2001.

Noddings, Nel. *Caring: A Feminine Approach to Ethics & Moral Education*. Berkeley; LosAngeles: University of California Press, 1986.

Nussbaum, Martha C. "Beyond the Social Contract: Toward Global Justice." (Lecture 1: Capabilites and Disabilities: Justice for Mentally Disabled Citizens.) *Tanner Lectures in Human Values*. Canberra: Australian National University, 12–14 November 2002.

Rawls, John. *Politischer Liberalismus*. Frankfurt a.M.: Suhrkamp, 2003.

Rommelspacher, Birgit. *Anerkennung und Ausgrenzung: Deutschland als multikulturelle Gesellschaft*. Frankfurt a.M.: Campus, 2002.

Rösner, Hans-Uwe. *Jenseits normalisierender Anerkennung: Reflexionen zum Verhältnis von Macht und Behinderung*. Frankfurt a.M.: Campus, 2002.

Sherwin, Susan. *No Longer Patient: Feminist Ethics & Health Care*. Philadelphia: Temple University Press, 1992.

Stemmer, Peter. *Handeln zugunsten anderer: eine moralphilosophische Untersuchung*. Berlin: De Gruyter, 2000.

Taylor, Charles. "Die Politik der Anerkennung." In Charles Taylor and Amy Gutmann (eds.) *Multikulturalismus und die Politik der Anerkennung*. Frankfurt a.M.: Fischer, 13–78, 1997. (Taylor, Charles. "The Politics of Recognition". In *Multiculturalism and 'The Politics of Recognition'*. Princeton: Princeton University Press, 25–74, 1992.)

Tronto, Joan C. *Moral Boundaries: A Political Argument for an Ethics of Care*. New York/London: Routledge, 1993.

Waldschmidt, Anne. "Die Flexibilisierung der 'Behinderung'. Anmerkungen aus normalismustheoretischer Sicht, unter besonderer Berücksichtigung der 'International Classification of Functioning, Disability and Health'." *Ethik in der Medizin* 15, 3, 191–202, 2003.

Wendell, Susan. *The Rejected Body: Feminist Philosophical Reflections on Disability*. New York: Routledge, 1996.

Wolbring, Gregor. *Folgen der Anwendung genetischer Diagnostik für behinderte Menschen*. Gutachten im Auftrag der Enquete-Kommission Recht und Ethik der modernen Medizin des Deutschen Bundestags, 2001.

Neuro-Prosthetics, the Extended Mind, and Respect for Persons with Disability

Joel Anderson

1 Introduction

In discussions of performance enhancement, as in applied ethics generally, it is tempting to think that we can answer the hard ethical questions by discovering boundary lines that lie in the subject matter itself. The hope, here, is that we could settle an array of thorny issues if only we could identify the fault-lines between, for example, therapy and enhancement, or between pharmacological identity-management and the restoration of the authentic self. And within the wider public debate, there are loud voices declaring that the boundary-lines are actually so obvious that only self-interested lawyers and out-of-touch intellectuals could miss them. Consider, for example, the following quote:

> Some people want to engineer their babies. They hope to buy them an edge in the lottery of life, to fix them up with special genes for extra intelligence or height or skin color or sexual orientation. How is this different from giving your kid piano lessons and extra tutoring, some of them ask? It's very much like the difference between cutting out junk food and injecting yourself with steroids. There is a line – and the vast majority of us can see it (Shanks 2005).

Especially in the face of such pronouncements, one of the most important contributions that philosophers can make is in guarding against precisely this tendency to think that the answers lie in *discovering* these boundary-lines in the phenomena themselves. My aim here is to make such a contribution to the debate over neuro-enhancement. My goal, then, is to question the pat answers.

In particular, I will be arguing that, in discussions about the ethical status of neuro-prostheses, there are deep conceptual and ethical problems with the implicit reliance on what I call the "Invasiveness Criterion", that is, the idea that what makes some neuro-prosthetic technologies problematic is that they violate a boundary between what is inside the person and what is outside. This underwrites, I believe, a widespread and influential intuition that we ought to be vigilant with regard to enhancement technologies that penetrate the body. Indeed, I believe that this focus on the inner-outer boundary helps explain much of the deep-seated suspicion regarding

J. Anderson
e-mail: joel.anderson@phil.uu.nl

M. Düwell et al. (eds.), *The Contingent Nature of Life*,

enhancement technology that involves genes and drugs, a suspicion that is reflected in the typically negative talk of "dope" and "pill-popping" or of "genetic *manipulation*". After all, since pharmaceuticals and gene therapies work only by being taken up within our bodily organs, they always involve a breach of the boundary of the body. There is, I think, a link in the minds of many between suspect forms of enhancement and the idea that the person is being invaded by something alien. But it becomes unclear to what extent the problem lies with the invasiveness or something else.

This is part of why I find it so fruitful to look at the case of neuro-prosthetic devices, which include now-familiar technology (such as cochlear implants) as well as more experimental or even still-hypothetical technology (such as brain-machine interfaces). For they can be integrated into the body to varying degrees, which makes it easier to analyze the role played by metaphysical assumptions, especially the idea that crossing the boundary of the skin represents something that is intrinsically problematic. Indeed, the prosthetic devices that seem to generate the least amount of moral unease seem to be precisely those – like artificial limbs, wheelchairs, and hearing aids – that seem to work more from the outside, as detachable supports or as tools. Perhaps this is because employing devices to do what we couldn't otherwise do is an essential part of the human condition. We are, after all, the "tool-using animals." But here again, the distinction between a "tool" and a prosthesis that is implanted into the body is more apparent than real. Or so I shall be arguing.

I begin by describing two largely hypothetical cases of neuro-prostheses used for non-therapeutic enhancement, one involving enhanced control of the hearing experience and the other, improved face recognition. These are, I believe, typical of the sort of cases – brain-machine interfaces – that tend to raise serious ethical concerns. My aim is to analyze the assumptions underlying this reaction. After setting aside important pragmatic concerns (about health risks, lack of informed consent, or social inequality of access and benefits) and then rejecting criticisms based on the "unnaturalness" of neuro-prostheses, I focus on two criteria that might be used to criticize neuro-prosthetic implants: the "Invasiveness Criterion" already mentioned, and the "Combined Criterion", which adds the idea that it is *non-therapeutic* invasions that are intrinsically problematic. Using variations on the two core cases, I argue that we ought to reject both criteria, on conceptual grounds (Sections 5 and 8) as well as ethical grounds (Sections 6 and 9). This suggests that we should look elsewhere for grounds on which to object to these neuro-enhancements, and in Section 10 I discuss several principles that may be informing objections to them, several of which are worth taking seriously. However, I refrain from either taking a position on which neuro-enhancements are acceptable or on which principles ought to guide us in making such assessments. My focus is rather on criticizing the widespread though tacit assumption that the problematic character of some cases can be read off the nature of prosthetic devices themselves, and particularly on whether they are implanted in the body.

2 Two Cases of Neuro-prostheses

In this Section I describe two cases of neuro-enhancement by means of a device. I will refer to such devices as "neuro-prostheses" to distinguish them from improvements in performance that are achieved genetically, pharmacologically, or developmentally (e.g., through training). But I shall be using the term "neuro-prosthesis" in a rather broad manner. I will not be restricting it to medical or other therapeutic purposes, nor will I be restricting it to cases in which a device replaces a specific body part that has been lost or damaged. This last point is particularly important. I am assuming that neuro-prostheses are intended to enable persons with regard to certain capabilities or forms of human functioning, and that these functions can be realized in different ways. In this sense, a blind person's cane is a neuro-prosthesis: it contributes to restoring a lost capacity for spatial orientation. Neuro-prostheses thus come in a wide variety of forms.

The case of Anna. Consider first the case of Anna, a lover of classical music and professional recording engineer. She has devoted her life to ensuring that music recordings capture the balanced sound of symphony orchestras in an optimal fashion. Doing this requires discerning an extremely complex acoustical array: she must determine how the sound is to be mixed, so as to capture the optimal balance of frequencies. Anna has extensive training in this, along with a certain degree of natural talent. But she is only human, and there are subtleties in the orchestra's sound that can only be picked up by a computer-assisted array of microphones. This suggests a possible (but fictitious) adaptation of existing hearing aid technology that would involve replacing her cochlea with much more sensitive artificial follicles and then hardwiring that to her ventral cochlear nucleus. In the scenario envisioned in this thought experiment, the practical upshot is that after several years of training, she is able to make much finer discriminations of pitch, and this enables her to produce the higher-quality recordings that she has always been striving for but could not produce.[1]

The case of Peter. In a related case, imagine someone who is tired of bumping into people he's met before and not being able to remember their names, or even be sure whether he's met them before. He has a special face-recognition system installed that works as follows.[2] A tiny camera in his eyeglasses sends images wirelessly to a small computer in his pocket, which uses fast, flexible software to recognize faces out of a database of acquaintances and generate an audio signal of the name that is then channeled directly to the brain via an interface with the auditory system, such that after several months of training up the system, he no longer even notices the difference between cases in which he is prompted for the name of someone he meets and cases in which he recalls the name without assistance. The vocalized names merge into the constant flow of perceptual input and cognitive processing that comprises human subjective experience. He can also effortlessly indicate to the computer that a face not in the system needs to be recorded for later updating. He now feels very much at ease at class reunions and conferences, situations that used to frustrate him enormously.

3 Objections Based on Pragmatic Concerns or Unnaturalness

What ought we to think of these neuro-enhancing brain-machine interfaces? The face recognition technology in particular might actually be so appealing that many would not seriously object, but that would be disingenuous for many, since it is in fact precisely the sort of brain-machine interface that otherwise generates almost visceral opposition. We are talking, after all, about holes being drilled into people's skulls to feed wires in. My target, in any case, is the view that neuro-prostheses of this sort are at least *troubling*.

In what follows, I shall be setting aside some of the potentially very significant pragmatic grounds for objecting to such devices. Clearly, for example, neuro-prosthetic implants raise concerns about health risks, increased vulnerability to abuse, or unfair advantage in competitive contexts. In addition, as in the case of cosmetic surgery, people may often not really be clear on the benefits and risks. These are important and relevant concerns. But my focus in the present context is with the question of whether there is a distinct line of reasoning that can be traced to the nature of the technology and its impact on the human, especially in terms of the violation of the boundary of the skin. All I assume is that many would still be opposed to this sort of neuro-enhancement even without these pragmatic worries.

Before turning to my main diagnosis, I need to comment briefly on one approach that I shall be setting aside, namely the approach based on the "violation of the natural", which is often linked to feelings of revulsion and abhorrence. On this view, which has been gaining quite a bit of attention from ethicists recently,[3] our feelings of revulsion serve as a last-ditch indication that something gone wrong, a violation of basic categories. As Leon Kass puts this, for example, in "The Wisdom of Repugnance", "In this age in which everything is held to be permissible so long as it is freely done, repugnance may be the only voice left that speaks up to defend the central core of our humanity" (Kass 1997: 20). Along these lines, one might object to the cases of Anna and Peter as repugnant. There may well be cases (necrophilia, perhaps) for which the only compelling grounds for rejecting it are that it generates a widely shared and deeply felt revulsion. But there are also plenty of reasons to be concerned about this line of argument, the most obvious being that people's "yuck"-responses over the centuries have all too often turned out to be morally problematic prejudices, such as reactions to interracial marriage.

For my purposes, however, there is a deeper problem with the whole idea that neuro-prostheses might be criticized as violations of nature. This is because the very idea that prostheses are unnatural relies some problematic assumptions about how the human body is implicated in the actions of agents. This is not simply the first-year philosophy student's objection that nothing is either normal or natural. The point is that we need to be cautious about making normative claims on the basis of the purported naturalness of how particular aspects of human agency can be embodied or instantiated. For, in cases where we are talking about some essential feature of human agency that we want to protect or encourage, what we are really talking about is not how many fingers one has or the precise mechanism by which one's serotonin levels are maintained but rather that one is able to shape the environment or that one's sensitivity to changes in circumstances is reflected in changes in emotion or mood. One reason for taking this broadly functionalist approach stems from a commitment to a conception of the human that is not prejudiced against persons with disabilities – those with fewer fingers, for example. The burden of proof ought to lie with those who would argue that what ought to matter, for a normative conception of human agency, is not functions and capabilities but rather their instantiation. But the other reason is that most features of the mind show evidence of "multiple realizability", such that there are different mechanisms by which they can be realized.[4] Thus, for example, we might see the capacity for shared attention as a key feature of the broader intersubjectivity distinctive of persons, but that does not mean that the mechanisms that instantiate that capacity have to be of a certain type.[5] Given this multiple realizability, it cannot be assumed that the ontology of the nervous system will draw the lines between *the* "natural" way of instantiating a capability and abnormal ways of doing so. In any such case, we face a choice between doing the normative line-drawing in terms of capability or physical realization, and given the risk of exclusionary bias associated with the latter, it seems good policy – both as a matter of metaphysics and ethics – to favor a characterization in terms of capability rather than in terms of a particular instantiation of that capability.

4 Boundary Violations

In the next five sections, I take up the idea that the wrongness of cases such as that of Anna and Peter lies, at least to a significant extent, in the fact that the physical integrity of the person is being violated. In addition to the remarks I've made so far, in specifying the scope of my argument, I should emphasize that the issue I'm focusing on is a matter of distortion that is traceable to the device as it is implanted within the body of the person. Thus, it is crucial that we distinguish criticisms of the enhancement of the capacity from the way in which it is achieved. As I said at the outset, this is one of the advantages of focusing on neuro-prostheses: because some neuro-prosthetic technology can function from outside the body, we can separate the permissibility of its enhancing effects from the permissibility of putting it inside the body. Put this way, the cases I'm interested in are those in which the neuro-prostheses are generally considered to be unproblematic in a handheld form, but not once implanted into the body. And the position I wish to challenge is the common view that, whereas implants raise serious concerns, one is at liberty to employ all the latest wireless technologies and other gadgets to improve my performance (as long as the technology is not dangerous to me or others, and we are not talking about a context of circumscribed competition where it would be against the spirit of the game to gain this advantage).

I believe that to the extent to which the remaining line of critique is actually distinctively about the invasiveness of the brain-machine interface, it is indefensible. And that it only appears compelling on the basis of other issues that are not necessarily linked to the fact that the device penetrates the barrier of the skin and skull. There are two main objections to make to the supposition underlying this critique, that is, the supposition that the skin-and-skull barrier is a relevant ethical watershed: it involves bad metaphysics, and it has unacceptable ethical implications, particularly with regard to the standing of persons with disabilities.

5 Conceptual Difficulties with the "Invasiveness Criterion": Bad Metaphysics

The difficulties with the Invasiveness Criterion start with the very notion that we can make a sharp distinction between what is "inside" or "outside" the person. Ordinarily, of course, we think of the skin as the relevant boundary between where I end and the rest of the world begins. But it turns out that there are good reasons to question the idea that my *person*, as something not to be transgressed or violated, coincides with my *body*. Some parts of my body, for example, may be of no concern to me. The tips of my fingernails that I am about to clip off, for example, are not essential to my person. Conversely – and most important for my purpose – there are aspects of who I am and what I can do that extend beyond the boundary of my skin.

This last point has been developed in a particularly compelling manner – and with special relevance for *neuro*-prostheses – in so-called "extended mind" approaches within the philosophy of mind.[6] On this view, the limits of what counts as an agent's "own cognitive system" can plausibly be thought of as a fluid and thus as determined not by any ontology of the human organism but rather by the social practices in which the agent participates. In light of this, it is hard to see why we should think that the boundaries of the human agent coincide with the physical boundary of an organism's skin and skull, especially given what I said earlier about multiple realizability, namely, that it cannot be assumed that the ontology of the nervous system will draw the lines between *the* "natural" way of instantiating a capability and abnormal ways of doing so.

To see the force of the "extended mind" perspective, consider a few examples that are more mundane than the cases of Peter and Anna. Take, first, the paradigm case of the extension of agency, that of a blind person's cane. What the reports from expert cane-users indicate is that the cane is not simply a tool employed by the hand, but it comes to be experienced as itself part of one integrated sensory organ. They don't feel the vibration of the cane and infer that there is an object in the way; they feel the object through the cane as much as they feel it through their arm (Merleau-Ponty 1962: 143).

Now consider a case of cognition, that of long division. If someone asks me what 2384 divided by 127 is, I realize immediately that I can't do it "in my head". I either reach for a calculator, or I grab pencil and paper. But as I work the arithmetic out on paper, what are we to say about *where* the cognitive system is located – within my brain, or also partly in what I am doing with the paper and pencil? Since I cannot do the arithmetic without the paper and pencil (or calculator), it seems clear that they are essential parts of whatever system it is that is doing the long division.

This point can be extended further, as David Chalmers and Andy Clark have done with their fictional example of "Otto":

> Otto suffers from Alzheimer's disease, and like many Alzheimer's patients, he relies on information in the environment to help structure his life. In particular, Otto carries a notebook around with him everywhere he goes. When he learns new information, he writes it down in his notebook. When he needs some old information, he looks it up. For Otto, his notebook plays the role usually played by a biological memory
>
> (Clark and Chalmers 1998).

For Otto, in other words, his notebook is as much a part of his person as the deteriorated neurons inside his skull. Indeed, tearing sheets out of his notebook may be a greater assault on his cognitive system than removing some of his brain tissue. Of course, as Clark and Chalmers are aware, there are significant differences between how one "consults one's memory", depending on whether one looks a notebook or *just remembers* – differences in portability, speed, and so on. But these differences do not map neatly onto the differences between "extended" cognitive systems and those inside the boundary of skin and skull. Sometimes it's quicker or more reliable to glance at my notebook than to rely on what's in my skull. What we need to realize is that it is these features of the cognitive system and not the boundary of the skin and skull per se that should be seen as the basis for the distinctions we might want to make. I'm not denying that we may want to make distinctions here, on the basis of our social practices and the value-commitments embodied in them. I am only denying that the question of what distinctions we ought to make is answered by given features of the skin-skull boundary.

Consider now three variations on the case of Anna: BMI-Anna (as described previously, with the brain-machine interface); PDA-Anna (using a handheld personal digital assistant); and DNA-Anna (whose performance stems from genetically rooted talent, not assistive devices). Consistent with the methodology I have adopted, what we need to ask is, What ethical differences between the three remain once one takes away differences in how they function. Assume that they perform indistinguishably well across the entire spectrum of relevant tasks, that they all pass a sort of "Turing test" relative to each other for this capacity:[7] if observers were to judge only on the basis of performance, not knowing which performance was caused by which version of Anna, there would be no way to tell the difference between them. Of course, if it is impossible for BMI-Anna to ever be as good as DNA-Anna, or if it is impossible for PDA-Anna ever to perform as well as BMI-Anna, then the embodiment will have ethically relevant implications. But then we don't have a case where the embodiment *itself* is what makes the difference, and that is what we are concerned with.

One might object that performance is not all that matters; that a sense of ownership is crucial, or a sense that the relevant actions flow from me as a person. But here the parallel point holds. In the absence of any principled grounds for thinking it impossible for it to seem as natural to do something by means of a tool than not, we can contemplate the ethical status of neuro-prostheses in which Anna's tool or implant is experienced transparently, in the same way that for some blind persons, it is the tip of their cane and not the tip of their fingers where their organ for touch ends.

The point for our purposes is that once we have the cases in which there are no differences in performance or experiential transparency between BMI-Anna, PDA-Anna, and DNA-Anna, then it is unclear what non-question-begging argument there could be for saying how BMI-Anna's implant is "inside" her person in a way that PDA-Anna's handheld device is not. But then, once it becomes clear that the distinction between inside and outside the body does not coincide with the distinction between inside and outside the person, it becomes unclear why the boundary of the skin should be seen as so ethically significant. And it becomes evident that the Invasiveness Criterion trades on some conceptual confusions. For it's not clear how the inside–outside distinction can even be formulated in a way that matters.

6 An Ethical Difficulty with the "Invasiveness Criterion"

In addition to these conceptual and metaphysical difficulties with the Invasiveness Criterion, there is an ethical concern worth mentioning. There are several ways in which to think about the idea that, as a matter of principle, we ought to view with suspicion any assistive devices that penetrate the skin or that involve an interface with the nervous system. One might base these reservations on the idea that, empirically, there are greater risks associated with such devices or the procedures associated with them. Thus, as in cases of genetically modified foods, one might think that a "precautionary principle" is appropriate, such that we should be cautious about introducing technology whose impact is not yet understood. The increased difficulty of reversing the implantation of devices – especially those, like cochlear implants, require a great deal of training – might be thought to strengthen these concerns.

But there is a concern here about the way in which these arguments work. To begin with, they appear to depend on the idea that we need non-rational taboos to protect us from ourselves. Otherwise it is hard to see why the invasiveness itself ought to be the relevant principle rather than various well-founded principles regarding the risks, irreversibility, etc. that may or may not be empirically linked with invasive devices. This implicit suggestion that we need to short-circuit people's decision-making in this way is problematically paternalistic. Moreover, the difficulty with taboos is the way in which they take on a justificatory life of their own. People start worrying about whether or not a device is invasive rather than whether it is effective, safe, and so on. And this can lead to a situation in which people are encouraged to focus on aspects that may not be important. Just as there are serious risks to mental and physical health resulting from widespread tendencies to demonize "popping" pills or visiting a "shrink", there is a risk that subtle (and not-so-subtle) taboos about neuro-prostheses may cloud people's thinking about what is really good for them. As with considerations regarding cosmetic surgery, we would be well advised to avoid ethical decision-making that trades on the same emotional reactions as freak shows. Enlightenment is called for.

7 Medical Therapy vs Non-Medical Enhancement: The "Combined Criterion"

In arguing against the idea that invasiveness alone should serve as a moral criterion, I have not mentioned one quite natural form that this objection can take, namely, that such invasive neuro-prostheses certainly ought to be permitted in the case of disease and disability, but not otherwise. Think of a case in which Anna is deaf and elects to get a cochlear implant. Or imagine that Peter's face-recognition prosthesis serves to relieve his prosopagnosia, a rare but severely debilitating condition in which persons can perceive and recognize details of their environment but cannot recognize the faces of even their closest friends and family.[8] If Peter's brain-machine interface is very effective at providing him with a subjectively transparent and objectively accurate means of recognizing faces via the audio clues channeled into his brain in association with computer analyzed video images, then it's hard to see why the penetration of the skill ought to matter.

This line of thinking suggests that what may have been missing in the discussion thus far is the distinction between forms of invasive neuro-prostheses that serve a medical purpose and those that are merely for cosmetic or non-therapeutic purposes. Perhaps, in other words, what makes certain neuro-prosthetic devices worrisome is that they are both non-therapeutic and invasive. Call this the "Combined Criterion."

Of course, if my arguments in the preceding sections are correct, the whole notion of invasiveness ought not to be treated as marking a genuine ethical boundary, and if that is true, it won't help to combine it with another consideration. But the argument against it is strengthened by showing that it is problematic even when qualified in a plausible manner. And however much the enhancement/therapy distinction may be in dispute, there is no denying the intuitive relevance of those invasions of the body that occur in a medical context and those that don't. Consider surgery. What renders it permissible for a surgeon to cut me open (or even to perform "minimally invasive" laparoscopic surgery) is not simply that I gave her permission. The permissibility lies in the whole package of therapeutic aims, practices of training, professional responsibilities, etc. Without this larger meaning-complex – and especially without the purpose of healing me – cutting me open is morally suspect, even if I consent. The claim I wish to consider is that a similar sort of argument can be made for the case of neuro-prostheses: that their invasiveness creates a presumption against them, except when they are part of a reasonable medical treatment program.

As with the Invasiveness Objection, there are both conceptual and ethical grounds for questioning the Combined Criterion. Conceptually, there are, of course, well-known difficulties with the therapy/enhancement distinction, and I review them quickly in the next section. But there are also serious ethical difficulties that I take up in Section 9 regarding not only the restriction of treatment option but also the implicitly denigrating effects of viewing neuro-prostheses as a necessary evil.

8 Conceptual Difficulties with the "Combined Criterion": Therapy and Enhancement

In the recent debates over "enhancement", the emerging consensus seems to be that the prospects for identifying a principled difference between enhancement and therapy are dim,[9] and that the really decisive considerations actually lie elsewhere, in the purposes of medicine,[10] the fair distribution of public health resources,[11] or the vision of human beings as part of a wondrous creation that calls for gratitude and appreciation.[12] Even there, however, problems remain, particularly regarding the prospect of distinguishing those cases in which what the performance-enhancing treatment addresses is a disability or not. For we are all disabled in the sense that there are things that we cannot do but that we could do with some assistance. Thus, the relevant question must be reformulated as having to do with the level of functioning to which we believe all persons have a claim-right. But that is a question that can only be formulated on the basis of the contextual specifics of what treatments are available, what resources are available, and how various treatments will enable various forms of functioning. This is a matter of determining which capacities are to be viewed as especially important and which deserve a lower priority. But these issues are not going to be answered by finding bright lines in the treatments themselves or in the forms of disability themselves. In short, these are important questions, but it is a confusion to think that they can be answered in a conceptual distinction between "treatment" and "enhancement".

9 Ethical Difficulties with the "Combined Criterion": Disrespect for the Differently (En)Abled

But suppose that there were some general way of saying which particular methods of treatment are permissible only as treatment and not as enhancement. I believe that *ethical* difficulties would remain, having to do with the respect that is owed to persons with disabilities who use prostheses. In a nutshell, the concern is that the Combined Criterion builds on and reinforces a tendency to view various forms of prosthetic devices as intrinsically disturbing, an attitude that is at odds with the respect and recognition legitimated claimed those who are, as I shall say, "differently (en)abled".

There are at least two ways in which this particular disrespect for persons with disability occurs. First, it is important that discussions of assistive technology appreciate the fact that many people who use prosthetic devices to restore lost function often do not view these devices as a misfortune, but rather as opening up new opportunities or at least as an equally valuable way of instantiating the function. This is true, for example, of some people and their (motorized) wheelchairs.[13] This revaluation of enabling tools is part of a larger effort within the disabilities movement to de-stigmatize prostheses. But insofar as the Combined Criterion is designed to provide a way of saying how a neuro-prosthesis that is intrinsically monstrous might nonetheless be permitted because it is part of therapy, it leaves intact the purported monstrosity of the device itself. This is the logic of a "necessary evil": it remains an evil. After all, according to the Combined Criterion, we are talking about something that would be an abomination in a non-disabled person. But to those differently (en)abled persons who view their prostheses as part of who they are, a failure to leave open the space to stake a claim to that de-stigmatized cyborg identity involves a lack of respect.

Second, the Combined Criterion has the clear implication that a person's prosthesis is "not really part of her". Respect for persons is not to be confused with respect for their bodies; rather the target of the respect is the full person, however that site of concernful agency is physically realized. Again, I am not saying that all users of prostheses do or ought to identify with their prostheses. Rather, this is a point about the metaphysics of the addressee of respect for persons. It needs to be at least an open possibility that in some cases, this is most appropriately understood as including various assistive devices. This is a rather straightforward implication of something that I take to be uncontroversial: that what we respect are not bodies but persons.

There are two further unwelcome ethical implications of the Combined Criterion for persons with disabilities that I wish to mention briefly. The first becomes a problem if one formulates the anti-enhancement principle in such a way that prostheses are *impermissible* in cases in which the level of functioning is raised above normal levels. This would make it obligatory to avoid prostheses that boost individuals' performance beyond normal levels, as if there were an obligation to make sure that a hearing device didn't allow someone to be able to hear better. It would be perverse to require that the person would have to choose the prosthesis that was not as effective.

Finally, one further problem of moral perception is that, since the Combined Criterion entails the invasiveness criterion, it still relies on the logic of abomination discussed earlier. And it is simply implausible to suggest that the fact that the prosthesis serves a therapeutic purpose will be sufficient to disengaged the disgust-reaction, since the gut reactions to the breaching of the skin boundary are hardly going to be fine-tuned enough to track distinctions between therapeutic and non-therapeutic uses of the same technology. If, by hypothesis, an aversion to invasions of the body is supposed to be part of the deep grammar of our ethical responses, say, upon seeing a computer cable going into someone's skull, it is unclear why we should think that that feeling would evaporate once we see that the person is a quadriplegic. And this is all the more complicated in the case of *cognitive* deficits, which are typically much less visible.

10 Conclusion: Post-metaphysical Concerns with Neuro-prostheses

Where does this leave us with the issue of how to evaluate various neuro-prostheses ethically? As I said at the beginning of Section 5, my focus here has been on the notion that there is something intrinsically problematic about implanting neuro-prostheses that would be unproblematic if they were outside the body. My argument has been that this boundary of the skin is a misleading distraction. But I still want to hold open the possibility for arguing on other grounds that there are grave concerns regarding the use of neuro-prostheses. Indeed, part of my motivation in developing this argument is to open up room for a discussion of enhancements that is free from these distracting metaphysics. And before concluding, I would simply like to illustrate this point by sketching two potential concerns that could be raised about some neuro-prostheses – whether they are inside the body or are "mere tools" – and that are distinct from more familiar concerns regarding the safety of devices, the quality of the consent given for incorporation of such technologies, the social effects of differential access, and the character of individuals' motives for using them. The first has to do with issues of the notion of first-personal authority, and the second has to do with interpersonal practices.

As I mentioned earlier, one concern that people have about neuro-prostheses is that they will cut the agent off from her actions, in the sense that the machinery involved diminishes an experience's immediacy, transparency, or naturalness. I suggested that there was no reason, in principle, why neuro-prostheses would have to interfere with the experience. And in the relatively familiar cases of tool use – getting used to bifocals or a new computer keyboard – it is relatively clear what the variation in immediacy or transparency involves. After a while, you just don't notice any more. But there is also the particularly complex type of immediacy or transparency associated with the first-personal authority of avowals, in which the immediacy of the expressive act is of a piece with the authority of what one is saying. Richard Moran has recently made this point in discussing the case of someone who accepts the truth of her psychotherapist's assertion that she feels betrayed by her family, but although she can report this fact about herself (as a third-person observation), she cannot avow it (as a first-person report on a more-or-less immediate sense she has of her emotional state) (Moran 2001: Section 3.3). In this sense, she can speak authoritatively *about herself* but not *for herself.*

In terms of *this* form of transparency, we can ask whether there might be a form of self-dissociation associated with avowal that might result from the kind of indirectness seemingly involved in, say, Peter's face-recognition implant. For even if we can image that Peter no longer notices that the information is prompted aurally, it is hard to shake the idea that he is just hearing voices rather than really recognizing people and recalling their names. Of course, as I have already said in discussing the "unnaturalness" objection, there are problems with the quick assumption that relying on prosthetic devices necessarily involves something "outside" the agent. There are plenty of cases in which people use various devices but experience it to be fully integrated into their thinking, acting, and feeling. And perhaps it is, in principle, just as possible to say, "My left wheel is really bothering me today" as it is to say, "My knee is really bothering me". And yet there does seem to be a real worry here, about having an alienated relationship with one's self.

Typically, these questions about dissociation are asked with regard to the effects of neuro-pharmacological interventions on authenticity.[14] And more work needs to be done to work out how this discussion could be applied to the case of devices. But in the space available to me here, I would like to suggest that there are important concerns about dissociation that can best be formulated in terms of a loss of self-trust. In extreme cases, individuals who have been victims of rape or torture or other forms of profound physical abuse lose the ability to take their own desires and feelings seriously as authoritative. This self-dissociation can be profoundly damaging to human agency.[15] And it might be the case that in some cases a cognitive prosthesis like Peter's could disrupt the relationship of avowal in a way that would have similar results. Just as victims of trauma lose the ability to avow their desires, someone who relied extensively on a cognitive prosthesis – to provide an alternative means of accessing memories or an alternative source of awareness of deficit – might no longer be in a position to avow certain feelings, desires, and beliefs – that is, to assert them with the kind of automatic first-person authority with which we ordinarily report our states of mind.

There are complex issues here, which I am really only broaching here. Ultimately, the question to be asked in individual cases is whether people with neuroprostheses are able to integrate them in such a way that that can be part of the system responsible for one's avowals. The central point I have been making in this paper is that whichever of these concerns turn out to be relevant, they will apply equally to prostheses inside the body and to tools outside the body.

Finally, I would like to mention a further class of considerations that might be raised, having to do with the ways in which neuro-enhancements may disrupt presuppositions of important social practices. If we expect people to be able to remember the faces of loved ones better than those of casual acquaintances, then various assistive devices might disrupt the social meaning of recognizing someone. Imagine Peter telling you that he recognized you only because he was prompted by his implant. It's not just that you'd be less impressed with his uncannily good memory; rather, you might well feel hurt or slighted, not to mention deceived. He might seem not to be as close a friend as you thought, despite his insisting that it makes no difference. His "recognition" of you just seems less personal. After all, he never forgets *anyone's* name.

Appreciating the force of this concern – and properly placing it – requires recognizing both the contingency of social practices and the potential for real loss regarding experiences that are available only within that practice. The first point is straightforward: social practices change, including practices involving cognitive tools. Fifteen years ago, if someone had asked me for my wife's office phone number and I didn't know it "by heart", it would certainly have been viewed a puzzling, or even a cause for concern – either about my absent-mindedness or about my marriage. Now, however, few of my colleagues (and *none* of my students!) would find it odd in the least if I had to look at my cell phone to say what the number was. Perhaps something like that could happen with the social meaning that is currently attached to recalling people's names. And it might entail a major improvement in the lives of people with memory impairments.[16]

But there is another side to this phenomenon. For it might be that there is a complex and deeply significant package of emotional responses, rewarding experiences, social meanings, and so on that are bound up with the social practices that are held together by the expectation that one remember one's friends and loved ones' names – such that it might be crucial to retain the practice, in order to retain the constellation of meaningful experiences that it makes possible.[17] There are difficult questions here, having to do with the primacy of current practices as providing continuity of social meaning, or rather a primacy of individual experiences within the practices, as motors of change.[18] But again, my central point is that these questions ought to be asked across the board with regard to assistive devices (whether implanted or not) and even more broadly, to other forms of enhancement, including education.

These are only a few of the concerns that can be raised about neuro-prosthetic devices. Much more work needs to be done to analyze them, with appropriate attention not only to the principles involved, but also to the concrete details of their specific technologies, their social context, and the widely varying ways in which they are used. These are big issues. Here, I have been primarily concerned with a rather modest point, although it is often overlooked. My point is not that the dangers associated with elective, implanted neuro-prostheses are exaggerated, but rather that the debate needs to be carried out in terms of features of neuro-enhancements that are also found in other assistive technologies, especially those outside the body. Only by running the debate in that way can we avoid the dual dangers of demonizing legitimate interests in enablement and trivializing the potential dangers of many enhancement technologies by saying they are "merely tools". We need to be even-handed. And a simplistic metaphysics of "inside vs outside" or "therapy vs enhancement" just gets in the way of thinking clearly about what the real concerns are regarding the neuro-enhancement technologies that will play an ever more powerful role in our lives.[19]

Notes

[1] I'm leaving aside here the questions of whether the subjective qualitative "feel" of the sound would be different for her. For a discussion of this issue of "qualia," see especially David Lewis's functionalist reply to Frank Jackson's "knowledge argument" in Lewis 1983: 130–132.

[2] After developing this thought experiment, I discovered that Bradley Rhodes (1997) describes a similar "remembrance agent" and that it is also discussed by Andy Clark (2003: Ch. 2).

[3] Renick 1998; Stout 1983; Kass 1997. Stout and Renick both draw on Mary Douglas's notion of the "liminal"; see Douglas 1966.

[4] The *locus classicus* for this pivotal argument against mind-brain identity theories is Putnam 1975.

[5] It might be that only certain mechanisms are able to realize various capabilities, but that is an empirical question.

[6] Central texts here are Clark and Chalmers 1998 and Clark 2003. Although it is beyond the scope of this paper to take up the debate over the extended mind, it is not uncontroversial.

[7] See Alan Turing (1950: 433–460). Turing proposed a way of operationalizing intelligence, not by specifying necessary and sufficient conditions for intelligence, but by focusing on whether in double-blind tests an entity behaves in a way that is indistinguishable (in terms of intelligence) from an intelligent human. My suggestion is that this point could be generalized to other cases of "capacity-equivalence".

[8] For a particularly interesting discussion of prosopagnosia, see Duchaine et al. 2003.

[9] See especially the now-classic texts by Parens (1998) and Juengst (1998). See also the essays in Schöne-Seifert et al. (forthcoming).

[10] For an excellent discussion of these issues, see Talbot forthcoming.

[11] On issues of distributive justice, see the contributions by Reinhard Merkel and by Achim Stephan and Saskia Nagel to Schöne-Seifert et al. (forthcoming); see also parallel discussions in Buchanan et al. 2000.
[12] As discussed by Erik Parens 2005.
[13] I am grateful to Tom Shakespeare for a discussion of related issues; see also Shakespeare and Watson 2002.
[14] Kramer 1993; Elliot 2003; DeGrazia 2005: esp. Ch. 5 ("Enhancement Technologies and Self-Creation") and Schmidt-Felzmann forthcoming.
[15] For further discussion and references, see Anderson and Honneth 2005.
[16] And, of course, there is also the possibility of intentionally controlling the information in Peter's neuro-prosthesis, so as to make the information about loved ones more quickly or fully available.
[17] This idea is a central theme in the work of Charles Taylor. See especially Taylor 1989.
[18] Indeed, I think that this dynamic provides a way of articulating Erik Parens's distinction between an ethic of "gratitude" or appreciation and one of "creativity" (Parens 2005).
[19] In preparing and revising this paper, I have benefited from suggestions and criticisms of audiences in St. Louis, Nijmegen, Delmenhorst, Utrecht, and Doorn, and from an anonymous reviewer for this publication.

References

Anderson, Joel and Honneth, Axel. "Autonomy, Vulnerability, Recognition, and Justice." In *Liberalism and the Challenges to Liberalism*. John Christman, and Joel Anderson (eds.). New York: Cambridge University Press, 127–49, 2005.

Buchanan, Allen; Brock, Daniel W.; Daniels, Norman and Wikler, Daniel. *From Chance to Choice: Genetics and Justice*. Cambridge: Cambridge University Press, 2000.

Clark, Andy and Chalmers, David. "The Extended Mind." *Analysis*, 58, 7–19, 1998.

Clark, Andy. *Natural-Born Cyborgs: Minds, Technologies, and the Future of Human Intelligence*. New York: Oxford, 2003.

DeGrazia, David. *Human Identity and Bioethics*. New York: Cambridge University Press, 2005.

Douglas, Mary. *Purity and Danger: An Analysis of Concepts of Pollution and Taboo*. London: Routledge, 1966.

Duchaine, Bradley; Parker, Holly and Nakayama, Ken. "Normal Emotion Recognition in a Prosopagnosic." *Perception*, 32, 827–38, 2003.

Elliot, Carl. *Better Than Well: American Medicine Meets the American Dream*. New York: Norton, 2003.

Juengst, Eric. "What Does Enhancement *Mean*?" In *Enhancing Human Traits: Ethical and Social Implications*. Erik Parens (ed.). Washington, D.C.: Georgetown University Press, 29–47, 1998.

Kass, Leon R. "The Wisdom of Repugnance: Why We Should Ban the Cloning of Humans." *The New Republic*, 216, 22, 17–26, June 1997.

Kramer, Peter. *Listening to Prozac*. New York: Penguin, 1993.

Lewis, David. "Knowing What It's Like." In *Philosophical Papers*. Vol. 1. New York: Oxford University Press, 1983.

Merleau-Ponty, Maurice. *Phenomenology of Perception*. London: Routledge, 143, 1962.

Moran, Richard. *Authority and Estrangement: An Essay on Self-Knowledge*. Princeton: Princeton University Press, 2001.

Parens, Erik. "Is Better Always Good? The Enhancement Project." In *Enhancing Human Traits: Ethical and Social Implications*. Erik Parens (ed.). Washington, D.C.: Georgetown University Press, 1–28, 1998.

Parens, Erik. "Authenticity and Ambivalence: Toward Understanding the Enhancement Debate." *Hastings Center Report*, 35, 34–41, 2005.

Putnam, Hilary. *Mind, Language, and Reality: Philosophical Papers*. Vol. 2. Cambridge: Cambridge University Press, 1975.

Renick, Timothy. "A Cabbit in Sheep's Clothing: Exploring the Sources of Our Moral Disquiet about Cloning." *The Annual of the Society of Christian Ethics*, 18, 259–274, 1998.

Rhodes, Bradley. "The Wearable Remembrance Agent: A System For Augmented Memory."*Personal Technologies Journal Special Issue on Wearable Computing*, 1, 4, 218–224, 1997.

Schmidt-Felzmann, Heike. "Authentizität bei psychopharmakologischem Enhancement." In *Neuro-Enhancement. Ethik vor neuen Herausforderungen*. Bettina Schöne-Seifert et al. (eds.). Paderborn: Mentis-Verlag, 2007.

Schöne-Seifert, Bettina; Johann, Ach; Uwe, Opolka and Davinia, Talbot (Hrsg.). *Neuro-Enhancement. Ethik vor neuen Herausforderungen*. Paderborn: Mentis-Verlag, 2007.

Shakespeare, Tom and Nicholas, Watson. "The Social Model of Disability: An Outdated Ideology?" *Research in Social Science and Disability*, 2, 9–28, 2002.

Shanks, Pete. "Of baseball and enhancement bondage." *San Francisco Chronicle*, B5, September 26, 2005.

Stout, Jeffrey. "Moral Abominations." *Soundings*, LXVI, 1, Spring 1983.

Talbot, Davinia. "Neuro-Enhancement: Die ärztliche Rolle." In *Neuro-Enhancement. Ethik vor neuen Herausforderungen*. Bettina Schöne-Seifert et al. (eds.). Paderborn: Mentis-Verlag, 2007.

Taylor, Charles. *Sources of the Self: The Making of Modern Identity*. Cambridge, Mass.: Harvard University Press, 1989.

Turing, Alan. "Computing Machinery and Intelligence." *Mind*, 50, 433–460, 1950.

Part V
Others' Views: Intercultural Perspectives

Normative Relations: East Asian Perspectives on Biomedicine and Bioethics

Gerhold K. Becker

The search for cultural commonalities in East Asian biomedicine and bioethics is overshadowed by the notorious debate about Asian values during the late 1980s and early 1990s. I agree with Amartya Sen that claiming a quintessential set of Asian values amounts to "generalizations of heroic simplicity" (Sen 1999: 15–29). Yet a more sympathetic reading is possible that would place the debate within the broader context of a search for a sense of identity, authenticity, and sovereignty that is shared by many across East Asia. One would need to distinguish several dimensions in the values discourse, in particular its political, economic, cultural, and moral dimensions.

1 Multi-dimensional East-Asian Moral Discourse

1.1 The Political Dimension of the Asian-values Debate

There can be little doubt that the so-called Asian-values debate was politically motivated. The notion of "incompatible values" was enshrined in the 1993 *Bangkok Declaration*, in which Asian nations attempted to translate their definition of cultural identity into political and economic practice. The *Declaration* states that standards of human rights override neither "Asian" values and state sovereignty nor requirements of economic development (Davis 1995: 205–209). Those values emphasize "the primacy of order over freedom, family and community interests over individual choice and economic progress over political expression" (Lim 1998: 27).

While the *Declaration* was equally directed against internal dissent (Chua 1995: 31; 187) and Western critique of authoritarian rule, it utilized and rekindled the strong anti-colonial sensitivity that extends well beyond the most outspoken representatives of Asian values, Singapore and Malaysia. This sensitivity is still a unifying bond for East and South-East Asian nations, stimulating them in their search for cultural authenticity and for traditional values that range from particular ways of life to areas of biomedicine and healthcare. It is also a significant motive of

G.K. Becker
e-mail: gkbecker@hkbu.edu.hk

M. Düwell et al. (eds.), *The Contingent Nature of Life*,

suspicion about the underlying reasons for the promotion of global legal instruments regulating medical and biotechnological research and practice.

Value differences are real and, as empirical comparative research confirms, cannot be denied, but they are not incompatible. In a survey on societal values, Asian respondents put societal harmony first and relegated freedom of expression towards the bottom, whereas American respondents ranked the latter first, closely followed by personal freedom and the rights of the individual (Hitchcock 1994: 26).

1.2 The Economic Dimension

Perception and application of biomedicine take place within specific socio-economic conditions. Naturally, such conditions vary considerably between countries and cover the full spectrum of diverse opportunities within a socio-economic space marked out at opposite ends by countries such as China and Japan. There are, however, commonalities and one revolves around the strong apprehension about the implications of asymmetrical globalization. Although talk about globalization tends to conflate genuine interdependence at the business level with a flattening of cultural diversity and value pluralism, it is exactly the practical difficulty to keep those dimensions of international cooperation separate that fosters greater awareness of matters of cultural authenticity. And this reflects another facet of the sensitivity to real or perceived threats of domination by outside forces. For East Asians, the dilemma seems to be that if you want affluence you have to adapt to the MacDonaldization of the world.

At the same time, modernization, understood as the wholesale transfer and application of modern science and technology to all sectors of society, is broadly endorsed. For the developing nations in the region, it is the most decisive factor towards achieving developmental parity with and thus full political independence from the West.

It seems inevitable that globalization and modernization produce tensions, in particular in the areas of biotechnology, medicine, and healthcare. Observers have blamed the erosion of traditional virtues in the medical profession on the amalgamation of medicine, research, and health-care with the capitalist, market-driven economic system. As a consequence of Meiji Japan's (1874), Republican China's (1914) and other countries' fascination with Western science and technology at the expense of traditional holistic medical practice, physicians in East-Asia are mainly Western-trained.[1] In addition, they bring along a bioethical orientation that originated from the specific conditions of the practice of medicine in the West, above all in the United States. Their loyalties are therefore frequently divided between scientific and therapeutic commitments, economic viability, and the moral demands of traditional culture. The huge differences of access to healthcare in urban and rural areas as well as a general shortage of doctors and nurses are, above all, caused by economic imbalances. As Pinit Ratanakul has pointed out, "in Thailand, for example, 62% of doctors and nurses are in Bangkok, where most of the country's hospitals are, while there are too few doctors and nurses in the provinces, where most of the people are. There are also too many hospitals in Bangkok and too few neighborhood clinics and public health centers in rural areas" (Ratanakul 2004: 1715). The situation is similar in China and many East Asian countries. In 1990, two hundred towns in the Philippines had no resident doctors, and only an estimated 32% of all qualified Filipino doctors and nurses practice their profession in their own country (ibid.: 1716).

1.3 The Cultural Dimension

In spite of Max Weber's claim that the unifying premise of East Asian world-views is their "indifference to the world" and an opposition to "practical rationalism of the West" (Weber 1920: 364; 266), Asian bioethicists find commonalities rather in a broadly shared fundamental naturalism that transcends strict distinctions between the natural and artificial and inculcates holistic views of human life within the greater natural order. Instead of setting the human life-world apart from nature, there is a general tendency to see it as a part of nature and integrated in the natural order. Polar complementarity and pure immanence instead of a dichotomy between the individual and culture or society are frequently taken to provide ideal-typical paradigms of such alternative models of rationality and ethics. It is the longing for a "metaphysics of presence, wholeness, and totality", which seems unavailable within Western traditions that stimulates this search for holistic conceptions of reality, "in which everything exists as part of Ultimate Reality, equal in degrees of being and reality, and fundamentally interrelated as in the luminous image of Indra's net" (Olds 1991: 21). The favorite Buddhist metaphor gained prominence in the Hua-yen school of China but continues to exert influence on current East Asian social and political thought. It can also illustrate the alternative conceptions of East Asian biomedicine, including those of disease and treatment, which are rarely seen as exclusively biological but always as embedded in complex holistic systems of belief and value (Ratanakul 2004: 1717).

1.4 The Moral versus the Ethical Dimension

The need for universal moral standards in the appraisal and regulation of developments in the life-sciences and beyond seems obvious and without alternative. Yet from an East Asian perspective, the prospects for the convergence of particular ethical world-views towards a common morality appear rather dim. Nevertheless, this may be more a sign of hope than despair, as it provides at least some breathing space before the vision of a new global moral order becomes (a stifling) reality. The fear is based on the intrinsic dynamics that transform universal morality into a global order that would erode traditional values and cut deep into the sovereignty of nation states.

With regard to biomedicine, the continuous entrenchment of fundamentalist positions across the religious, political, and cultural spectrum was reflected in the UN debate on human cloning. The fledgling Asian nation states with multiple ethnicities saw their (political) integrity threatened by what they suspect is the resurgence of colonialism under the disguise of supposedly universal moral values and their global implementation through legal instruments. The decision by the UN to put the *Cloning Declaration* to a vote instead of seeking consensus is a case in point. Particularly countries from East Asia including China, Korea, Japan and Singapore were unanimous in their disapproval of both process and content.[2] It would therefore be too simplistic and even counter-productive to brush such critique aside as mere political posturing or immature moral thinking as it clearly reflects the values listed already in the Singaporean *White Paper*, which stirred up the debate and affirmed consensus, tolerance, and harmony as typical Asian values. Although it appears likely that in the long term some sort of consensus on globally shared fundamental moral norms will emerge, there will always be room for divergent ethical views. Instead of aiming at the discovery of "ultimate truth" and the trumping of disagreement, "the path of conversational restraint" would be a more promising route to "reasonably responding to [their] continuing moral disagreement" (Ackerman 1989: 19). Consensus on the universal content of moral duties does not necessarily imply moral unanimity in the justification of such content. "Why", one might ask with Charles Larmore, "can we not affirm a set of duties binding on all without supposing they must be justifiable to all?" (Larmore 1996: 57).

2 Family-centered Biomedicine

I break off here to embark on a different path that may lead towards more specific insights at the micro level. I propose to find in the *family* one of the most distinctive value-factors shaping East-Asian societies and providing a central focus for ethical traditions. The family's defining social role holds a prominent place on all lists of Asian values and it is the pervasive model of order. While its theoretical underpinning varies according to particular ethical traditions, its practical significance for daily life suggests a core set of moral beliefs that can be broadly shared. Although it seems an overstatement to characterize the East Asian ethos as "family-centred harmonious self-suppression" (Hattori 2003), the moral authority of the family remains strong and is hardly questioned.

In a similar approach, bioethicists from the Philippines have recently sought to construe an alternative Asian version of bioethics that would have its experiential basis in a moral phenomenology of socially lived values revolving around the family and thus be representative of the specific "spirit" of Filipino culture (Tan Alora and Lumitao 2001). The organization of all areas of social life is modeled after the familial community, which holds final authority in decision-making for its members. Grounded in the "socio-moral responsibilities" of familial authority and its overriding concern for harmony, Filipino bioethics similarly is supposed to assume an unabashed paternalistic attitude that leaves little space for individual choice and dissenting views. Moral agents are expected to have internalized the dominant social order and to submit to authority. As the authority is represented by individuals and extends in concentric circles from parents and elders to employers, the state and the Church, Filipino culture has been portrayed as person-oriented insofar as persons "take precedence over abstract, impersonal issues or ideas" or "concerns for social justice or honesty." The medical world that provides the culture-specific parameters within which the alternative model of bioethics unfolds is a direct extension of the larger social world and its unquestioned hierarchy of traditional authorities.

Similar claims have been put forward with regard to other East and South East Asian countries. The Buddha confirmed traditional Indian teachings on the significance of the family and emphasized family relationships. He exhorted the laity to maintain close family ties and to respect the dignity of the family as a social unit (Saddhatissa 1987: 117).

The moral authority of the family has been most comprehensively conceptualized within Confucianism and is, in principle, acknowledged even today in all countries with strong Confucian traditions (China, incl. Taiwan and Hong Kong, Korea, Japan, Singapore). Wei-ming Tu has argued that for all of Confucian East Asia the family "as the basic unit of society is the locus from which core values are transmitted. The dyadic relationships within the family, differentiated by age, gender, authority, status, and hierarchy, provide a richly textured natural environment for learning the proper way of being human" (Tu 2002: 55–77). Confucians regard the family as the "natural habitat of humans", the necessary and the most desirable "environment for mutual support and personal growth" (Tu 1985a: 123).

Yet the moral authority of the Chinese family must not be idealized and one should not gloss over the severe tensions and grave injustices associated with it, particularly regarding the effects of its strict hierarchy and paternal authority on the subjugation of women. Sketching out the ideal-type of the Chinese family may, however, facilitate the identification of cultural factors that characterize the specific conditions within which biomedicine has to operate in countries under Confucian influence and beyond.

In Confucian moral and political theory the family is central and recognized as an entity with social properties substantially richer than those of its members. In contemporary bioethical discourse, the moral role of the family is therefore frequently interpreted as contrasting sharply with rights-based conceptions of human life, individual autonomy, and human dignity, and thus as providing the conceptual basis for the construction of alternative versions of ethics. The theoretically most ambitious among such projects have been launched from within contemporary Confucianism by ethicists in mainland China, Taiwan and Hong Kong. It has been argued that this version of Confucianism is representative of a specific tradition of philosophical thought and a cultural perspective that is "generally endorsed by the Chinese and embodied in their way of life and practices" (Chan 1999: 213). On its account, the moral conception of the family includes at least the following three morally significant components.

The Confucian vision of the ideal family and in consequence of the community rejects the notion of mutually disinterested and independent individuals. Instead it centres in a conception of human relatedness and relational agency. The human being is conceived above all as "constituted by a web of unique role relations and by the way concrete responsibilities are performed" (Tao 1999: 576, 578). Ontologically, the person is grounded in "the Heaven-Earth-man triad", and thus "inseparably connected to and mutually dependent on the transcendent Heaven, the natural environment, and above all the community in which the person lives and from which the person derives his or her identity" (Hui 1999a: 159; see also Hui 2000). Moral status is a direct function of the capacity for compassion and sympathy, rooted in a moral community of social persons. Confucians are, however, aware of the tension between the ontological basis of personhood and human dignity, which Mencius (孟子) found in the "four seeds"[3] (*duan* (端) "sprouts") of morality that distinguish every human being from birth on, and the social constitution of the relational self in its specific roles and their moral demands.[4]

According to the Analects, "filiality and brotherliness are the bases of humanity" and "being filial and brotherly is the initial step towards realizing one's humanity" (Tu 1985b: 123). As the *Classic of Filial Piety* (孝經 *Xiao Jing*) stated, "filial responsibility is the root of (all) virtue, and (the stem) out of which grows (all moral) teaching." The *Doctrine of the Mean* (中庸 *Zhongyong*) states that filial piety "means being good at continuing the purposes of one's predecessors and maintaining their ways" (Ames and Hall (2001): 81).

Filial piety (*xiao* 孝) then can be defined as the respectful and obliging attitude towards the parents and as the specific virtue that cultivates and organizes moral relations in family life (Roetz 1993: 53). In traditional Chinese society it not only served to ensure that "the elderly were properly taken care of by their families" but functioned also as a means "to regulate the behavior between family members or even those with the wider society" (Chow 1992: 125). Based on the twin principles of love by gradation and care by extension, the goal is to extend moral concern from those who are close to those who are unrelated (Tao 1999: 581).

Incidentally, the moral significance of filial piety (responsibility) is not restricted to Confucianism but has a firm place also in Buddhism. Although it played a more significant role in Chinese Buddhism, it is not unique to it but existed in India from early times (Jan 1991: 35). According to the Mangala Sutra[5] parental love "can never be compensated even if one were to carry one's parents on the shoulder without putting them down for a hundred or a thousand years" (Hallisey 2002: 246; 245). The gratitude to parents for their love is exemplified by numerous acts of respect and care embodied in the concrete, personal relationships of daily life, including bathing them and providing them with food and drink.

Familial relationships are socially expressed in rituals (禮 *li*) in which they are simultaneously re-enforced and integrated into the all-pervasive natural order. The origin of rituals in ancestral worship[6] binds them back to this order and re-enacts the natural order in familial contexts (Ho 1992). By establishing regulative familial and social relations, directing actions, and shaping institutions, rituals inculcate moral practice as they extend from behavioral patterns of daily interactions and familial celebrations to specific rites and socio-political systems. The individual's virtue of humaneness (*ren* 仁) is a direct function of the internalization of rituals, which prescribe human action and enable human self-cultivation. Rituals are instrumental in achieving "the moral vision needed to appreciate that the family constitutes the normal way of human life – a social reality necessary for the full development of cardinal human virtues and full human flourishing" (Fan and Li 2004). Thus the ever-deepening and broadening awareness of the presence of the other in one's self-cultivation is a distinctive feature of Confucian ritualization (Tu 1985a: 114).

3 Implications for Biomedicine and Bioethics

3.1 Shared Decision-making

One of the most obvious implications of the Confucian conception of the family in biomedicine is the emphasis on harmonious decision-making between patient, family, and physician, and the stress on benevolence and utility. In contrast to patient-centered health care settings, familial interdependence implies that health issues are not exclusively left to the individual but treated as family matters. As Ruiping Fan has argued persuasively, this applies particularly in cases where informed consent to treatment options is required. The family, in cooperation with the physician, will decide collectively without necessarily concurring with the patient's explicit wishes. As the family takes responsibility for the patient even when competent, there is "virtually no room" for advance directives or durable powers of attorney (Fan 2002). In addition, in a Confucian family setting any advance directive as a formal statement put in writing by one of its members would suggest a level of distrust that would render it ineffective, particularly as familial authority could either regard it as invalid or disregard it altogether.[7]

The scenario is largely identical in all Confucian countries, in spite of recent legislation seeking to strengthen the position of the individual. It has been reported that in Taiwan more than ninety percent of surgical consent forms are signed by family members regardless of the patient's status of competence (Yang 2003: 100). Although empirical evidence indicates a strong desire of patients for information about their conditions, recent studies confirm the traditional practice of disclosure to family members rather than to patients. Considering the emphasis of nonmaleficence and beneficence and the relative power of the family, there is a well-recognized suggestion that, in Asian culture, informing cancer patients about their diagnosis and prognosis should be modified according to the family's opinion.

In Japan, the organ transplantation act of 1997 endorsed the traditional position of the family and provided for a veto right even against the explicit wishes of the donor (Bagheri 2003: 135). In Buddhist countries, too, the concept of patient autonomy is challenged as merely individualistic and in conflict with the fundamental principles of interdependence of all beings and of continuous transformation.

It seems a common characteristic of medical practice in the cultural setting of East Asia that disclosure of diagnosis and prognosis is likely to be modified according to the family's opinion. The family, in cooperation with the physician, decides how much information should be shared with the patient about his or her health conditions or the prospects of disease. If the diagnosis concerns terminal illness, the virtue of humaneness or benevolence would demand to keep such disturbing news hidden from the patient as long as possible. As a consequence of this strategy, the patient's questions to the attending physician will usually be redirected to the family, and the family is thought justified to even employ deception in its strategy to protect the patient's emotional well-being (see also Morioka 1995).

3.2 Relational Self and Moral Status

The family as the most basic form of communal life provides the moral space of the defining relationships at the early stages of human development. Although there is no consensus on its implications for moral status and personhood, the concept of the relational self puts all emphasis on the process of socialization within the bounds of the family rather than on ontological characteristics. Frequently, this line of thought is backed up with reference to a famous remark by Xunzi who simply stated that "human life begins at birth and ends in death" (Qiu 2000: 332; see also Qiu 2004: 1697). The widely accepted practice of abortion throughout Chinese history (Cong 2003) lends support to this view, particularly since, according to Paul Unschuld, it has "evoked little, if any, concern among Confucian thinkers" (Unschuld 1995: 465–469; see also Bao 1999; Cong 2003: 252).

As to contemporary Confucian bioethics, one extreme position claims that moral status is a strict function of family membership and its subsequent transferal of rights and responsibilities. It is assumed to be the mother's prerogative to confer membership in the moral community to the child that she is bearing, and that "nobody else has a right that overrides hers. If she does not want to confer this right to her future child, she has the right to abort, though within certain limitations that arise from the growth of the fetus" and its increasing potential for social community (Lee 1999: 175–176).

In contrast, moderates have argued that classical Confucianism, particularly of Mencian provenance, is able to accommodate rights-prone views on human dignity and human worth that are independent of role commitments. As the human community forms an integral part of the larger Heavenly order, parental relationships too are subordinated to its ultimate authority.[8] Yet even on this interpretation it is generally acknowledged that full moral status does not apply, or at least not equally, to all prenatal phases of human life but is relative to human development and familial recognition.

It appears that similar perceptions of the moral status of prenatal life are shared across East Asia and not restricted to Confucianism. With regard to traditional Filipino society it has been argued that even the notion of the "fetus" and its moral implications are based on biomedical categories which have no indigenous cultural equivalent: "hence, a phenomenon biomedicine may readily describe as 'abortion' need not be morally controversial at all" (Sy 1998: 105).

While the much discussed Japanese tolerance for, and ritualization of, abortion is mainly rooted in Mahayana traditions of Buddhism, which apparently adopt a more lenient attitude towards it than the overwhelmingly antiabortionist Theravada Buddhism, it should be noted that from the perspective of Buddhist metaphysics the ontological status of the human embryo is not of major significance. Though there is no unanimity on the implications of the classical Buddhist teachings for contemporary bioethical dilemmas such as abortion, the dominant view seems to suggest that prior to birth human life does not yet exist as the fetus lacks the embodiment of all five aggregates (skandhas) comprised of corporeality, sensation, perception, unconscious formation and consciousness. In addition and more importantly, Buddhist ethics is less focused on concerns for the victims of medical interventions than on the agents and the karmic implications of their actions. As Mettanando Bhikkhu put it: "The Buddhist approach to medical ethics is oriented around the person performing the actions, i.e. the doctor's behavior rather than protective legislation and right of the patient as a 'consumer'. This is because the karma of the medical treatment belongs to the doctor" (Mettanando 1991: 202). This seems also reflected in the belief of Japanese Buddhists that abortion may be a "sorrowful necessity", that can be amended by offerings to the aborted bought at the temples and thought to facilitate a more propitious rebirth (Hughes and Keown 1995).

Similar tensions between traditional Buddhist beliefs and actual practice can be found in Thailand where Theravada Buddhism is the official religion to which 95 percent of the population adhere. Although abortion is still illegal and rejected by most Thais on the grounds that it breaches the First Precept of Buddhism which prohibits killing any living being (and, by extension, unborn human life), the high rates of illegal abortions suggest that even within the parameters of a traditionally Buddhist society, socio-economic factors gain increasing influence on ethical decision-making, and that abortion is regarded as a supplementary means of fertility control. It seems that policies aiming at the reduction of population growth together with rapid modernization played a major role among the factors contributing to the rise in the number of illegal abortions performed in Thailand.[9] Accordingly, Buddhist bioethicists in Thailand have begun to focus more on the social and economic conditions of life, and in particular of health care and issues of fair distribution of resources, than on the ontological status of the human embryo (Ratanakul 1998).

3.3 *Filial Responsibility and Biomedicine*

Mencius' statement that of the three unfilial acts "to have no posterity is the greatest of them all"[10] is frequently invoked as the moral basis for the liberal use of reproductive technology in Chinese society and thus illustrates its enduring influence in modern biomedicine. Respect for filial obligations is a strong motive for infertile Chinese couples to resort to the whole array of means of artificial reproduction. Nevertheless, here too divisions run deep among Confucian bioethicists. While the use of donor eggs and the employment of surrogate mothers usually raises concerns about parenthood and the preservation of natural, familial relationships, for some such concerns are secondary to the filial duties of continuing the family (Hui 1999b: 133). For others, they mark the line that separates legitimate from illegitimate means of reproduction as the following example from Hong Kong illustrates. During the public consultation on the proposed legislation on reproductive technology, Confucian bioethicists demanded that surrogacy and reproductive cloning be banned. They saw in the former a violation of filial relations and an assault on the concept of relational personhood (Tao and Chan 1997; Lo 1996), in the latter (as a-sexual reproduction) the distortion of the natural order, the destruction of parental relationships, and the illicit objectification of the reproductive act.

3.4 Scientific and Spiritual Dimensions of Biomedicine

It is a characteristic of most East Asian countries that traditional and modern medicine are harmoniously practised side by side. Depending on the types of illness, one attends either the practitioners of traditional medicine or the physician whose clinic walls proudly display the certificates of his studies at Western medical schools. And frequently, one sees both. The great respect for and the keen interest in modern science and technology that is typical of all East Asian countries is also reflected in their attitude towards modern medicine. The tension between the two medical paradigms emerges, however, more tangibly in views about disease aetiology than in alternative strategies of therapy. In spite of differences in cultural traditions, the Thai-Buddhist attitude seems largely representative for East Asia in general. While 60 percent of the respondents to a survey saw no conflict between science and religion, 54 percent thought that scientific explanations of disease were not in congruence with Buddhist (karmic) beliefs (Ratanakul 1993).

There is a general perception that modern medical diagnosis does not reveal the true root-causes but only scratches the surface of disease and thus points merely at another set of symptoms, including those at the genetic level. Accordingly, modern scientific medicine has been charged with neglecting the spiritual and holistic dimension both in its diagnostics and, more importantly, in its therapeutic efforts. A fruitful and long-term development of medicine would require creative cooperation between modern and traditional medicine so as to render it simultaneously "holistic, scientific, and humane" (Ratanakul 1998).

The well-known problems surrounding organ transplantation and brain stem death in various countries in East Asia indicate such an intersection between the paradigms of traditional and modern medicine and their differing spiritual and scientific bases. In all countries under the influence of Confucianism, cultural factors have been blamed for the fact that rates of organ donation are considerably lower than in Western countries.[11] Transplantation medicine challenges not only the traditional Chinese view of the human body but also ritualistic and filial duties owed to the deceased. Surveys about public attitudes towards organ donation in Hong Kong, for example, have consistently confirmed that there is a desire to preserve a "complete body" after death and a belief that cadaveric organ donation will not only mutilate the body but also constitute a violation of filial piety (Chan et al. 1990; Li 2000). The reluctance in the general public to accept brain stem death as a reliable criterion of the cessation of life has also been linked to the same set of moral beliefs and ritual requirements. Even stronger concerns have been raised in Japan and fuelled a debate stretching more than two decades and culminating in the organ transplantation act of 1997. Yet even this act did little to calm the strong emotions on both sides. As a result, the rate of organ donations dropped even further.

3.5 Society: Family Writ Large

The recent introduction of legislation and statutory, regulative bodies in all countries with Confucian background confirms the unquestioned authority of society to define individual roles within the moral framework of traditional responsibilities and rights. The restriction of reproductive technologies to legally married couples and the prohibition of gender selection for social reasons in Hong Kong law are cases in point. Gender selection reflects a manifest desire across all Confucian societies for male children, and also a Confucian tradition that is less concerned about issues of moral status in the early stages of life. Relevant legislation from China, Korea, Japan to Singapore exhibits a utilitarian bend that allows the use of "spare" embryos for research, takes a favorable position on the creation of human embryos for research purposes, and advocates therapeutic cloning.

4 New Challenges to Traditional Values

As the effects of globalization are gaining momentum, traditional value systems have come under considerable pressure from shifting social parameters and the economic forces of the market. The trend towards the nuclear family is not restricted to the West but has affected East Asia as well. Whereas measures of population control in mainland China are unique and have cut deeply into traditional conceptions of familial values, the nuclear family is the preferred and self-chosen option in Taiwan, Hong Kong, and Singapore, despite strong discouragement by government. The results are felt in particular by the elderly and the sick. Filial responsibility has been one of the first casualties of rapid societal transformation as it depends on a socio-economic order that is favorable to the development of extended families within a supportive clan-system. All countries with strong Confucian influence witness symptoms of an erosion of the moral authority that filial responsibility commanded in traditional Chinese society. Suicide patterns in Asian societies reveal that the elderly "are killing themselves at rates up to five times higher than their own younger generations and eight times higher than their Western counterparts" (Hu 1995: 202). Most vulnerable among the elderly are women. This suggests that as the traditional, Confucian-based ethics of care is being undermined, additional moral resources are needed to curb the rise in abuse, neglect, and ill treatment of the elderly. As the effect of filial piety depends largely on prevailing socio-economic conditions, it rather seems to define a moral ideal of familial relationships than describe daily reality (Ikels 1980).

Additional conflicts arise between filial duties of care and the adverse effect of life-prolonging medical technology on the quality of life. Traditionally, the preservation and protection of the parents' life was the most basic of all duties and the first of filial obligations. Yet in market-driven clinical settings economic factors increasingly influence the interpretation of the filial imperative as traditional parameters of familial decision-making unravel. This situation dramatically aggravates both the patient's vulnerability and familial responsibility. It is difficult to imagine how families in dire economic conditions, in rural China or elsewhere, should not be tempted to influence the physician's treatment options so as to serve their collective interests. Empirical evidence seems to confirm that medical advice in the best interest of the patient is likely not to be followed when the family faces financial hardship that makes necessary medical treatment nearly unaffordable.

In discharging its moral duties, the Confucian model of the family depends to a considerable degree on a co-operation with the physician that is equally based on mutual trust and the physician's moral integrity. While ideally the family collectively decides what is in the best interest of the patient and the physician treats patients as if they were members of his own family without regard to financial interests, the economic conditions of modern health care systems require additional safeguards, preferably through legislation, lest the traditionally elevated authority of the physician should be instrumentalized for personal gain at the expense of family and patient alike. As a Chinese bioethicist observed, China is currently experiencing the "darkest period of the doctor-patient relationship in its history" as more and more patients sue hospitals and doctors and a general crisis of trust has emerged between patients and physicians (Cong 2003: 249).

Various indicators suggest that the model of harmonious familial decision-making is increasingly questioned in favor of a greater recognition of patient autonomy. This seems particularly relevant with regard to informing terminally ill patients about their conditions. Recent empirical research from Taiwan, Hong Kong, and Japan seems to confirm a common trend towards more direct patient involvement in medical decision-making. While Hong Kong physicians, like their mainland counterparts, used to reveal terminal illness only to the family, they are now more likely to respond truthfully to patients' explicit requests for information even against the express wishes of the family. Attempts have been made to reconcile such tendencies with normative Confucian views on the family under a conception of "moderate familism" characterized by an inclusive process of joint decision making between the patient and the family (Chan 2004). Similar developments are emerging in other Confucian countries. Researchers in Taiwan observed strong preferences among patients to be informed about their conditions prior to the family. Studies in Japan revealed positive attitudes in patients towards participation in medical decision-making based on truthful information from health-care professionals.[12] Cross-cultural research in clinical settings in Japan and in the US revealed a similar tendency towards greater patient involvement in disclosure strategies. While the study suggests that "Japanese physicians are moving away from a rigid policy of nondisclosure to a policy of selective disclosure", their general practice continues to be more restricted than that of their Western counterparts (Elwyn et al. 1998).

Notes

[1] The following observation from Thailand seems paradigmatic for the whole region: "Many Thai doctors who are practicing modern medicine were trained in America and/or Europe and they brought with them to Thailand not only knowledge of modern medicine but also some of its accompanying values such as the free market ideology which values wealth over persons and human needs. If this tendency continues, modern medicine will create an inequitable society where only the rich get quality health care" (Ratanakul 1998: 98–99).

[2] Fifty-ninth General Assembly of the United Nations' Sixth Committee, Declaration on Human Cloning, 18 February 2005.

[3] See Hansen 1992: 397.

[4] In Mencius's theory of human nature (*Mencius* 2A: 6), every human being is endowed by Nature with four potential moral virtues at birth. These are the seeds of compassion, shame, modesty, and right and wrong which are equally possessed by all. They are also the "four beginnings" or "four moral possibilities" of the virtues of humaneness, righteousness, propriety and wisdom. The dignity of humans is grounded in an inborn moral sense, which accords them a nobility based upon their moral worth, instead of their worldly ranks. See Bloom 1998.

[5] See Mettanando 2004.

[6] This is still visible in the compound character, which is an ideograph connoting the presentation of sacrifices (shi) to ancestral spirits at an altar to them (li) (Ames and Hall 2001: 69).

[7] Similar claims have been made for Japan by Ohi (1998) and Tuchida (1998).

[8] According to Ruiping Fan, human beings are placed in a system of *natural* relationships, which make normative claims on them, and interventions such as reproductive cloning would destroy this relationship and violate human dignity. See Fan and Ruiping 1998: 73–93; 193–196.

[9] For 1993, the Ministry of Public Health estimated that as many as 80,000 illegal abortions were performed, see Lerdmaleewong and Francis (1998).

[10] Apparently, the first unfilial act was blind obedience to parents and causing them to fall into immorality, and the second to serve in the government when the parents are old and poor. See Chan 1973: 75.

[11] Over the years, the organ donation rate has been steady at about 1–2 per million population (compared with a donation rate of 19/million in Australia, 16–20/million in USA, and 18/million in most European countries with the exception of Spain where the rate is even 22/million).

[12] Results from an empirical study of disclosure attitudes among Tokyo residents suggest that "the greater proportion of respondents wants or permits disclosure of cancer diagnosis and prognosis." See Miyata, Tachimori, Takahashi, Saito and Kai 2004: 7.

References

Ackerman, Bruce. "Why Dialogue?" *The Journal of Philosophy*, 86, 5–22, 1989.

Ames, Roger T. and Hall, David L. *Focusing the Familiar*. Honolulu: University of Hawaii Press, 2001.

Bagheri, Alireza. "Can the 'Japan Organ Transplantation Law' Promote Organ Procurement for the Brain Dead?" In *Asian Bioethics in the 21st Century*. Sang-yong Song, Young-Mo Koo, and Darryl R.J. Macer (eds.). Christchurch, Tsukuba: Eubios Ethics Institute, 133–137, 2003.

Bao, Nie-Jing. "The Problem of Coerced Abortion in China and Related Ethical Issues." *Cambridge Quarterly of Healthcare Ethics*, 8, 463–479, 1999.

Bloom, Irene. "Mencius and Human Rights." In *Confucianism and Human Rights*. Wm. Theodore De Bary and Wei-ming Tu (eds.). New York: Columbia Press, 94–116, 1998.
Chan, A.Y.; Tse, M.H. and Cheung, R. "Public Attitudes Toward Kidney Donation in Hong Kong." *Dialysis Transplant*, 19, 242–257, 1990.
Chan, Ho-mun. "Informed Consent Hong Kong Style: An Instance of Moderate Familism." *Journal of Medicine and Philosophy*, 29, 2, 195–206, 2004.
Chan, Joseph. "A Confucian Perspective on Human Rights for Contemporary China." In *The East Asian Challenge for Human Rights*. Joanne R. Bauer and Daniel A. Bell (eds.). Cambridge: Cambridge University Press, 212–237, 1999.
Chan, Wing-tsit. *A Source Book in Chinese Philosophy*. Princeton: Princeton University Press, 1973.
Chow, Nelson. "Family Care of the Elderly in Hong Kong." In *Family Care of the Elderly: Social and Cultural Changes*. Jordan I. Kosberg (ed.). Newbury Park: Sage Publications, 123–138, 1992.
Chua, Beng-Huat. *Communitarian Ideology and Democracy in Singapore*. London: Routledge, 1995.
Cong, Ya-li. "Bioethics in China." In *The Annals of Bioethics: Regional Perspectives in Bioethics*. John F. Peppin and Mark J. Cherry (eds.). Lisse: Swets & Zeitlinger, 239–260, 2003.
Davis, Michael C. (ed.). *Human Rights and Chinese Values*. Hong Kong: Oxford University Press, 1995.
Elwyn, Todd S.; Fetters, Michael D.; Gorenflo, Daniel W. and Tsuda, Tsukasa. "Cancer Disclosure in Japan: Historical Comparisons, Current Practices." *Social Science & Medicine*, 46, 9, 1151–1163, 1998.
Fan, Ruiping. "Human Cloning and Human Dignity: Pluralist Society and the Confucian Moral Community." *Chinese & International Philosophy of Medicine*, 1, 3, 73–94, 1998.
Fan, Ruiping. "Reconsidering Surrogate Decision Making: Aristotelianism and Confucianism on Ideal Human Relations." *Philosophy East & West*, 52, 346–372, 2002.
Fan, Ruiping and Li, Benfu. "Truth Telling in Medicine: The Confucian View." *Journal of Medicine and Philosophy*, 29, 2, 179–193, 2004.
Hallisey, Charles. "Auspicious Things." In *Religions of Asia in Practice*. Donald S. Lopez, Jr. (ed.). Princeton: Princeton University Press, 237–251, 2002.
Hansen, Chad. *A Daoist Theory of Chinese Thought: A Philosophical Interpretation*. New York, Oxford: Oxford University Press, 1992.
Hattori, Kenji. "East Asian Family and Biomedical Ethics." In *Asian Bioethics in the 21st Century*. Sang-yong Song, Young-Mo Koo, and Darryl R.J. Macer (eds.). Christchurch, Tsukuba: Eubios Ethics Institute, 229–231, 2003.
Hitchcock, David. *Asian Values and the United States: How Much Conflict?* Washington, D.C.: The Center for Strategic and International Studies, 1994.
Ho, Ping-ti. "Original Ritual (Yuan Li)." *The 21st Century*, 11, 102–110, 1992.
Hu, Yow-Hwey. "Elderly Suicide Risk in Family Contexts: A Critique of the Asian Family Care Model." *Journal of Cross-Cultural Gerontology*, 10, 199–217, 1995.
Hughes, James J. and Keown, Damien. "Buddhism and Medical Ethics: A Bibliographic Introduction." *Journal of Buddhist Ethics*, 2, 105–124, 1995.
Hui, Edwin. "A Confucian Ethic of Medical Futility." In *Confucian Bioethics*. Ruiping Fan (ed.). Dordrecht: Kluwer, 127–163, 1999a.
Hui, Edwin. "Chinese Health Care Ethics." In *A Cross-cultural Dialogue on Health Care Ethics*. Harold Coward and Pinit Ratanakul (eds.). Waterloo: Wilfried Laurier University Press, 128–138, 1999b.
Hui, Edwin. "Jen and Perichoresis: The Confucian and Christian Bases of the Relational Person." In *The Moral Status of Personhood: Perspectives on Bioethics*. VIBS Special Series Studies in Applied Ethics. Gerhold K. Becker (ed.). Rodopi: Amsterdam/Atlanta, 95–117, 2000.
Ikels, Charlotte. "The Coming of Age in Chinese Society: Traditional Patterns and Contemporary Hong Kong." In *Aging in Culture and Society*. Christine L. Fry (ed.). New York: J. F. Bergin Publishers, 80–100, 1980.

Jan, Yün-hua. "The Role of Filial Piety in Chinese Buddhism: A Reassessment." In *Buddhist Ethics and Modern Society*. Charles Wei-hsun Fu and Sandra A. Wawrytko (eds.). New York: Greenwood Press, 6–39, 1991.

Larmore, Charles. *The Morals of Modernity*. Cambridge: Cambridge University Press, 1996.

Lee, Shui-chuen. "A Confucian Perspective on Human Genetics." In *Chinese Scientist and Responsibility: Ethical Issues of Human Genetic in Chinese and International Context*. Vol. 34. Ole Döring (ed.). Hamburg: Mitteilungen des Instituts für Asienkunde, 187–198, 1999.

Lerdmaleewong, Malee and Francis, Caroline. "Abortion in Thailand: A Feminist Perspective." *Journal of Buddhist Ethics*, 5, 22–48, 1998.

Li, Philip Kam Tao. "Public Attitudes Towards Organ Donation in Hong Kong." *The Hong Kong Medical Diary*, 5, 4, 8–9, 2000.

Lim, Linda Y.C. "Whose 'Model' Failed? Implications of the Asian Economic Crisis." *The Washington Quarterly*, 21, 25–36, 1998.

Lo, Ping-cheung. "Ethical Reflections on Artificial Reproduction Policies in Hong Kong." In *Ethics in Business and Society: Chinese and Western Perspectives*. Becker, Gerhold K. (ed.). Berlin: Springer, 1996.

Mettanando, Bhikkhu. "Buddhist Ethics in the Practice of Medicine." In *Buddhist Ethics and Modern Society.* Charles Wei-hsun Fu and Sandra A. Wawrytko (eds.). New York: Greenwood Press, 195–213, 1991.

Mettanando, Bhikkhu. "Buddha Through the Lens of Mangalasutta. Ideal Model of Systematic Engagement of Buddhism for the Emerging Civil Society of Thailand." In *The Role of Civic Religions in Emerging Thai Civil Society.* Imtiyaz Yusuf (ed.). Bangkok: Assumption University, 51–59, 2004.

Miyata, Hiroaki; Tachimori, Hisateru; Takahashi, Miyako; Saito, Tami and Kai, Ichiro. "Disclosure of Cancer Diagnosis and Prognosis: A Survey of the General Public's Attitudes Toward Doctors and Family Holding Discretionary Powers." *BMC Medical Ethics*, 5, 7, 2004. http://ww.biomedcentral.com/1472-6939/5/7.

Morioka, Masahiro. "Bioethics and Japanese Culture." *Eubios Journal of Asian and International Bioethics*, 5, 87–91, 1995.

Ohi, Gen. "Advance Directives and the Japanese Ethos." In *Advance Directives and Surrogate Decision Making in Health Care*. Hans-Martin Sass, Robert M. Veatch and Rihito Kimura (eds.). Baltimore: Johns Hopkins University Press, 175–186, 1998.

Olds, Linda E. "Chinese Metaphors of Interrelatedness: Re-Imaging Body, Nature, and the Feminine." *Contemporary Philosophy*, 13, 8, 16–22, 1991.

Qiu, Renzong. "Medical Ethics and Chinese Culture." In *Cross-cultural Perspectives in Medical Ethics*. 2nd ed. Robert M. Veatch (ed.). Boston: Jones & Bartlett Publishers, 318–337, 2000.

Qiu, Renzong. "Contemporary China" *Encyclopedia of Bioethics*. 3rd ed. Vol. III. Stephen G. Post (ed.). New York: Macmillan; Thomson, 1693–1700, 2004.

Ratanakul, Pinit. "A Survey of Thai Buddhist Attitudes Towards Science and Genetics." In *Intractable Neurological Disorders, Human Genome Research and Society: Proceedings of the Third International Bioethics Seminar in Fukui, 19–21 November, 1993*. Norio Fujiki and Darryl R.J. Macer (eds.). Christchurch, Tsukuba: Eubios Ethics Institute, 199–202, 1994.

Ratanakul, Pinit. "Bioethics in Thailand." In *Bioethics in Asia*. Norio Fujiki and Darryl R.J. Macer (eds.). Christchurch, Tsukuba: Eubios Ethics Institute, 98–99, 1998.

Ratanakul, Pinit. "Medical Ethics, History of South and East Asia." In *Encyclopedia of Bioethics*. 3rd ed. Vol. III. Stephen G. Post (ed.). New York: Macmillan; Thompson, 2004.

Roetz, Heiner. *Confucian Ethics of the Axial Age*. Albany: State University of New York Press, 1993.

Saddhatissa, Hammalawa. *Buddhist Ethics: The Path to Nirvana*. London: Wisdom Publications, 1987.

Sen, Amartya. "Economics, Business Principles, and Moral Sentiments." In *International Business Ethics: Challenges and Approaches*. Georges Enderle (ed.). Notre Dame: University of Notre Dame Press, 15–29, 1999.

Sy, Peter A. "Doing Bioethics in the Philippines: Challenges and Intersections of Culture(s) and Medicine(s)." In *Bioethics in Asia*. Norio Fujiki and Darryl R.J. Macer (eds.). Christchurch, Tsukuba: Eubios Ethics Institute, 103–106, 1998.

Tan Alora, Angeles and Lumitao, Josephine M. (eds.). *Beyond a Western Bioethics: Voices from the Developing World*. Washington, D.C.: Georgetown University Press, 2001.

Tao, Julia and Chan, Ho-mun. "Should Hong Kong Government Ban All Forms of Surrogate Arrangements – Some Ethical Considerations." (in Chinese). *Values and Society*. Vol. I. K.P. Yu (ed.). Beijing: China Social Sciences Publishing House, 137–155, 1997.

Tao, Julia. "Does it Really Care? The Harvard Report on Health Care Reform for Hong Kong." *Journal of Medicine and Philosophy*, 24, 6, 571–590, 1999.

Tu, Wei-ming. *Confucian Thought. Selfhood as Creative Transformation*. Albany: State University of New York Press, 1985a.

Tu, Wei-ming. "Selfhood and Otherness: The Father-Son Relationship in Confucian Thought." In *Confucian Thought. Selfhood as Creative Transformation*. Wei-ming Tu (ed.). Albany: State University of New York Press, 113–130, 1985b.

Tu, Wei-ming. "Multiple Modernities – Implications of the Rise of 'Confucian' East Asia." In *Chinese Ethics in a Global Context: Moral Bases of Contemporary Societies*. Karl-Heinz Pohl and Anselm Mueller (eds.). Leiden/Boston: Brill, 55–77, 2002.

Tsuchida, Tomoaki. "A Differing Perspective on Advance Directives." In *Advance Directives and Surrogate Decision Making in Health Care*. Hans-Martin Sass, Robert M. Veatch, and Rihito Kimura (eds.). Baltimore: Johns Hopkins University Press, 209–221, 1998.

Unschuld, Paul. "Confucianism." *Encyclopedia of Bioethics*. 2nd ed. Vol. I. Warren T. Reich (ed.). New York: Simon & Schuster MacMillan, 465–469, 1995.

Weber, Max. *Gesammelte Aufsätze zur Religionssoziologie*. Vol. I/II. Tübingen: Mohr/Siebeck, 1920.

Yang, Hsiu-I. "Bad Living than Good Death? A Cultural Analysis of Family Paternalism in Death and Dying in Taiwan." In *Asian Bioethics in the 21st Century*. Sang-yong Song, Young-Mo Koo, and Darryl R.J. Macer (eds.). Christchurch, Tsukuba: Eubios Ethics Institute, 99–107, 2003.

Limits of Human Existence According to China's Bioethics

Reproductive Medicine and Human Embryo Research[1]

Ole Döring

1 The Sky is the Limit?

An often-quoted motif from classical Chinese philosophy in contemporary Chinese bioethics is taken from the *Zhongyong*. This is one of the early Confucian 'Four Books' that dominated orthodox scholarship and the curriculum for higher education in Imperial China since the Song dynasty. The following passage highlights the practical relationship between an agent and the effects of his actions on other entities in general ethical terms.

> Only those who are absolutely sincere can fully develop their nature.
> If they can fully develop their nature, they can fully develop the nature of others.
> If they can fully develop the nature of others, they can fully develop the nature of things.
> If they can fully develop the nature of things, they can assist in the transforming and nourishing process of Heaven and Earth.
> If they can assist in the transforming and nourishing process of Heaven and Earth, they can thus form a trinity with Heaven and Earth.[2]

There is concrete bioethical relevance here. First, there is nothing intrinsically wrong with changing nature's course, because this is part of the human mission, to 'assist' nature ('Heaven and Earth') in its development, on condition that the practice be understood properly. For bioethics the quotation implies a moral context when practice is evaluated in terms of social capability and meaning. Of course, this line of reasoning cannot directly refer to or be reduced to biological science. It is concerned with ethics rather than biology, with practice rather than technology. However, since biology as a science belongs within the explanatory and normative scope of practice in general, the reasoning can be regarded as applicable in this modern context as well, as contemporary Confucian bioethicists have shown. It pointedly contends that,

- action that affects others must be legitimized,
- legitimacy depends on personal virtue (sincerity, empathy) of the actor,

O. Döring
e-mail: ole.doering@ruhr-uni-bochum.de

M. Düwell et al. (eds.), *The Contingent Nature of Life*,

- virtue depends on development of human nature,
- the charge of legitimacy is becoming more urgent apace with the increasing range and quality of human causation or power.

In this light, normative judgments about the limits of human existence, together with the related practice, register as individual acts which elude classification in generalizing ontological or scientific terms. Putting it briefly, to qualify as legitimate judgment about or act upon another entity, be it a human embryo or a comatose patient, that judgment or act cannot be based on scientific grounds solely. It must essentially be based on *the actor's personal virtue*, and on a proper account of the practical status of the affected entity. That is to say, reference to protocols, manuals, codes of ethics or legal works cannot substitute an individual's reasoning and decision making. The actor must proceed on the basis of his/her own virtue and moral responsibility.[3]

An exemplary evaluation will begin with an assessment of the real 'case', in terms of its individual characteristics and the social, (or otherwise relational), meaningfulness of the affected entity, or human being, respectively. Its development will be based on the real experience of social relatedness,[4] facilitated through shared human nature (*xing, si duan*) and empathy (*shu*). Hence, for example, the pregnant woman is privileged to decide, in moral terms, whether her unborn child should be aborted, for as long as she is the only social relation. Moreover, Lee concludes that entities without a potential to develop social relations, such as germ cells, cannot be assessed as if holding an independent moral status (Lee 2002: 175).

2 Cultural Limits of Respect for Human Existence?

In recent publications, Chinese life scientists and ethicists have claimed strong cultural grounds for a liberal environment for research on human beings in China. For example, the director of the Centre for Regenerative Biology at the University of Connecticut Storrs, Yang Xiangzhong, propagates a leading role for China in the life sciences, claiming scientific and cultural exceptionality. In his vision, China is still an 'embryonic nation', yet with a potential to become a world leader in embryo research.

Yang, whose institute has gained prominence by reason of successful experiments in cloning a second-generation cloned cow, in 2004, argues that,

> Therapeutic cloning, stem-cell studies and other research areas that use animal or human embryos are controversial and raise religious and ethical questions . . .
> As a result, many Western governments are weary of such research.
> These issues have led to unsupportive policies for cloning-related research, and the high costs of clinical trials for any proteins developed using this technology have forced many scientists and commercial companies to abandon promising research and to lose out on potentially profitable products.
> China has a cultural environment with fewer moral objections to the use of embryonic stem cells than many Western countries, and . . . it could take a leading role in this field . . .. China has probably the most environment for embryo research in the world . . .. In addition, the

> relatively easy access to human material, including embryonic and foetal tissues, in China is a huge advantage for researchers ….
> Collaborations with China are becoming very attractive to researchers based in the West. While Western researchers focus on animal models, partners at the new Chinese stem-cell research centres could focus on human models.
>
> (Yang 2004)

Yang suggests that in view of China's cultural characteristics, 'these technologies offer unprecedented research and commercialization opportunities for China' (*ibid.*: 210).

In a similar vein, other East-Asian countries are being described as less restrictive about unborn human life for reasons of religious orientation. For example, South Korea, especially after Dr Woo Suk Hwang's claimed cloning feat, is assessed through this lens in scientific publications. 'It is therefore not surprising that it was scientists from predominantly Buddhist South Korea that first cloned human embryos, rather than biologists in the USA or Europe' (Frazzetto 2004).

To those familiar with religious plurality and secular society in South Korea, this straightforward association of bio-policies and Buddhism may be astonishing (Joung 2004; Keown 1995; Schlieter 2004). Cultural accuracy does not seem to be in high demand these days.

The Korean cloning experiments were published in *Science* and drew immense public attention. The editorial of this same issue expressed concern beyond comparative religion or respect for culture. Phrased in a language of ethical humility, yet bordering on relativism, it says,

> The Korean success reminds us that stem cell research, along with its therapeutic promise, is under way in countries with various cultural and religious traditions. Our domestic moral terrain is not readily exportable: US politicians can't make the rules for everyone, and they don't have a special claim to the ethical high ground. And of course, political decisions in the United States may carry real penalties for its own scientific enterprise. Harvard's Doug Melton, a leader in stem-cell biology whose institution has just made a major commitment to it, says it this way: 'Look, life is short. I don't want spend the rest of mine reading about exciting advances in my field that can only be achieved in another country'.
>
> (Kennedy 2004)

As a consequence, some leading universities have publicly pondered how to outsmart existing restrictions within their own country, in order to keep the competitive edge (Cook 2004). Regional differences in quality and degree of ethical regulation of human embryo research incite the quest for loopholes and opportunities, serving as arguments to 'liberalize' research conditions, domestically and internationally. In an almost ironic twist, Dr Hwang once threatened that he would leave Korea with his laboratory and settle in another Asian country, or in England, should the forthcoming bioethics law happen to produce unwanted restrictions (Dreifus 2004). In the meantime, this particular concern has been mended (Joung 2004; Keun-min 2005). Apparently, the cultural rhetoric is part of a campaign of competitive downgrading of ethical standards in the name of biotechnological progress.

While *Science* played the Korean card, competing *Nature* magazine took substantial interest in 'embryonic' China. It promoted biomedical progress in special editions written in Chinese, and launched a series of strategic conferences, such as the *Nature Forum* 2004, 'China-California Connection: A Biomedical Alliance', on 30 March 2004, in San Diego.[5]

Two years earlier, in May 2002, Beijing had hosted China's first international conference on stem-cell research, celebrating the return of the first generation of Chinese experts from US partner institutions, with a strong network of high scientific and economic stakes. A shortage of ethic discussion on the program of the Beijing symposium was noticed. 'The scientists in the audience were not interested in the ethical dimension of their research.... Naturally, researchers that have invested much energy and family fortune into their careers are not likely to jeopardize their future by including bioethical considerations into their research practice' (Sleeboom 2002).

Obviously, reference to culture, which can be more or less accurate, in these cases can be seen as a disguise for vital stakes, such as in competitions among researchers or big publishing houses. More often than not, culture is used as a magic stick to shun rational analysis and reason-guided discourse, which would uncover mundane interest. Thus, 'culture' can be used as an ideological pattern in the fabric of the protective gear of biomedical researchers' proclaimed humanitarian mission. This observation opens a properly critical perspective on cultural arguments. The concerned observer finds it difficult to rely on the assumed authority of cultural self-interpretation.

3 Down to Earth: Regulations

It is generally difficult to assess the ongoing development in bioethics in China. Articles published in distinguished Chinese and international media do not fit easily into a coherent portrait. It is understood that Chinese life scientists have been engaging in human cloning (Mann 2003). Embryologists have transferred human cell nuclei into rabbit eggs (Cohen 2002; Weiss 2002). Research involving destruction of human embryos for the derivation of stem cells is taking place with no apparent public debate (Dennis 2002). Chinese and American reproductive doctors under the leadership of James Grifo have performed a medical experiment on a Chinese woman in Guangzhou, trying the method of somatic cell nuclear transfer (SCNT) (Zhang et al. 2003). This procedure could not be performed in the United States or European countries because of considerations of medical risks and ethical concerns (Weiss 2003).

On the other hand, official statements from China suggest a conservative and restrictive rather than a liberal policy against 'therapeutic' cloning. The Ministry of Foreign Affairs (2003) stated that human reproductive cloning is a 'tremendous threat to the dignity of mankind and may probably give rise to serious social, ethic, moral, religious and legal problems.' It warned that 'The Chinese Government is resolutely opposed to cloning human beings and will not permit any experiment of cloning human beings.' Given this discrepancy between policy and practice, greater effort than in the past can be expected for effective monitoring and examination of therapeutic cloning and other sensitive procedures in China.

As to bio-political control, regulations governing the practice of clinical biomedicine are comparatively more restrictive than those for research. On 1 October 2003, three new administrative regulations concerning reproductive medicine came into effect revising outdated regulations from 2001, they define ethical principles for assisted reproductive technology and human sperm-bank management (cf. Döring 2004b) and explicitly forbid human cloning. The regulations prohibit, in a clinical context, use of the technique of human egg-nucleus transfer for infertility treatment. But they do not cover basic research in vitro, which is under the authority of the Ministry of Science and Technology (MoST) (Leggett 2003).

The creation of human embryos is regulated in detail. Super stimulation – versus ordinary therapeutic stimulation – of ovaries is forbidden, and all procedures depend on informed consent from the donor. For women younger than 35, only 3 embryos may be implanted. Embryos are created for the sole purpose of procreation, but 'leftovers' may be donated to medical research, upon expressed wish of the donors. Commercial dealing is strictly banned (Döring 2004b).

According to these regulations, for example, the Grifo experiments are now prohibited, even though they took place in a renowned Chinese clinic. Besides safety concerns the main ethical objection is that involvement of biomaterial from more than two parents would interfere with accepted concepts of parenthood and family.

Accordingly, the birth of China's first 'fourth generation test-tube baby', announced in Wuhan in February 2004, will remain a singular exception (Li 2004). In fact, although the 2001 version of these Ministry of Health (MoH) regulations did not specifically ban human egg-nucleus transfer, a common moral assumption was that blurring of the 'natural' germ lines of individuals or species would not be acceptable in any clinical setting. Clearly, the positivistic principle 'if an action is not illegal, it is by definition legal' does not apply in China. The fact that policy making lags behind scientific and economic development, in terms of the entire legal and social infrastructure, cannot be interpreted as an expression of cultural values.

Initiatives from Chinese researchers in the life sciences and bio-ethicists have been designed in order to reduce ambiguity and enhance ethics in practice. In 2001, two proposals for scientifically and ethically satisfactory regulations on human embryonic stem-cell research were submitted to China's legislators.

First, an interdisciplinary advisory group from Beijing submitted a draft entitled 'Ethical principles and management proposals on human embryonic stem-cell research' to the two ministries (MoH and MoST) (Döring 2003). The document highlights principles of general respect for human life at all stages, informed consent, safety, and effectiveness of treatment and procedures. Biomedical research should be encouraged, but 'any form of gamete, embryo or foetal tissue trade' is to be banned. The document proposes standardized procedures, professional qualifications, and IRB reviews for all institutions that are involved in human embryonic stem cell research.

In 2001 a local bioethics committee in Shanghai submitted 'Ethical guidelines for human embryo stem-cell research'.[6] The gist of the two documents is quite similar. They share a general esteem of human life; emphasize informed consent, confidentiality, and voluntary donation, set a fourteen-day deadline for the permissible destruction of an embryo, and ban re-implantation of an embryo from research into a human uterus. Both documents also reject cloning for reproductive purposes but accept it for therapies.

In comparison, the Shanghai draft reflects issues of risk control, slightly contrasting the view of the Beijing draft. Interestingly, the first published version of the Shanghai guidelines (adopted 16 October 2001) permits cross-species recombinant experiments (Article 13.5), whereas the revision (of 20 August 2002) has deleted the clause stating that 'fundamental research may be permitted'.[7]

Furthermore, Shanghai allows 'human–animal cell fusion' if it is used for basic non-clinical research only. Any combination of human cells with animal cells for clinical purposes – e.g. for implantation into the human body – is prohibited (Article 14.4). Consequently, the creation of cross-species hybrids is allowed as long as they remain in vitro. Thus the proposed guidelines support Shanghai's local researchers who are engaged in such projects. In the absence of relevant national legislation, it is both noteworthy and consistent that the Shanghai proposal is cited formally as ethical reference in scientific publications, even though they are not accepted by MoST or MoH (Chen et al. 2003).

It should be noted that the Shanghai and Beijing guidelines, in line with all other related regulations of the last seven years, explicitly prohibit coercion of women into becoming pregnant and then choosing abortion, or into manipulating the method and time of abortion. These prohibitions obviously address a current (mal)practice which violates the requirements of informed consent. As far as ethical priorities are concerned, the guidelines seem to pay relatively greater attention to donors than to the protection of early human lives. Rather, improvement of the practice of informed consent and patient protection is balanced with the freedom of research.

China has accelerated and refined her bioethical regime, starting from a general and vague prohibition of human cloning in 1998. On 13 January 2004 came into effect the 'Ethical guiding principles for research on human embryonic stem cells' issued by the highest national authorities on the subject, MoST and MoH. These principles prohibit any research aimed at human reproductive cloning and confirm that 'it shall be prohibited to hybridize human germ cells with germ cells of any other species.' At the same time, they expressly permit research on 'surplus embryos' following in vitro fertilization (IVF), on foetal cells and on cells created by 'somatic cell nuclear transfer'. However, no embryo used for research may be cultured for more than fourteen days, and the sale of 'human gametes, fertilized eggs, embryos and foetal tissues' is prohibited.

Human embryos used in research are typically produced in IVF clinics close to the research establishment. However, the source of human embryos is often described as neither reliable nor constant. One reason for this is that acquisition of human oocytes is difficult. Tissue from aborted foetuses is widely used in a number of the research establishments. In spite of a new licencing system for IVF clinics, established since the fall of 2001, no reliable countrywide data are available about the number and quality of procured embryos or oocytes, or the donor's profile.

Information gathered in several interviews suggests that the availability of tissue for research depends on the reputation of the respective clinic and research unit.

China's IVF policy illustrates that attention is being paid to the potential social impact of biomedicine rather than to embryo protection. The revised MoH regulations of October 2003 forbid doctors to help single women get pregnant through assisted reproductive technology such as embryo transfer and artificial insemination. In a regional policy trial, the authorities of Jilin (Liaoning province) in 2002 gave single women the right to get pregnant via assisted reproductive technology. This initiative stirred moral debate about morality, the purpose of family and sexual activity. In effect, this debate was decided in favour of conservative moralists. The national Health Ministry overruled the Jilin regulations.

The situation is summarized in a recent report from Great Britain, 'the Chinese authorities are anxious to establish a regulatory and ethical framework in relation to human embryonic and foetal stem cells that reflects emerging international standards. It was not . . . to determine how fully such standards are accepted and reflected in practice outside the centres of international excellence' (Du et al. 2004).

In fact, in the bioethical field China is working towards a recognizably liberal European framework of regulation, in several cases actually based on the British House of Lords select committee's recommendations (Munro 1988).

4 Moral Demarcations

However, there are discernible Chinese peculiarities to be accounted for. In general terms, the relevant documents of bioethics reveal two lines of moral demarcation. I want to use this differentiation as heuristic reference frame for cross-cultural comparison and as a method to organizing the discourse on normative issues.

(1) First, a 'Chinese Rubicon' defines the beginning of a human's worthiness of protection. The transfer of an embryo from the petri dish into the uterus seems to demarcate the line between research and medical or invasive treatment. Manipulation in vitro is permitted, but implantation into the female system remains a taboo. The use of cloning technology for human reproduction is most unlikely to be endorsed in China.

This Chinese Rubicon is based on a strong notion of natural *purity* and *dignity*, which can, beyond ideological conservatism, be traced back in the history of Chinese philosophy to the neo-Confucian 'Principles School' (*Lixue*), developed in response to invading Buddhism during the Imperial Tang and Song Dynasties (Munro 1988). Reference to other philosophical sources of naturalism are made by popular Confucians, who associate it with the Han-dynastic amalgamation of cosmological, social-moral and political concepts, especially a sexualized interpretation of Yin and Yang (Qiu 2003).

In this light, 'assisting nature' through biomedical or other means includes a quest for moral purification of humanity *and* the world. It stretches medicine on the scale of a moral-teleological axis, while taking certain (controversial) *natural* features as undisputedly given. Action that is driven by deviating interests cannot be accepted. Both aspects, moral motivation and natural constitution, are interconnected. The natural constitution of an embryo may be altered if certain conditions apply, but an altered embryo may not become part of the causal chain of the social network or an individual's human social-biological system. Hybridization and other forms of manipulation must remain within the enclosure of the dish, even if meant for therapeutic purposes (Abbott and Cyranoski 2001).

This *Imperative of Purity* has already gained regulatory force in practice. This is expressed in the prohibition of nuclear transfer or ooplasma transplantation in reproductive medicine.

(2) The second line of moral demarcation according to Chinese bioethical regulations has a *legal* form. It is subject to adjustment due to political or social processes. This 'Chinese Limes' in bioethics is defined in terms of the social-moral dimensions that constitute the human being. As soon as a human is born into the social environment, the full power of legal protection begins to apply. This is the onset of a gradual development of the individual's social career, in the course of which nobody may be manipulated or killed (legal exceptions apply in cases such as the death penalty). According to this view, an embryo is only a human *life* but not a societal entity during the early phases of development when there is no psycho-emotional and physical relationship with a mother. Hence it can be seen as commodity for 'high-ranking medical purposes'.

I shall not discuss the relation between the two demarcations, nor their potential impact on different ethical concepts. Nor will I analyse their roots in China's culture and society. These are important hermeneutic tasks beyond the scope of this paper. From a cultural perspective, the demarcations cannot be played out against each other. And certainly, neither of them can be neglected. Together they contribute to the fabric of cultural context.

Considering the essentially pragmatic character of these policy regulations, they should not be expected to convey particular moral stakes or to express deeper cultural reflection. For example, primary concern for the protection of women and the security of scientists are highlighted, though one would hesitate to regard these as particularly strong points of Chinese culture. However, it should be considered that these norms still transport the signature of a deeper cultural purpose. Chinese society provides a strong value basis resting on Confucian, Buddhist, Taoist and secular humanistic ideas. There is notable debate about issues such as cloning and reproductive medicine. Except for a small number of expert forums, however, discussions take place in a largely scattered manner. Due to systematic political constraint there is no self-sustaining process of controversial civil debate (not to mention a *discourse*). This makes it more difficult to assess the value profile and conceptual landscape of China's relevant moral configuration.

The primary question is not about human dignity and the demand that it be respected. The question is *whether and in what sense* a certain entity is to be regarded as a human being in its own right. An influential group of conservative scholars[8] who often refer to Confucianism, for example holds that humans accumulate their moral status through social relationship and merits. In the absence of such qualities, someone is not regarded as a human being with full protection rights. This applies to early human life forms as well as to cases of adults who lose their dignity through moral deprivation.

5 Two Positions

Personal opinions of leading researchers are valuable sources for assessing regulations for the treatment of human embryos in practice. The elite of Chinese embryologists and fertility doctors is involved in the making of laws and the related counselling. And they directly account for the effect on the respective practice. I sketch two such views.

Dr Sheng Huizhen is a prominent embryologist at Shanghai's Xinhua hospital. According to her, experiments that hybridize human and rabbit's cells are morally acceptable for genuinely ethical reasons. Combination of genetic material from a human nucleus and animal mitochondria technically generates a *new species*. She believes that the biological differences between this species and a human can be neglected for the purpose of scientific modelling. Ethical objections do not apply here, because such an entity has no potential to develop to maturity or to procreate.[9] This procedure promises morally inexpensive advances in basic research. Sengh expresses the hope that in the long run these experiments are going to reduce the need to use human embryos for stem-cell research. From a biochemical point of view, Sheng opines that the fusion of sperm and egg cells marks the only plausible starting point for the beginning of a new human life, and it is only than that the issue of protection arises.

A different moral perspective is expressed by China's most renowned fertility doctor, Dr Lu Guangxiu. She engages in embryo research, cloning and stem-cell research in a clinical setting at Changsha's Xiangya Second Medical Hospital. Dr Lu holds that during the first fortnight of existence, an embryo is tissue matter. It should be compared with blood or body cells in this regard. Being a carrier of parental genes with the potential to become a full human entity and, at the same time, as part of a woman's body, it cannot be utilised at will. However, the principles of avoiding harm and protecting self-determination and personal integrity of the woman or couple overrule other moral concerns.

Dr Lu regards the imperative of protecting female donors as an insurmountable and unobjectionable obstacle towards the procurement of human eggs as a basic resource for an entire line of research. She firmly rejects any activity or clinical policy that would actively promote donation of eggs.[10] In China such a model is inapplicable on cultural grounds. Consequently, Lu criticizes the controversial cloning experiments of Dr Hwang in South Korea.

6 What Constitutes the Limits of Human Existence?

The foundations and limitations of the value or worthiness of protection of a human being remain heterogeneous, depending on the meaning of *humanity* and the chosen moral or ethical approach. There is no cultural consensus in China on this matter.

There is a tradition of co-existence of diverse philosophical, religious and political opinions and resulting dispute.

Common sense does not help to harmonize the disagreement either. Customary reference to the age of a newly born as *yi sui* (one year) suggests impregnation as the beginning of existence; on the other hand, many people regard humanity as a matter of *merit* acquired through accumulated 'good' social practice. Moreover, 'pre-natal education' has become a mushrooming fashion, (relevant literature can be found in imperial libraries, cf. Dikötter 1995). Traditional Chinese and imported manuals from the USA promote techniques such as diet, prenatal exposure to music, culture or conversation and *feng shui* (geomancy). On the other hand, abortion is still the most prevalent means of birth prevention.

From a perspective of philosophical Confucianism, the limits of human existence should be assessed in the light of two interrelated moral dimensions. First, it reflects the agent's moral character, in particular the capacity to make moral judgments about another entity (such as a human being at the end or at the beginning of life). Second, it explores the nature of this entity, through the relational connection between agent and object. This latter perspective would allow for different material judgments about the moral and practical characteristics of an entity, concerning both the (social and legal) Limes and the (moral) Rubicon. The first issue, from this interpretation of a Confucian view, affects the nature, theory and methods of ethics and of bioethics in particular, without primary interest or effort invested in such formal categorizations. So far, this asymmetric constellation in the cultural and philosophical debate has hardly been addressed in bioethics, disregarding as it does the ethical and methodological impact.

Notably, on the one hand this Confucian approach can illustrate the problem of moral judgment in cases that do not correspond to (any possible) moral experience. On the other hand, it carries considerable potential for an emerging, not yet mature, critique of action (*Handlungskritik*), as it had continuously been envisaged but never was theoretically elaborated in full by proponents of Confucian ethics.

The Confucian approach does not offer clear-cut definitions of the limits of human existence in bioethical terms. However, this hesitation, ambiguity or reflected modesty can be turned into a theoretical asset and interpreted as expressing a systematic ethical point. That is, only the accomplished moral character (*junzi*) is in a position to understand practice properly. However, it is impossible to establish independent and objective standards for the evaluation of such a figure. Thus it becomes a regulative idea. There is no alternative source of insight, no special qualification, be it scientific or another, which would corroborate high-claiming practical judgment in sensitive issues and, at the same time, contradict the judgment or practice of a *junzi*.

The riddle of normative orientation through asserted model characters is a genuine thread in Confucian philosophy and has contributed to a regular scheme in the description of Chinese societies (cf. Bakken 1994). It cannot be dissolved or disregarded without dismissing essentials of this school. One established strategy in response to this challenge is to focus on practice as the constructive, (that is, the synthesizing), process of *sustaining the practical tension* between moral intuitions and related accomplishments. It keeps the debate alert to the constant demand to avoid hubris and to counter totalitarian claims or relativistic in-group morality on the part of specific communities. According to historical evidence, high-claiming moral communities tend to invite or support paternalistic, authoritarian and oppressive regimes. A balanced Confucian approach would become a source of criticism rather than a pretender to 'truth'.

Obviously, even in its present shape, this nascent Confucian approach tends to turn the conventional language and perspective of mainstream bioethics upside down. As I see it, it can serve to adjust purpose and commitment in bioethics to concerns of humanity. For example, a Confucian aspiration offers a conceptual reference frame that would accommodate Kantian maxim ethics, narrative, feminist and anthropological avenues to bioethics. Moreover, it exhibits a distinct normative profile, challenging legalistic, utilitarian, hedonistic, materialistic, ideological and positivistic frameworks. This would be consonant with its aspirations since the times of the classics.

The demand for responsibility, sincerity and moral development on the part of the actor places the moral burden on the individual actor. At the societal level, this requires a set of basic preconditions so that the demands can in fact be met in practice. It supports a culture of moral experience and communication, focusing on comprehensive education and life-long learning. Obviously, present-day China has no such environment. Hence there is reason to be cautious about Confucianism as a key to interpret Chinese bioethics. Rather, it is recommended as a *critical* framework.

To sum up, Confucian ethics focuses on individual encouragement and procedural advice on how to assess the limits of human existence. It takes no particular interest in legal form or ontological status, but on the moral substance of practice, holding the mirror to reflect each individual's motivation and performance on the full scale of practice. Consequently, any action in the medical context is regarded as a challenge, because it interferes with the course of nature and humanity. Thus it assumes a practice that can only be legitimate when intervention takes place in terms of 'assisting Heaven and Earth'. Positive ethical standards cannot substitute the individual's responsibility or judgment. This is a very high claim, on a par with the Kantian description of the 'good will' – and as impossible to be positively established.

By emphasizing the part of the causal initiator (i.e. the actor), the issue of the moral status of human beings is neither solved nor addressed directly. In effect, the 'embryo matter' is brought to the level of individual social relation clusters, and taken out of the limited charge of the public and the experts. Thus it leaves room for a plurality of moral practices without inviting ethical relativism.

Probably the most uncertain element of this approach is, How can it be integrated in a system that is built on general, positive legal norms? Confucianism attempts to prevent practice from creating situations of juridical concern in the first place. It does not offer juridical decisions. It frames our institutions in terms of meaning and purpose. Thus it does not directly challenge or contradict any given normative system.

Moreover, from this perspective, the specific quarrels over biomedical criteria for human dignity appear to be of secondary importance, though not marginal.

7 Demarcation and Perplexity or Common Ground

It is the purpose of the above argument to counter expectations of an 'alien' moral culture in China, which would be fundamentally at odds with moral views in European and North American countries, while maintaining some peculiarities of the Chinese moral position. The moral landscape between the Chinese *Limes* and *Rubicon* is wide and shattered. Just as it is in Europe, the ethical and legal frameworks for bioethics in China are influenced by interests with historical and political contingencies (cf. in detail: Döring 2004a).

Accordingly, the explanatory merit of culturalistic concepts regarding the interconnectedness of culture and ethics is limited. Moral dissent is to a significant extent culturally immanent. What matters is how peaceful debate and cohabitation can be organized and reflected, in a manner that sustains local, regional and global flourishing. In other words: the challenge is to cultivate diversity and make it fruitful.

Considering the adaptability of international bioethical regulations in China, no major obstacles should be expected from the side of China's politics and administration, as far as technical issues and standards are concerned. The relevant Chinese policies can be understood as responses to the pragmatic demand for harmonization of international standards. In view of the practical implementation and monitoring of bio policy, however, concern is in place. Owing to the well-known obstacles in developing a modern civil society with a state and culture of law in China, the present situation inside the clinics and laboratories as well as the related exchange of resources and information requires particular scrutiny and alertness. Full assurance of respect of the powerful towards each one's basic rights, born or unborn, is still not necessarily guaranteed.

Finally, it should be noted that the limited focus of this paper is no substitute for a deeper cultural and social-political analysis. It indicates some major tendencies in the minds of leading figures in the life sciences and bioethics in China. Observers should appreciate that the currently discussed ethical standards are embedded in the dual purpose of facilitating the development of life sciences by increasing their regular performance and raising the level of acceptance within the public. These attempts respond to the demand to regulate the life sciences after scandals and reports about irregular experiments have irritated the population, while the transforming society continues to be in flux on all levels.

If we want to learn more about the 'Limits of Human Existence According to China's Bioethics' on a deeper level, I recommend serious engagement in related cross-cultural research projects, based on mutual respect. The agenda is on the table.

Notes

[1] This paper is a substantially revised translation of my 2005b article.
[2] *Zhongyong*, Chapter 22. Translation is taken from Chan 1963: 107–108. It is quoted like this, e.g., by philosopher Lee Shui-chuen (Li Ruiquan) (1999).
[3] Cf. a more elaborated analysis in my 2005a article.

[4] This theme is elaborated in Tao (2002).
[5] http://www.kintera.org/site/pp.asp?c=eiJVJ5ORF& b=10904.
[6] Ethics Committee of the Chinese National Human Genome Center at Shanghai (2004).
[7] Cf. Bioethics Committee, Southern China National Human Gene Research Center (2001).
[8] They share a strong esteem for the ties of family and community, restrictive opinions on sexual morality, paternalistic political ideas, and a sense of Chinese cultural pride.
[9] Personal communication Dr Sheng in Shanghai, 25 May 2004.
[10] Personal communication Lu Guangxiu in Changsha, 22 May 2004.

References

Abbott, Alison and Cyranoski, David. "China Plans 'Hybrid' Embryonic Stem Cells." *Nature*, 413, 339, 2001.

Bakken, Börge. *The Exemplary Society: Human Improvement, Social Control and the Dangers of Modernity in China*. Oslo: Department of Sociology, University of Oslo, 1994.

Bioethics Committee, Southern China National Human Gene Research Center. "Proposed Ethical Guidelines for Human Embryo Stem Cell Research." *Zhongguo yixue lunlixue* (Chinese Medicine and Philosophy) 6, 8–9, 2001.

Chan, Wing-tsit. *A Source Book in Chinese Philosophy*. Princeton: Princeton University Press, 1963.

Chen, Ying; He, Zhi Xu; Lu, Ailian et al. "Embryonic Stem Cells Generated by Nuclear Transfer of Human Somatic Nuclei into Rabbit Oocytes." *Cell Research*, 13, 251–64, 2003. http://www.cell-research.com/20034/2003-116/2003-4-05-ShengHZ.htm.

Cook, Gareth. "Stem Cell Center Eyed at Harvard? Researchers Seek to Bypass US Restrictions." *Boston Globe*, 29 February 2004. http://www.boston.com/news/local/articles/2004/02/29/stem_cell_center_eyed_at_harvard?mode=PF

Cohen, Philip. "Dozens of Human Embryos Cloned in China." *NewScientist.com*, 6 March 2002.

Dennis, Carina. "Stem Cells Rise in the East." *Nature*, 419, 334–336, 2002.

Dikötter, Frank. *Sex, Culture and Modernity in China*. London: Hurst, 1995.

Döring, Ole. "China's Struggle for Practical Regulations in Medical Ethics." *Nature Reviews Genetics*, 4, 233–239, 2003.

Döring, Ole. *Chinas Bioethik verstehen. Ergebnisse, Analysen und Überlegungen aus einem Forschungsprojekt zur kulturell aufgeklärten Bioethik*. Hamburg: Abera, 2004a.

Döring, Ole. "Chinese Researchers Promote Biomedical Regulations: What are the Motives of the Biopolitical Dawn in China and Where are they Heading?" *Kennedy Institute of Ethics Journal*, 14, 1, 39–46, 2004b.

Döring, Ole. "Der Embryo in China: eine Frage des Charakters?" In *Der Zugriff auf den Embryo. Ethische, rechtliche und kulturvergleichende Aspekte der Reproduktionsmedizin*. Reihe Medizin-Ethik-Recht Band 5. Fuat S. Oduncu, Katrin Platzer, and Wolfram Henn (eds). Göttingen: Vandenhoeck & Ruprecht, 126–145, 2005a.

Döring, Ole. "Reproduktionsmedizin und traditionelle Werte im modernen China." In *Biomedizin im Kontext. Beiträge aus dem Institut Mensch, Ethik und Wissenschaft 3*. Sigrid Graumann und Katrin Grüber (eds). Münster: LIT-Verlag, 2005b.

Dreifus, Claudia. "2 Friends, 242 Eggs and a Breakthrough: A Conversation with Woo Suk Hwang and Shin Yong Moon." *New York Times*, 17 February 2004. http://www.nytimes.com/2004/02/17/ science/17CONV.html?pagewanted=all

Du, Jiansheng; Johnston, Geoff; Minger, Stephen et al. "DTI, Global Watch Mission Report: Stem Cell Mission to China, Singapore and South Korea." September 2004. http://www.oti/globalwatchonline.com/online_pdfs/36206MR.pdf

Ethics Committee of the Chinese National Human Genome Center at Shanghai. "Ethical Guidelines for Human Embryonic Stem Cell Research." *Kennedy Institute of Ethics Journal*, 14, 1, 47–54, 2004.

Frazzetto, Giovanni. "Embryos, Cells and God: Different Religious Beliefs have Little Consensus on Controversial Issues such as Cloning and Stem-Cell Research." *EMBO Reports*, 5, 6, 553–555, 2004.
Joung, Phillan. *Ethische Probleme der selektiven Abtreibung: Die Diskussion in Südkorea*. Bochum: Zentrum für Medizinische Ethik, 2004.
Joung, Phillan. "Forscher in der Rolle der Schurken. Das südkoreanische Klonexperiment – ein Lehrstück bioethischen Debattierens." *Neue Zürcher Zeitung*, 27 March 2004.
Bae, Keun-min. "Bioethics Law May Face Court Review." *The Korea Times*, 6 January 2005.
Kennedy, Donald. "Editorial: Stem Cells, Redux." *Science*, 303, 5664, 1581, 2004.
Keown, Damien. *Buddhism and Bioethics*. New York: St. Martins Press, 1995.
Lee, Shui-chuen (Li Ruiquan). "A Confucian Perspective on Human Genetics." In *Chinese Scientists and Responsibility*. Ole Döring (ed). Hamburg: Mitteilungen des Instituts für Asienkunde nr. 314, 187–198, 1999.
Lee, Shui-chuen. "A Confucian Assessment of 'Personhood'." In *Advances in Chinese Medical Ethics: Chinese and International Perspectives*. Ole Döring and Chen Renbiao (eds). Hamburg: Mitteilungen des Instituts für Asienkunde. No. 355, 167–177, 2002.
Leggett, Karby. "China Has Tightened Genetics Regulation – Rules Ban Human Cloning. Moves Could Quiet Critics of Freewheeling Research." *Asian Wall Street Journal*, A1, 13 October 2003.
Li, Liu. *Science and Technology Daily (Keji ribao)*, 13 January 2004. http://www.softcom.net/webnews/wed/bo/Uchina-testtube.Rh3E_EFQ.html.
Mann, Charles C. "The First Cloning Superpower: Inside China's Race to Become the Clone Capital of the world." *Wired Magazine*, 11 January 2003.
Munro, Don. *Images of Human Nature: A Sung Portrait*. Princeton: Princeton University Press, 1988.
Qiu, Renzong. "Klonen in der biomedizinischen Forschung und Reproduktion: Ethische und rechtliche Zwänge – Eine chinesische Perspektive." In *Klonen in biomedizinischer Forschung und Reproduktion*. Ludger Honnefelder and Dirk Lanzerath (eds). Bonn: Universitäts-Verlag, 329–342, 2003.
Schlieter, Jens. "Some Aspects of the Buddhist Assessment of Human Cloning." In *Human Dignity and Human Cloning*. Rüdiger Wolfrum and Silja Vöneky (eds). Dordrecht; Boston; Lancaster: Nijhof; Brill Academic Publishers, 2004.
Sleeboom, Margaret. *IIAS Newsletter*. 29 November 2002, 49. http://www.iias.nl/iiasn/29/IIASNL29_49.pdf
Tao, Julia Po-wah, Lai. "Global Bioethics, Global Dialogue: Introduction." In *Cross-Cultural Perspectives on the (Im)Possibility of Global Bioethics*. Tao, Julia Lai Po-wah et al. (eds). Dordrecht: Kluwer, 2002.
Weiss, Rick. "Stem Cells in Human Blood are Reported Potential Help in Tissue Repair, Regeneration Cited." *Washington Post*, A08, 7 March 2002.
Weiss, Rick. "U.S.-Banned Fertility Method Tried in China: Woman Became Pregnant Through Egg Transfer Technique but Lost All Three Fetuses." *Washington Post*, A10, 14 October 2003.
Yang, Xiangzhong. "An Embryonic Nation." *Nature*, 428, 210–212, 2004.
Zhang, Jianhong et al. "Pregnancy Derived from Human Nuclear Transfer." *Fertility and Sterility*, 80, Suppl. 3, 56, 2003.

There is the World, and there is the Map of the World

The Ethics of Basic Research

Laurie Zoloth

It is a commonplace in bioethics, nearly a ritual act, to begin with the words: "Humanity is facing a crossroads." I have done it myself, and it is fair enough to say such a thing, unless, of course, one reflects on history. Then, we could begin in this way: "Humans live at crosses in the road. It is where that species can be most commonly found; it is where it is happiest." We are creatures that love surprises, that survive failure, that have an odd capacity to dream things up. It has seemed to the collective us that this is the case in science since the time that the word "scientist" entered the vocabulary in the mid-1800s, but it has been true far longer: humanity has stood at many crossroads in which biological changes and decisions about how to interact with the environment (Eat it? Kill it? Plant things? Burn it? Flee? Move north across the ice bridge?) have made irreversible changes in what it meant to be human. We are no different in the gravity of this moment. At stake is an essential question: May we make the world? If the response is yes, then how do we retain our humanity, our freedom, and our capacity for justice? And if the answer is no, then the questions remain.

Let me begin with stories:

1 Story Number One: Autumn 1986

Majit Murandi was born far too soon, to a devout Hindu couple who had come to America to go to graduate school. He was their first child, and when he came into the world at 27 weeks, struggling to breathe, we, his doctors and his nurses, struggled to decide how to care for him – and whether to care for him. In 1986, the chances for Majit were bleak. We did not have the drugs to mature his lungs, the ventilator that kept him alive was, by current standards, primitive, and it scarred his lungs with each small, ragged machine-breath it pumped into him. He looked at us, black raisin-eyed boy, and curled his soft hand, tiny stars, around his mother's fingers, and his parents were smitten. They were serious people: educated northern Indians, with

L. Zoloth
e-mail: lzoloth@northwestern.edu

M. Düwell et al. (eds.), *The Contingent Nature of Life*,

a large family that telephoned from India in the middle of our night shift. Majit's mother came every single day, to oil his body, to tie red string[1] on his hands, her mechanical engineering book balanced on her lamp, her yellow *sari* glowing in the eerie light of the open table with the blue heat lamps. Majit did fine, actually, in relative terms. His brain intact, he grew slowly. He was just terribly small and his muscles too weak to move his lungs, and by the time they were strong enough – when he was beginning to reach out and to smile, and to like the color red – he was six months old and his scarred lungs on the X-ray looked like ground glass. He needed new lungs. He lived in the way babies and their families live in the neonatal intensive care unit, in a big-boy crib in the quiet, bright corner of the room, until he died. That night, I begged the funeral director, and he agreed, to look the other way for a few hours so that the family could secretly take their child's body home and wash him and dress him in a white dress, adorn him traditionally, and give him back to the funeral home covered in red rose petals. We had done the very best for him. Majit was given the pinnacle of our best efforts in medicine in 1986, especially since his family were kind, good cooks, and always present in the unit. We used well over a million dollars of the HMO[2] funds. He and his family taught neonatal medicine to six rounds of medical students and several dozen medical residents, and, since his family had asked, we came and stood at his funeral, precisely the way we stood there at his bedside, witnesses, but nearly helpless.

2 Story Number Two – Winter 2004

To live as an observer of science, being a teacher of bioethics, is like living in Chicago in February. As I write, the snow is falling and the world was still waiting, as skeptics know, the long time Chicago has to wait until spring; but when I pick the children up after work, something subtle has changed in the road – the light has shifted just a bit, its first slant toward summer.

The news in health care is grim in the winter of 2004. There is avian flu, and cholera, mad cows, SARS,[3] and ricin,[4] whole health care systems in Africa stagger under the burden of AIDS and malaria, resistant TB lurks in east Russia, and 43 million Americans lack decent access to health care. But the announcement is made by a South Korean team that they have made a stem cell[5] line of human cloned cells, and it is the kind of sudden, subtle, shift of light. In a lab in Korea, 242 human eggs donated by 16 women volunteers are coaxed through an intricate, delicate process toward becoming blastocysts,[6] and of the 20 that result, one is turned into the first successful human stem cell line. The achievement is significant, appearing in one of the leading peer-reviewed journals, *Science*, for it makes good on one of the first necessities of stem cell therapeutics – that tissue could be made that is a direct genetic match for patients in need of transplants. It is important as well for basic research in stem cell biology, for it offers a window into understanding how cells grow, develop, and change – the basis for all of human disease. It is only one more step on the very slow, iterative road of genetic medicine, only one step closer toward a tantalizing goal, on a path in which each tentative step matters a great deal. Dr. Hwang and Dr. Moon, and the technicians and graduate students that came with them, sit quietly behind the table in Seattle. They tell the press – huge numbers of press, from all over the world – that they have cloned one embryo, made one set of human stem cells, from ova and cumulous cells taken from 242 women volunteers. It is only one line, in one lab, in a small country that Americans rarely discuss, and it is, in the words of Jim Battey, director of the stem cell center at NIH (National Institutes of Health), only a "small incremental technical advance." Yet it is the first story on every major network and on the front page of every urban newspaper in the world. Why? What were we witnessing at this time? Indeed, what is presented will turn out to be largely a hoax, but why were we so deeply taken in by this event?

We are watching medicine, which increasingly means – and this is still a good thing – a broad political and social debate about biotech occurring all over the world. The Koreans, too, had debated stem cells and cloning and passed a law in 2003 that set up restrictions on human-assisted reproductive cloning, allowed human-assisted regenerative cloning for research and therapy, and allowed funding for a stem cell research center. The debate in Korea is different than in America. Buddhism and Confucianism are important faith traditions in which embryos are not seen as stable/sacred, for there is a belief in reincarnation and each embryo, each being, contains within it all possibilities, many endings and beginnings. Further, there is a cultural impermissibility of cadaveric or living organ donation (only 30 were done in 2003); hence, there was a strong popular desire to create alternate research in tissue transplantation. Conflicts of interest theoretically, were dealt with directly: a patent for the techniques was obtained, but the researchers repeatedly insisted that they had no "commercial intention," that the goal was basic research. Principal investigators, we were told, cannot make money from inventions. They could, we later understood, become wildly famous.

Why would this research have been such important research? Because for all of the promise of stem cell therapies, the major obstacle for serious use will be how to make stem cell tissues match up with sick or injured people. Like any sort of human transplant, the big problem with stem cells is that the tissue or organ will be swiftly and painfully rejected, and cloned tissue might just make a nearly perfect match, so it is one of several possible roads toward this goal. We are years away from any human use, but this first serious, peer-reviewed experiment seemed a critical proof of the concept, and with more research and more careful study, we thought we might be heading toward fundamental changes in how we treat human health. It would have been vitally important because it seems to open the door to an avenue of basic research – how to study the ways human diseases are expressed at the very earliest stages.

Also, in part, the case was important because of its impact on our sense of the new. Science seemed now global; pluralistic views contended for meaning, in a world famously called "flat" it seemed a triumph of the freedom of inquiry, and that many religions could be fully engaged in the fray. The science was merely iterative, technical, and well practiced. But it was world-stoppingly important in that it seemed to have achieved a wide approval, even enthusiasm (something that seems even to be true when the work was revealed as a fraud, so deep was the hope it engendered). It had captured a sense of consensus that seemed sincere: built governmentally, a controversy that could be mediated, public support of and pride in work, clarity about non-commercial interest, the conflict of interest/appearance of conflict avoided, and finally, it seemed coherent, scientific, and humble – the very humility of the researchers made it so difficult to believe later, in their duplicity.

The reaction was swift. The Vatican statement called the work "murder" and "akin to Nazi science." Others called it "frankly immoral" and the beginning of the slippery slope. There was a noticeable repositioning of political forces over both the applications of science (and fears of the future) and over basic science itself (forbidden knowledge).

But surely it is not just the science in this story that drove the ethics debate. Rather, it is the fundamental understanding, and destabilization of complexity itself: that the etiology of disease is best explained now, not by humors, or germs, or even environment, but by the complexities of genetics, protein interaction, cell signaling pathways, and events in time and space that trigger how this happens within the organism. This idea that one's fate is neither entirely in one's control, nor subject to morality or desire, challenges what we believe we are and what we believe about ourselves in nature.

Our challenge, and this means each of us – scientist, citizen, congregant, critic, and enthusiast – will be how to live decently and fairly in a complex world of difficult moral choices. Can stem cell research yield therapies that could help millions who now suffer? Will it yield cures for diabetes, Parkinson's, spinal cord injury? Who can yet know? That is the thing about stories in which one cannot turn to the back of the book. If this research were able to help even some, that might be light enough in the wintering world. Stem cell research and cloning for such research seems to be only a possibility now; still, even after the failure of Korea, scientists yearn to make it a fact. Where that road might turn us is of course uncertain, but what is certain is that despite the failure of the Koreans, no one is yet ready to abandon the road itself.

The calling of medicine and research science is one that urges physicians and researchers to see the full humanity of each patient – no matter how small, how vulnerable, how expendable. It is a new century now, and we stand at the edge of a new terrain in medicine. 27-week-old babies seem relatively mature now that we are discussing the chances for the 22-week-old ones. We have nasal CPAP,[7] surfactant treatment,[8] steroids, and far better equipment for helping these children. We understand far more, we are driven by research possibilities, and all along (1987–2004) we have been learning the complex code of genetics; we mapped the human genome. When I read about human embryonic stem cell research and see such cells that have been directed to become lung tissue, I never fail to think about Majit, born too soon for any of this. I remember the ipseity,[9] the self that I saw emerge in his dark eyes, the self that learned to turn over, to lift his head and reach for the world. The research imperative is largely driven by such haunting memories – of children with cancer, or children with diabetes, or cystic fibrosis. If you ask many of the researchers in the emerging fields of biotechnology, they will tell you a story like the one about

Majit, about a patient met in training, or a family member. They will tell you of the suffering of the world, and its fundamental injustice.

Yet, when bioethicists speak of biotechnology, we hardly begin in the brokenness of the present. We nearly always begin with an account of the future.

Consider the following, from one of our most thoughtful critics, Dr. Leon Kass:

> Let me begin by offering a toast to biomedical science and biotechnology: May they live and be well. And may our children and grandchildren continue to reap their ever tastier fruit – but without succumbing to their seductive promises of a perfect, better-than-human future, in which we shall be as gods, ageless and blissful. As nearly everyone appreciates, we live near the beginning of the golden age of biotechnology. For the most part, we should be very glad that we do.... Yet, not withstanding these blessings... we have also seen more than enough to make us anxious and concerned. For we recognize that the powers made possible by biomedical science can be used for non-therapeutic or ignoble purposes, serving ends that range from the frivolous and disquieting to the offensive and pernicious. These powers are available as instruments of bioterrorism (e.g. genetically engineered drug-resistant bacteria or drugs that obliterate memory); as agents of social control (e.g. drugs to tame rowdies, or fertility-blockers for welfare recipients); and as means of trying to improve or perfect our bodies and minds and those of our children (e.g. genetically engineered super-muscles or drugs to improve memory). Anticipating possible threats to our security, freedom, and even our very humanity, many people are increasingly worried about where biotechnology is taking us.... We are concerned that our society might be harmed, and that we ourselves might be diminished, indeed in ways that could undermine the highest and richest possibilities of human life
>
> (Kass 2003).

There is much to say about this long quote. I wanted to give Dr. Kass a fair, non-sound-bite chance to have his concern heard, including the retelling of the "Genesis" story, the recasting of the researcher as the evil tempter, and the eschatological[10] anxiety at the heart of his argument. But Dr. Kass is not alone: Charles Rubin,[11] notes that "the cutting edge of modern science and technology has moved, in its aim, beyond the relief of man's estate to the elimination of human beings." Francis Fukuyama (2003) fears that we have created a post-human future that will leave us soulless, stripped of the essential nature that makes us free.[12] Gil Meilaender[13] is concerned that it is suffering that makes us able to be compassionate, and that the relief of suffering is only one among many competing and noble goals. Bill McKibben (2004) and Carl Elliott (1998) speaking from a leftist perspective argue along the same lines: that in using biotechnology, we risk inauthenticity.

This idea, that human future will be, can be, made trans-human, is, to be certain, also too often put forward by the researchers themselves, or by bioethicist enthusiasts. The hopes for research are rather amazing, and the strides toward a certain kind of mastery, a certain kind of certainty, are powerful ones. The classic account of a medical research breakthrough gone bad is called "hyped" research, and anyone close to the so-far falteringly few clinical trials in genetic intervention or cancer can understand how press reports of cures can fall disappointingly short.

Scientists can add fuel to the conflagration at the margins of medicine by speaking of religion as an absurd fantasy, or dismissing all fears as Luddism.[14] Pharmaceutical companies can construct elaborate accounts of research that can be only a minimal advance over standard, effective, and genetic treatments – understandable from a business angle, but a distortion from a purely scientific perspective, and understood as deceptive by many in the field.

What a paradox that it is on this terrain, the terrain of medical research, once the very core of how we understood ourselves as fully human, that we find the very notion of humanity to be challenged. What should be the response of the compassionate, humanistic physician to the new biotechnological possibilities? Is the pursuit of healing perilously close to a tragic misunderstanding of our role, leading us into a posthuman future we cannot control? How can the duties and responsibilities of medicine act as our guide in this complex future, and how can the freedom of research be balanced with the core ethics of medicine itself?

Conferences such as Davos and Doorn are a tribute to science – our best hopes, and our deepest trepidations – and any serious exploration of the history of science and its meaning must be frank about its mutable and shifting knowledge base. What a paradox that the search for the actual description of how the world works is so fraught with remarkable unknowability. Molecular genetics, and nanotechnology[15] in particular, are developments of an idea at the heart of science that the world is real, and at the atomic level perfect, constructed, finally actual, even if we cannot fully understand it, even if it cannot be fully interpreted. Medical research aims at far more than basic research. Like bioethics, great medical science actually has both a descriptive and a normative component; it implies a direction and a meaning and a goal.

As I see it, the free inquiry of research science can be understood as a sort of free speech. It is protected by the larger social polity, and it has to be responsive to the larger civic discourse and to the meaning of the moral gesture of medicine. If medicine's future lies in genetics knowledge, how will such terrain shape our view of the self? If medicine's future lies in transgression of boundaries understood as natural, how will we reconstruct a robust sense of morality and of a connection to the past narrative?

We live in the world as we find it, but medicine is, in a sense, about the world as we imagine it could be. The task of the next century in medicine will be a complex and difficult freedom, for with emerging, transformative powers will come serious and vexing challenges. Creating a duty-based response in research as well as in medicine will be needed if the calling at the heart of medicine continues to guide the work of the physician.

Just as we stand on the edge of a rigorous new understanding of genomics, molecular biology, and regenerative medicine, with the first actual clinical trials being suggested, bioethics has turned its attention to the fear of a future entirely out of our control. At the moment in which human health is constantly improving, when lifespan is longer that at any point in history, and when the emerging health problem is not starvation but obesity, we are beset with fear. A fearful polity and a fearful response to new biotechnology are a commonplace. What is at stake when moral philosophers write about a post-human future and seek to convince science policymakers to stop basic research as a genre of forbidden knowledge? What is attendant on the slippery slope argument, and why does such language resonate so clearly with the classic language of eschatology?

The fear that is so clearly expressed by many in bioethics is a fear of a future end time, in which what has come is a dystopia, but an odd one: Let us return to Dr. Kass:

> The last and most seductive of these disquieting prospects – the use of biotechnical powers to pursue perfection, both of body and of mind – is perhaps the most neglected topic in public and professional bioethics. Yet it is... the deepest source of public anxieties about biotechnology, represented in the concern about "man playing God." Or about the Brave New World, or a "post-human future." It raises the weightiest questions of bioethics, touching on the ends and goals of the biomedical enterprise... and the intrinsic threat of dehumanization – not the old crude power to kill the creature made in God's image, but the new science-based power to remake him after our own fantasies.

Since these are such serious charges, let me respond seriously, and to a large extent historically because the objections to medical research might be described in historical terms, noting the features of classic tension between modernity and traditionalism existing in the debate.

The objections of the critics are largely based on six arguments:

1. Human bodies are inviolate from the moment of conception. Our moral status, which is to say our bodies, and increasingly our DNA, is our self, and to touch, alter, even an early embryo in the first moment after that DNA is assembled is impermissible. Our dignity requires this intactness, our ipseity is this identity. This moral status means that, most of all, destroying human embryos is always wrong, and also that any deliberate approach to the DNA is wrong as well.
2. Nature – human nature and the nature of the green and living world – is fixed, for it has borders that cannot be broached without violation. Nature is normative, and morally good, if left free, and it will express itself in a primal harmony that our use of it, and our machines, threatens.
3. Suffering is the main thing that defines our species' being. Suffering and its noble acceptance are the great teachers of our need and of our love. Without suffering, we would become soulless. Death defines us, and this technology is intended to, and might lead to, immortality.
4. Slopes are slippery. If we use a technology for good, there will be no way to stop it from being used for larger and progressively more evil or more trivial purposes.
5. This finitude, this death, is what leads each of us to our aspirations. And it is these "fine aspirations acted upon [that] is itself the core of happiness." Real happiness is not being really happy, but is achieving one's fine aspirations.
6. The mix of marketplace and medicine is troubling, and the very success implied by genetic medicine – its widespread applications – should alert us to its danger.

Not everyone makes every claim; some emerge politically from the left, some from the right. I will make two claims about these arguments. First, all of them are more than trivially correct, and any sensible person could agree with many of these statements. The trouble begins here in their extremity when taken to their logical conclusion. (Yes, slopes are slippery, but they are not impossibly slippery, just to give one example.)

Second, all of these are profoundly religious statements. They are statements of faith, worldview, and eschatology, not moral arguments. As such, they will not – cannot – be entirely agreed upon in a pluralistic democracy. Like many faith claims in our world, they tell us, in a sense, that the end is near, unless. . . .

What response can be given? In the interest of time, let me make only two of the many.

2.1 Living Under the Fallen Sky

First, is the sky really falling? Is the world about to be fundamentally changed by genetic medicine? This is, I believe, the fundamental problem, and it is a folly *à deux*, shared by both extreme sides in the debate. Both sides actually believe that the future is a certainty, that we are at an eschatological moment in human history, and this is an error. The contemporary literature about biotechnology – from Kass to BIO (Biotechnology Industry Organization) – reflects a rather touching sense that the spectacular future can be named and known. But it cannot be fully known. There are two logical problems in thinking about the future. First, is it empirically possible to know it? Will we actually and in a widespread way ever use, say, cloning (to take a popular panic), and will cloning ever really have an effect on people's reproductive practices? Even IVF, widely used to be sure, is hardly a threat to the family or the practice of erotic sexuality. Are we seriously debating the good or evil of immortality or agelessness as if this was an actual concern?

Second, we cannot know the social impact of technologies that will occur against a changing horizon of other temporal and technological changes. It was not long ago that living as a healthy 80-year-old was unprecedented, and this has in one generation swiftly become attainable for many. And the defining suffering for all of human history – infectious disease – can now be addressed in nearly every case, but there is no empirical evidence that we are living without our souls. Is the world empirically cheapened by IVF babies, or even Botox or steroids? It is odder, perhaps, and complex, but it is hardly the matrix.

May we make the world? On some level, how can we not? What would it mean to turn away from learning how to sequence mammalian genomes? To cease drug development because a drug might be used for evil intent?

This is perhaps a religious matter, as is all manner of discussion about the end of time. For many Jewish traditions, the sky fell long ago when the Temple of Jerusalem was destroyed. The idea of the sacred canopy, in the sense that Peter Berger (1990) intends it, has surely fallen in pieces after the Holocaust, and lies like shattered glass, everywhere. People walk around a world that is difficult, not an Eden. The snake is just a snake, nature morally neutral; people intended to create by picking up the shattered pieces, the jigsaw puzzle of the fallen sky, trying to figure it out. From a Christian perspective, as Jean Belke Elshtain,[16] reminds us, the tree at the center of the moral universe is the Cross, but for the Jewish philosopher, it is the Torah, Law, justice, the pact of solidarity and witness that is *Etz Chaim*, the tree of life, at the center of the lost world.

Utopia, the "world to come," is, in a sense, one that we must make. In the words of the Jewish philosopher Levinas, utopia is a state where one can be chosen to be called upon by the needs of other people. For Levinas, authenticity has little to do with some aspect of being. Rather, "the authenticity of the I" is this listening on the part of the first one called, this attention to the other without subrogation . . . the possibility of sacrifice as the meaning of the human endeavor. It is this sort of authenticity that concerns this Jewish theology, and it is the very meaning of utopia – justice and transcendence, love for the other, and the willingness to live within a code of law that still contains within it the possibilities of mercy.

It is no accident that Levinas argues that free people are ones who "before all loans, have debts," meaning that duties are prior to rights. Moreover, duties are specific: we owe something to the neighbor, we are responsible. We are chosen and unique in this responsibility, and in this responsibility we want peace, justice, and reason. This is the way to think about the meaning of being human. If one thinks that the world as it is, is morally correct, and that change is not only dangerous but a seduction, a destabilizing one, threatening to alter our species' being, that must be, in this account, protected, singular, a thing entirely belonging to us, then the sort of thing that stem cell research represents is terrifying, for it deconstructs the world into pieces and suggests that it can be re-imagined, and re-imagined largely for the need of the other. In a sense, it is a close account of the moral gesture of medicine.

I would argue that it is not suffering that defines us, it is healing. The moral gesture of healing is the precise non-animal thing. It is not our pain, which I share with every mouse, it is the fact that in the medical school in which we teach ethics, we teach doctors and nurses to work in blood and feces and bone for the need of the other alone, the fact that the dying man and the laboring woman command our complete attention. It is not achieving my fine aspiration that makes me human, or even reading the *Iliad*; it is my willingness to sacrifice my very breath to resuscitate if need be, or to wash the back of the indigent stranger if that is who I am given to care for, that allows me to become human. Without our duties, and without our love, we are just words on paper.

We are turned to by the neighbor because the world is not-yet-good and we can change it, because medicine is powerful and can be wise, and our neighbor understands that. We – and by this I mean especially but not exclusively medicine as a profession – must be interrupted in our story line to answer to our debt to him and her.

I might perhaps be more persuaded by the claims of the goodness of suffering if that claim was consistently put forward by parents of children with dreadful diseases, or AIDS orphans, or burn patients – people who might be considered experts on the topic – rather than abstractly by professors, I might add.

2.2 A Familiar Discussion

Is this a unique time in human history? Is our world-making ability unprecedented? Hardly. At many times in human history, the sense that large forces clashed over the power to control the world was apparent. One critical time was the 9th through the 12th centuries, but the conflict also appeared in the Renaissance, and in early modernity as well. Let us turn to the medieval period, and to the work of medievalists. Jonathan Cohen,[17] Richard Landis, and others have argued that mysticism, millennialism, and calls for a return to the natural order often emerge just at the moment that science forges new relationships of knowledge and power. Landis, who studies millennialism, notes the comparison between our period and the ninth century. We can see that the literature of opposition to science is largely based on resonant tropes, images and metaphors particularly, but not uniquely, shaped in the early medieval period.

Hence the debates about these are given force by secular argument but are deeply driven by faith commitments and arguments of religion. What are examples? They include: fear about the role of women and a call for the return of women to the inner, private world of the family; certainty that suffering is both unavoidable and redemptive (that our suffering will make us pure); fundamentalism in faith; and a taking on of symbols of origin so as to be more authentic, such as costumes of origin; a sense of living in an "end time"; and claims that plagues, especially plagues driven by external persons, endanger the polity. There is a fear of contagion and intrusion. There is an aesthetic that emphasizes blood, a feature of art of the medieval period, and this can be seen in the popular attention to the film *The Passion*.

We are faced, globally with a deep yearning for the past, for the present can be a terribly uncertain place. There is a sense that medieval tropes are true or essential ones, that we live in a spirit-haunted world, with an affection for clarity between good and evil, that there are aliens, or witches, and pure natural beings that are linked to nature, as in the *Lord of the Rings* or the tales of Harry Potter. Old prejudices re-emerge, and fundamentalism is a powerful force politically and religiously. In such a period, there is a struggle with phenomenological knowledge – science, observation, empirical wisdom, and received knowledge from religious leaders.

In the West, such fears are often the subject of intellectual discourse, just discourse, but in the developing world in 2003 and 2004, such theories had dramatic consequences. In Nigeria, polio vaccine was refused out of fear that it would cause female sterility, and in Zimbabwe, grain was refused in a famine out of fear of genetic engineering. In America, the President's Bioethics Advisory Council raised questions based on a notion that an instinctual moral repugnance, or the deep-seated knowledge of faiths, ought to guide moral discernment – but what are we to think of moral repugnance when it means that babies are not given polio vaccinations, or women are not allowed to read, or five-year-olds go to bed malnourished? In bioethics, the basic Heideggerian ideas – that truth can be recovered from a pre-literal past, that authenticity is formulated by one strong "I," and that suffering and death are the fulcrum points of essential humanity – have formed a strong argument.

Yet rationality, science, and what came to be understood as "modernity" have always been in collision with such ideas. Modernity is uneasy with received knowledge, and seeks to learn from a process of trial and error, of open knowledge and phenomenology, in which the observation of the world, its description, its failure to comply with theory was long noted, by Galileo, by Vesalius,[18] and others. Linked to open knowledge and to a positive legal code, not one hinted at by "the yuck factor," is an equality preceding positive law. Science is carried out with the idea of progressive optimism, with its intervention to change fate, with an affection for machines, and with an attention to units of time. Natural philosophy, that is, science, is based on cause and effect and direct experience. This is why medical research was always radically violative of both the natural order, that is, the naturalness of suffering, and of death; think of the theft of bodies so that cadavers could be studied.

How can such claims of the critics of medical research be best evaluated? What normative course is suggested by this understanding of science? In the American stem cell research debate, in which the slippery slope/eschatological arguments have a powerful hold, we have slowed the clock in public funding, albeit not all of it, and the NIH cell lines are steadily being put to use, but private funding is also supporting serious research. Very recently, Harvard and HHMI (the Howard Hughes Medical Institute), funded 17 new cell lines that have been freely distributed to 50 labs internationally. In Belgium, China, India, Israel, South Korea, the UK, California, and New Jersey in the US, regenerative cloning is permitted, and this even when the science is still tentative. As the report in *Science* that South Korean researchers have made a stem cell line of human cloned cells, shows us, all science is now fundamentally global. Different faith traditions – Buddhist, Hindu, Sikh, Muslim, Jewish, as well as Christian sensibilities – will need to be considered now, and in most of these, the duty to heal the sick and the need for free scientific inquiry will be the primary considerations in this work. For many whose religion now prohibits any use of the early embryo, no matter how it is created, much of this research will be impermissible. But other will argue that this opens the door to a critical research direction. As each member of the clergy, as each lawmaker, reads this report, each must think: how do we balance the many competing moral appeals?

Many agree fully with these arguments I have enumerated, and they argue that the future is something that should worry us. The fear of technology, clones, or catastrophe plots directly with the general sense of fear that permeates our lives. In fact, bioethicists have turned to giving advice about new subjects in biology, such as neuroscience and its impact on happiness and memory. In so many matters of medical research, for example in neuroscience, in nanotechnology, the technology is already deeply and persistently feared, each with its own set of science fiction literatures, movies of terrible futures, and plucky natives who resist the evil scientists and other moral terrors. In contention in this debate are the same arguments, and once again they are about happiness and the meaning of the lives of children. So my final story is about that.

3 Story Number Three – Speaking of Actual Children

We are in Wisconsin in the US Midwest, driving home from a school retreat. It is raining and we cannot stop at the horseback riding place where I wanted to give my children a sense of self in the natural world. This being, after all, the natural world, it is pouring rain, and the horses are, sensibly enough, eating hay in the barn, with no plans to tote Californians around the woods for our authentic woods experience. But this being the Wisconsin Dells,[19] we are a few yards from the Kalahari Water Park, which we are told on enormous signs is the world's largest water park, "just like the other one down the road," notes my son. It is now 2004, I am a 53-year-old philosopher, and we have come to this: my children are begging to go to the water park, a place they have wanted to visit ever since it began to be advertised relentlessly on TV. I decide to go, as this is a one-time opportunity and they know it. I hate everything about it, and I wish I had Leon Kass at my side just at that moment. You have to pay, and once you pay you are given a plastic wristband with a number which you must strap to your wrist, there are thousands of people, and it is entirely fake. It is also very scary, I think. In fact, I begin to regard the entire *scène* with genuine moral repugnance.

As you emerge from the changing room where you put on clothing suitable for water activities, you enter a huge, football-stadium-size, enclosed, fake jungle scene. It has huge tree-like structures; structures that permit people to slide down from four stories high, gleaming with water; red plastic and blue plastic waterfalls, deep caves, fake beaches with fake waves on the fake beaches; and a fake river with fake rapids that are not rapid at all. But then, oddly, slowly, I see it differently. I look at the people and they are happy. They are not fakely happy, they are playing. We are all primates at play, messing around, we are climbing the big ladders, we are getting into the boats. The 53-year-old philosopher goes hurtling down the terrifying yellow slide, a rush to her apparent death, in the Heidegger manner. But we do not die, and the kids laugh and want to do it again. It is a thing to see. Thousands of regular, ordinary people, and we are near to naked in bathing suits, our vulnerable flawed selves, just as Levinas would want, without the pretense of clothing; who knows which is the philosopher, which is the lawyer, which is the trucker, and which is the women who cleans the office at night? There are things for people in wheelchairs to do, and there are very old people, and very young babies, and we are all floating in the fakest place one can imagine, and it is also the most natural of places because it is a place human primates have made to play with our babies and to teach them to be brave in trees and in water, to swim and to jump. It is happiness in itself.

I realize that for most of us – for most of us when we are being parents, and not making a turn at bioethics – this is okay, it is enough. I think of Leon Kass and how he has also written about the moral disdain he has for people who eat ice cream in public as they walk, and I look at the fathers feeding four-year-olds, and the messy little ice-creamy Americans, and I think, "Yes, these are my people; I am one of them." These folks do not want to make perfect babies. The fee at the water park is steep enough, for one thing, and they do not want to do anything unsafe, or cruel, or inauthentic either, and this is too often not appreciated. I think that, like Majit's parents, who just wanted their child to live, most parents like sharing a good joke, playing chase games, or telling a story. Will there be people who use their wealth to make evil or injustice, of course, just as on that Sunday, far from the Wisconsin Dells, far from the water park, far from the fake Kalahari, which is, by the way, an

African desert, there were people using their money to trick their way into Harvard, or to obtain elaborate surgery – and it did not matter.

There are people in the real Africa too and, frankly, we fail them in ways even more tragic when we write article after article about the horrors of perfection, rather than the need to vaccinate for polio so we can, yes, perfectly, eliminate naturally occurring polio from the world. The way that most of us really make the world, even when we use plastic, or aspirin, or anything on the continuum that can be understood as biotechnology (and that begins with ice cream cones), is as natural as the crow outside my window now, who is building its intricate, messy nest using the hair clips my daughter has dropped, the blue ones.

I am a watcher of science, of crows, of medicine, and what I see – I have learned – I often first worry about, but this would be wrong, partially. I remember that a turn to the classic texts of the Jewish tradition reclaims a praxis of duty to "the world to come," and this often means the world made for our children. "The heavens belong to the Lord, but the earth belongs to the children of men," not to us but to our children, and in these classic texts there is a suggestion that a far more optimistic future, a far more complex and contingent future, and a future that is still profoundly human awaits us if we make it. It is, in the terms of Levinas, "the adventure of possible holiness," who reminds us to ask not *"am I righteous?" But "is it righteous to be?"*

4 Normative theory: May We Make the World? Quick Answer: Yes

And here is the slower answer: we must, for the world is an unfinished place indeed, most surely in terms of the need for justice we yearn for. In this work, everyone will have much to do, and in this work, we must act as teachers and healers. Here are thing that bioethicists must do. First, try harder to, if not love, at the least trust, our neighbors as ourselves. Remembering that the failure of science is only a part of the sentence about the future; we have also to have some humility about our history of failure in philosophy and faith traditions. We ought to remember that our neighbor, the scientist, is largely driven by the same sort of duties that any scholar learns to love – to speak the truth, and to show a new path across interesting terrain. Further, we need to work patiently to remind others that it is not rights but duties that make us human, and that the duty to heal is even prior to freedom.

I can suggest four things that we might think could be agreed on. We cannot reasonably agree to agree on everything, and thus we need to find a way to allow faithful differences, such as when life begins, to be respected.

- Basic bench research ought to be like free speech – explored without coercion, or undue restraint, or conflicts of interest engendered by undue marketplace pressures.
- Such research is a public good and needs to be done with peer review, public oversight, and public funding and amidst open debate in which all sides are heard.

- Medical research is a moral gesture that must be aimed at the alleviation of suffering, done with justice with attention to full protection and consent of human subjects, and human gametes must be treated with respect. Ethicists can insist on reasonable responses from the science community.
- We can insist on a voluntary ban on all use of this technology for human reproductive cloning, which has been done already in every science community around the world. As in Asilomar,[20] this ban has been initiated by the science community, with regulatory mechanisms in place, by the US National Academy of the Sciences and other academies.
- We can ask that efforts are maintained to delineate the difference between hope and difficulty of therapeutic interventions, and ethicists should not hype our promises of dystopia[21] either. In this, we must work with the press and the clinicians to be honest about limits and finitude in medical care, be serious about what we will do to carve out a niche on that slippery slope and defend the limits that are agreed upon, be clear about the nature and meaning of suffering so it is not romanticized, and act in such a way that the duty to heal is primary in every gesture, every word.
- Finally, there must be a real conversation about the inevitability of two things we will encounter as the world becomes more known: error and evil. We do not have a sophisticated theory of either, neither do we have a strong normative response about both the possibility of real evil when it is done in the name of science, by rogues or by unjust marketing process, or a fair way to respond to honest mistakes.

Human use of genetic medicine has been heralded and dreaded in equal measure, and until now has only been a tantalizing theory. The events of the last years, of each quiet unheralded and heralded experiment in the lab, whether it fails or not, each clinical trial, make the theory arch toward the real.

Critical urgencies are always upon us. First, we can see that science is a global concern. Research is international, and will be, as is genomic research, the result of a fully international research effort. Scientists from many faiths and traditions will do the work, and citizens with many competing moral appeals will watch and worry and hope as it progresses. No one religion and no one moral authority can claim to be the final arbiter, for the serious differences in how we see the moral status of human embryos, how we think health care should be fairly distributed, what we argue that women can do with their bodies, and how scientific research can be reasonably regulated will vary. Second, extreme fears and over-exuberant promises will both need to be thoughtfully stilled as the small steps in this science are considered and reviewed. There is a thoughtful middle ground on which to stand. The future is not now. We have a great deal of work to do to shoulder the weight of the, literally, not-yet-born future. But that is what moral imagination is for. I can imagine the next step, I know that there is a journey across this territory in a real world, with real burdens, and that what we do – because it is what we need – is that we are mapmakers, drawing our approximate art, in only two dimensions. We have at least this: we can aspire to become who we must be to be the bearers of the weight, the ones who draw mountains.

5 Credits

Baruch Brody, Leroy Walters, Jonathon Moreno, Francis Kamm, Jean Bethke Elshtain, Emmanuel Levinas, Richard Landis, Jonathan Cohen, Leon Kass, Andy Lustig, Elisabeth Blackburn, Briget Hogan, Gil Milander, Carl Elliott, Art Caplan, Carl Sagan, Jim Battey, Len Zon, Doug Melton, HHMI, and Northwestern University.

Notes

[1] For protection against the "evil eye."

[2] HMO: Health Maintenance Organization, a form of prepaid health insurance common in the US. HMOs provide comprehensive health care to voluntarily enrolled individuals and families by medical professionals on their staff and member hospitals with limited referral to outside medical care.

[3] Severe acute respiratory syndrome.

[4] Ricin is a poisonous toxin that can be made from the waste left over from processing castor beans. It is one of the most toxic and easily produced plant toxins. Ricin has been used as a bioterrorist weapon.

[5] Any precursor cell; cells that can produce cells that are able to differentiate into other cell types.

[6] The embryo at the time of its implantation into the uterine wall.

[7] CPAP stands for "Continuous Positive Airway Pressure." CPAP is an airway treatment used for infants with respiratory distress. CPAP therapy delivers a constant, stable pressure to an infant's airways, facilitating breathing and intake of oxygen.

[8] Respiratory distress syndrome occurs almost exclusively in premature newborns because, generally, production of surfactant begins after about 34 weeks of pregnancy; the more premature the newborn, the greater the likelihood that respiratory distress syndrome will develop after birth. Premature infants on ventilators may be at risk of dying because they cannot produce a natural surfactant (a mixture of lipids and proteins) which acts as a wetting agent and lines the surface of the air sacs, where it lowers the surface tension, allowing the air sacs to stay open and stopping the lungs from collapsing from surface tension during the inflation of the lungs. The use of man-made surfactant lets the air sacs of the infant's lungs open with less pressure and allows breathing with less stress.

[9] Selfhood, individual identity, individuality.

[10] Eschatological: relating to the ultimate destiny of mankind and the world.

[11] Associate Professor of Political Science, Duquesne University Graduate Center for Social and Public Policy.

[12] Fukuyama, Francis. *Our Posthuman Future: Consequences of the Biotechnology Revolution*. Picador, 2003.

[13] Professor of Christian Ethics at Valparaiso University, associate editor of the *Journal of Religious Ethics*, and member of the President's Council on Bioethics.

[14] Ned Ludd and colleagues, the "Luddites" were a group of early nineteenth century workmen who destroyed stocking frames in an attempt to prevent the use of labor-saving machinery by destroying it, in the belief that such machinery would diminish employment. Luddite and Luddism became terms referring to one who opposes technical or technological change.

[15] Nanotechnolgy: the manipulation of materials on an atomic or molecular scale.

[16] Professor, University of Chicago, and co-chair, Pew Forum on Religion and Public Life, Elshtain is a political philosopher who has striven to show the connections between our political and our ethical convictions.

[17] An Israeli-born scholar, Dr. Jonathan Cohen directs the Hebrew Union College-University of Cincinnati Center for the Study of Contemporary Moral Problems. An assistant professor of Talmud and Halachic literature, he is responsible for coordinating the M.A. and Ph.D. programs that focus on Jewish law and ethics.
[18] Physician 1514–1564.
[19] A holiday vacation area.
[20] In 1973, a few days after Dr. Herbert Boyer and Dr. Cohen described their successful attempt to recombine DNA from one organism with that of another, a group of scientists responsible for some of the breakthroughs in molecular biology sent a letter to the US National Academy of Sciences (NAS) and the widely read journal *Science*, calling for a self-imposed moratorium on certain scientific experiments using recombinant DNA technology. The scientists temporarily halted their research and publicly asked others to do the same. Even though they had a clear view of their work's extraordinary potential for good and no evidence of any harm, they were uncertain of the risks some types of experiments posed. They suggested that an international group of scientists from various disciplines meet, share up-to-date information, and decide how the global scientific community should proceed. International scientists in this exceptionally competitive field complied with this request to halt certain research. In February 1975, 150 scientists from 13 countries, along with attorneys, government officials and 16 members of the press, met at the Asilomar Conference Center in Pacific Grove, California, to discuss recombinant DNA research, consider whether to lift the voluntary moratorium and, if so, establish strict conditions under which the research could proceed safely. The conference attendees replaced the moratorium with a complex set of rules for conducting certain kinds of laboratory work with recombinant DNA, but disallowed other experiments until more was known. The final report of the Asilomar Conference was to the NAS in April 1975, and a conference summary was published in *Science* and the academy *Proceedings* on June 6, 1975. Known officially as the International Congress on Recombinant DNA Molecules but remembered simply as "Asilomar," that meeting was widely hailed as a landmark of social responsibility and self-governance by scientists. At no other time has the international scientific community voluntarily ceased the pursuit of knowledge before any problems occurred, imposed regulations on itself, and been so open with the public.
[21] An imaginary place or state in which the condition of life is extremely bad, for example from deprivation, oppression, or terror.

References

Berger, Peter L. *The Sacred Canopy: Elements of a Sociological Theory of Religion*. New York: Knopf, 1990.
Elliott, Carl. *Bioethics, Culture, and Identity: A Philosophical Disease*. New York: Routledge, 1998.
Fukuyama, Francis. *Our Posthuman Future: Consequences of the Biotechnology Revolution*. New York: Picador, 2003.
Kass, Leon. "Ageless Bodies, Happy Souls." *The New Atlantis*, 1, 9–28, Spring 2003.
McKibben, Bill. *Enough: Staying Human in an Engineered Age*. New York: Owl Books, 2004.

Reflections on Human Dignity and the Israeli Cloning Debate

Carmel Shalev

1 Introduction

In 1997, the birth of Dolly the sheep sparked an intense public debate on human cloning that resulted in a wave of legislation aiming to ban the practice. International instruments, such as the Universal Declaration of the Human Genome and Human Rights, the European Convention on Bio-Medicine and Human Rights, and the Charter of Fundamental Rights of the European Union, consider reproductive cloning to be a violation of human dignity. In Israel, too, soon after Dolly's birth, the Knesset enacted a law that prohibited both cloning and genetic manipulation of eggs and sperm. However, the ban was a temporary one, for five years, in order to allow time for further reflection. The public debate that took place, in 2004, around the extension of the moratorium revealed the system of values that characterizes Israel's official position on the issue of cloning.[1]

Human dignity is entrenched in Basic Law: Human Dignity and Liberty, as a fundamental value of a Jewish democracy. Thus it is a vital principle in Israel's constitutional law, and has multiple meanings in its jurisprudence. Nonetheless, the prohibition of reproductive cloning in Israeli law does not rest on this value, but on general principles of ethics in medical research. This paper describes the legislative debate in Israel about the prohibition of human cloning, and analyses the underlying cultural values of a Jewish democracy, that explain why cloning is not considered to be a violation of human dignity. A major factor is the influence of Jewish religious law (*halakha*), which attaches high importance towards healing and reproduction, on the one hand, and low importance to the status of the early human embryo, on the other hand. However, these themes also resonate with a secular culture of fertility and pronatalism, a social reverence for science and acceptance of technology, and a constitutional tradition of liberty and freedom.

C. Shalev
e-mail: cshalev@012.net.il

M. Düwell et al. (eds.), *The Contingent Nature of Life*,

2 The International Arena

The question of cloning had been a subject of discussion in bioethics literature since the 1970s,[2] but it carried a certain flavor of science fiction. The birth of Dolly seemed to make human cloning an immanent reality, and the response was a gut reaction of moral repulsion and repugnance, articulated as an affront to human dignity. The first expression of this instinctive moral objection is found in the Universal Declaration on the Human Genome and Human Rights, adopted by the UNESCO General Conference in 1997. Article 11 states: "Practices which are contrary to human dignity, such as reproductive cloning of human beings, shall not be permitted". During the drafting of this international instrument by UNESCO's International Bioethics Committee, the Israeli member, Professor Michel Revel, had objected to specification of reproductive cloning as being a practice that is contrary to human dignity. Professor Revel was the chair of the bioethics advisory committee to the Israel Academy of Sciences, and later also became a member of the National Helsinki Committee for Genetic Research in Human Beings,[3] and he had major influence in the formulation and articulation of Israel's position on cloning.

The objection to cloning on grounds of human dignity was reiterated in European legislation. The Additional Protocol on the Prohibition of Cloning Human Beings (1998) to the European Convention on Bio-Medicine and Human Rights, states in Article 1: "Any intervention seeking to create a human being genetically identical to another human being, whether living or dead, is prohibited." This statement is explained in the preamble to the instrument in words that echo Kant's categorical imperative: "the instrumentalisation of human beings through the deliberate creation of genetically identical human beings is contrary to human dignity and thus constitutes a misuse of biology and medicine". In addition, the Charter of Fundamental Rights of the European Union, 2000 opens with a chapter entitled "Dignity", which declares in Article 1 that "human dignity is inviolable" and also goes on, in Article 3, to prohibit the reproductive cloning of human beings (together with other practices in the area of biology and medicine, such as eugenic selection and trading in human body parts).

European countries also took the lead in an attempt to bring about an international prohibition of cloning. In 2000, Germany and France undertook an initiative in the United Nations to bring about the drafting of an international convention banning reproductive cloning. Despite what appeared to be initial unanimous support for an international ban, the initiative did not result in a convention, because of disagreement about whether the ban should be limited to reproductive cloning or whether it should extend also to so-called "therapeutic" cloning. Several countries, including the United States of America, wished to prohibit development of cloning techniques by means of embryo research even for non-reproductive medical purposes, including stem cell research. Eventually, in 2005, the General Assembly resolved to adopt the United Nations Declaration on Human Cloning, which called upon member states "to prohibit **all forms** of human cloning, inasmuch as they are incompatible with human dignity and the protection of human life" [emphasis added – CS].[4] Many member states voted against the resolution, including France, and many abstained.[5]

Israel's position in the UN debate was formed by considerations of diplomacy. Initially, it joined what appeared to be a universal consensus to impose a legally enforceable ban on the practice, but not on grounds of human dignity. The first statement of Israel's delegation to the Sixth Committee working group [hereafter – the statement], where the initiative was being discussed, reveals that in fact Israel was in favor of only a temporary ban, and did not find any principled objection to cloning.[6] However, as the debate went on, Israel voted at least once together with the USA, on a matter of procedure, and ultimately abstained from the UN Declaration, even though a vote against would have been more consistent with its internal law and policy.

The statement acknowledged that the values of human dignity and the integrity of the human being must be valued above all other considerations, in order to avoid any misuse of biological and scientific advances. However, it suggested that the notion of human dignity should be construed as an imperative to view the individual not as a genetic profile, but as a unique and irreplaceable human being.[7] This seems to imply that cloning would be contrary to human dignity only if one adhered to a determinist belief that humans, like objects, can be duplicated – a belief which should be rejected. But because of the current dangers of research "at present, one thing is clear, that human reproductive cloning is absolutely unethical". Therefore, cloning should be banned temporarily, because of general principles of ethics in biomedical research. In other words, cloning might not be an intrinsic violation of human dignity, but it would be wise to proceed with caution to explore the moral, legal, social and scientific issues of reproductive cloning, and to reflect further upon their implications for human dignity.

The statement did not express outright a position that disagreed altogether with the condemnation of cloning on grounds of human dignity, because Israel wanted to be part of the consensus, and not to isolate itself. But it took what it considered to be a "unique" stand in proposing that a moratorium be imposed, rather than a permanent ban,[8] as had Israel's parliament.

3 Israel's Anti-Cloning Law

In 1999, the Knesset enacted a statute entitled The Prohibition of Genetic Intervention (Human Cloning and Genetic Manipulation of Reproductive Cells) Law, 5759-1999 [hereinafter – the Law].[9] The legislative initiative came from a member of parliament and was received in government circles with some alarm at the prospect of irrational populist restrictions on science (Prainsack 2003: 71–74, 152–56). It was, however, tabled as a bill by the parliamentary committee for scientific research and development. The commentary to the bill[10] referred to the stormy public debate throughout the world, following the birth of Dolly, and to emerging positions against cloning in other countries and in the international arena. In light of the fact that public debate had not yet resulted in a uniform position, the bill proposed and the law set a moratorium for a period of five years, "in order to examine the moral, legal, social and scientific aspects ... and the implications of such on human dignity."[11]

In its substance, the Law prohibited, in general terms, human cloning and the genetic manipulation of eggs and sperms ("germ line gene therapy").[12] The National Helsinki Committee was appointed as an Advisory Committee and charged with the responsibility to follow developments and submit annual reports to the minister of health. The commentary explained that this would guarantee periodic review of the need for the law, so that it would not be a stumbling block in face of scientific progress.

The Law did not distinguish between reproductive cloning and research into cloning techniques for other purposes. Simply read, it prohibited any intervention for the purpose of human cloning, defined as "the creation of a complete human being, chromosomally and genetically absolutely identical to another person or fetus, living or dead".[13] However, an opinion that came out of the office of the legal advisor to the Ministry of Health construed this to mean that the law was intended to prevent cloning of a "*complete*" human being, and did not make any provision with regard to processes not intended to result in birth. In other words, she read the law as prohibiting reproductive cloning and allowing non-reproductive cloning (basic research into cloning, and eventually – "therapeutic" cloning). Hence, in the absence of an express prohibition, there was no legal obstacle to approving stem cell research by means of cloning embryos, subject, of course, to ethics review and approval of the National Helsinki Committee.[14] When a proposal involving stem cell research by means of a nuclear transfer (cloning) technique came before the Committee, it refused to approve the study because of serious ethical shortcomings, particularly in regard to the procurement of the eggs needed for research. But it also decided, by a majority vote, in reliance on the opinion of the legal advisor, that it would be willing in principle to approve such research.

The Law had mandated the Advisory Committee to report annually to the minister of health on scientific, legal and ethical developments in the area of human genetics, but its first report was submitted only in December 2003, just one month before the moratorium was to expire, during the course of the parliamentary debate on its extension. The debate on the extension of the Law was once again a parliamentary initiative, taken by the chair of the Knesset Science and Technology Committee, Mel Polishuk-Bloch. Her position was that the prohibition on reproductive cloning should become permanent. In response the government submitted a bill[15] in which it proposed another temporary ban to allow for further deliberation on the formulation of policy, taking into consideration the value of "freedom of scientific research for the advancement of medicine".[16]

The Advisory Committee had indeed recommended that the moratorium be extended for an additional five-year period, rather than a permanent ban. Having examined the state of the art, it concluded that results of cloning experiments in animals show a low rate of success, with many embryos suffering from defects and dying in the course of the pregnancy or at birth. There were also risks for the carrying mothers. Therefore it considered human cloning to be currently dangerous and unsafe, and not to be allowed. At the same time, it recommended to permit stem cell research and cloning research for non-reproductive purposes, subject – in addition to the standard Helsinki ethics review – to a monitoring mechanism, and to the regulation of egg donations for research.[17]

The essence of the government bill proposed to clarify that there was a distinction between reproductive and therapeutic cloning, by specifying that the legal prohibition applies to the implantation of a cloned embryo in a womb. In other words, reproductive cloning would continue to be prohibited – at least for the next five years. The commentary explained that "the ethical rules governing scientific research in human beings do not currently allow experiments designed to bring about the birth of a person by means of cloning, so long as the dangers entailed in them remain without a solution." However, the prohibition would not apply to cloning for therapeutic purposes, including basic research into early stage embryonic development and cell differentiation processes, in light of the potential benefit to prevent and cure severe diseases and to provide organs for transplantation.[18] Aside from this, the bill proposed only one additional amendment, to increase the severity of the penalty clause.

It should be noted that Israeli law is notable for its liberalism and leniency compared with legislation in other countries, including those that were at the forefront of the development of reproductive technology, such as Britain and Australia. Although there is prior ethics review by the National Helsinki Committee, in accordance with the universally accepted standards that apply to all medical research involving human subjects, the Committee's discretion is unlimited by any statute. This means that, embryo research (including embryonic stem cell research) is neither prohibited nor regulated, as in other legal systems that permit it only within certain limits, subject to restrictions and to penalties in the range of ten years of imprisonment. For example, in some countries research is allowed on surplus IVF embryos, but the creation of an embryo for the sole purpose of research is not allowed. In others, creating embryos other than by fusion of human eggs and sperm (i.e., mixing human and animal gametes) is not allowed. Most countries enacted in legislation the rule that researchers have always abided by, that research embryos may not be developed beyond 14 days. Likewise, paying for research eggs, sperm and embryos is not allowed, and there are control mechanisms for the import and export of such. None of these limitations on substantive issues are embedded in Israeli law. Nor were they discussed in the parliamentary debate, which focused on the formalistic question of whether the prohibition of cloning should be temporary or permanent.

In the parliamentary committee discussions, the government took a hard-line position that freedom of science and medical research was the paramount consideration. Were it not for the currently unsafe state of the art, reproductive cloning could be regarded as carrying foreseeable benefits. The temporary ban on reproductive cloning was justified by pragmatic considerations of the dangers to fetuses and carrying mothers entailed in the current state of the art, rather than by any principled objection stemming from the notion of human dignity.

Eventually, the government had its way, and the law was extended for five more years, until March 1, 2009.[19] Besides clarifying that the prohibition against the cloning of a human embryo applies only when undertaken for the purpose of reproduction, the amended law also gave the Advisory Committee supervisory powers, increased the criminal penalty from two to four years imprisonment, and added the option of a fine (in the amount of approximately US $250,000) to counter the "significant economic incentives"[20] to clone a human being despite the ban.

4 Internal Debate

We have seen so far, that Israel's position on cloning favors freedom of science and fails to find any intrinsic moral objection, including on grounds of human dignity. This applies to reproductive cloning as much as to non-reproductive cloning. The temporary prohibition of reproductive cloning does not rest on grounds of human dignity, but results from the balancing of liberal principles of freedom of science and medical research, on the one hand, with ethical standards of safety in research with human beings, on the other hand. Reproductive cloning is currently unsafe and it would be unconscionable today to bring about the birth of a cloned human being. Yet, if the cloning technique were perfected and shown to be safe and efficacious, there would be no principled objection to the use of reproductive cloning for medical purposes.

This position brought together a broad coalition of government representatives, scientists, rabbis and ethicists in a uniform front, which presented the issue as a matter of defending and protecting science from unreasonable fears of progress. Three public committees endorsed the government bill – the Law's Advisory Committee, the bioethics advisory committee of the National Academy of Sciences, and an internal Ministry of Health Committee on the Status of the Embryo [hereinafter – the MoH Committee]. The protocols of the latter are most revealing of the internal governmental deliberations.[21]

The MoH Committee was appointed against the background of the rapid developments in embryonic stem cell research and genetic cloning, as well as in imaging and biological techniques for prenatal embryonic diagnosis, and began its work at the beginning of 2003. The mandate was "to examine the need for legislation that would regulate the application of the fundamental values of human dignity relevant to the matter at hand." It seems that the Committee met only twice and did not conclude its mandate, for two probable reasons: firstly, a change in government that replaced the appointing minister with another person; and secondly, an overlap in its mandate with the statutory mandate of the Law's Advisory Committee. Be that as it may, the minister requested that this Committee start its deliberations with the subject of cloning, since the Law was to expire by the end of the year, and the protocol of its second meeting indicates that the international context was a matter of some discussion.

Professor Michel Revel expressed the view that what was going on in the UN debate was a blow to science: "What is going on today in the United Nations is very sad. There is a virtual crusade to prohibit reproductive cloning through an international convention and to call it a crime against humanity. This is in contempt of the very notion of a 'crime against humanity', and in my view the State of Israel should say so.... One should not close the door to the possibility of applying an entire area of very important and interesting research in the field of future medicine."[22]

Professor Asa Kasher, a prominent philosopher and leading ethicist, perceived the international arena as being much influenced by Germany, whose policy was informed by Catholic views, on the one hand, and by the trauma of Nazism, on the other: "I think we have a special position in this matter. We, the Jews from Israel, are the only ones in the world who can come to the Germans and tell them, you are exaggerating.... Because you have reached an improper balance, and are obstructing millions of people who could benefit from the results of cloning research. We should create an official Israeli position that explains and explicates itself, and that will be the proper place to which the whole world will grow nearer slowly – the proper position, the balanced position."[23]

The basis of this unique position lies in the special Jewish religious approach to the human embryo. Professor Avraham Steinberg, a physician and leading authority on *halakhic* bio-ethics, emphasized the permissibility of embryo research, including cloning: "Cloning, whether reproductive or therapeutic, is permitted at a basic fundamental moral level. The moral Israeli position says this is a right process. I have no argument against the Catholics, but they cannot dictate to me what my position ought to be. If I think that from a moral perspective there is a difference between a pre-implantation embryo, an embryo, a fetus, and an infant, that there is a moral gradation, then I want to say this is my position."[24]

The sense of taking pride in the moral courage to hold to a unique position in face of a contrary world, also comes through a position paper submitted to the Science and Technology Committee by the heads of all of the three abovementioned public committees and the chief scientist of the Ministry of Health [hereinafter – "the Position Paper"]:[25]

> Above all, a decision of the parliamentary committee to prohibit genetic duplication permanently rather than temporarily would negate the positive distinction of the State of Israel, in both its own eyes and the eyes of others. In our own eyes, we are supposed to act as a Jewish and democratic state. The values of Israel as a Jewish democratic state require it to encourage all research that might have therapeutic benefit, within the reasonable and accepted limits of medical research. Israel has developed a position that is sometimes different from the Christian cultural foundation of Europe and other countries, but is suited to the Jewish cultural foundation.

These statements do not go without comment, and there are dissenting voices both on the MoH committee and in the parliamentary debate. There is a reservation that what appears to be a stand of moral integrity is in fact a lack of humility, and is disrespectful of lessons learned by humankind at large from the trauma of the Holocaust. There is concern about Israel's reputation in the world, and particularly in the scientific community, for being too permissive, lacking self-restraint, and failing to observe accepted international standards. There is also indignation at the implication that Israel might have a unique position on the value of human dignity, which is revered as a universal precept.

However, the protocols of the parliamentary debate reveal that there was little tolerance for minority views, which were predominantly female voices. Pressure was brought to silence dissent, and opposition voices were labeled as anti-science or feminist. MK Polishuk-Bloch, the chair of the Committee, said at one point: "I myself felt more than once how people tried to shut the mouths of those who think differently. I have never seen in the meetings of the Science Committee so much passion and pressure exerted by a particular group on another group as in this matter.... And I repeat for the umpteenth time that there is no inkling of harm to research."[26]

5 Government Policy

The Position Paper is a short two-page document, and more emotive than analytical. The first page sets forth the consensus of the three public committees and claims also to represent the common understanding of the bioethics community in Israel. The premise of the consensus is the faith that progress in medical science is always of some benefit to certain individuals, and that it would be socially irresponsible to suppress an entire area of medical research, especially in the area of infertility. Cloning has several foreseeable benefits. Research in cloning techniques could produce important results for the scientific understanding of the early stages of embryonic development. Likewise, there are the highly trumpeted possibilities of using cloned stem cells for regenerative medicine, that is, for the production of tissue and organs for transplantation in sick persons. Moreover, even reproductive cloning as such could have beneficial therapeutic applications, in isolated individual cases of singular medical need.

The second page explicates the reasoning that underlies the government's position, primarily in terms of Jewish religious law (*halakha*) values:

> From the perspective of Jewish *halakha*, central value is given to saving lives (*pikuah nefesh*) (including curing infertility and genetic disease), and the central perception is that man is a partner in the act of creation, to improve the world (*tikkun*) and correct defects in nature. It is not perchance that Israel is a leader, numerically, in the field of *in vitro* fertilization, because helping infertile couples is considered important, as a matter of *pikuah nefesh*.

Although it is made very clear from the start that there is unanimous agreement that reproductive cloning should be banned for the time being because of the dangers, the Paper says that there is no principled rabbinical objection to reproductive cloning in the treatment of infertility. It opposes this position to that of the Catholic church, not in terms of the graded moral status of the embryo and the permissibility of medical research in fertilized eggs, but by rejecting the notion of genetic determinism, which it supposes to underlie the Catholic view:

> The Catholic church views the fertilized egg as a complete human being, and therefore considers that a person's destiny is pre-determined in one's genetic constitution at the moment

of fertilization. Judaism, on the other hand, acknowledges that "all is foreseen and free will is given" (*Pirkei Avot*, 3:15) – that is, a person remains responsible for his actions, since the status of personhood is not dictated from the start but is rather acquired in his development. For this and other reasons, the rabbis and *halakhic* authorities do not find any prohibition of cloning where it has therapeutic benefit for reproduction.

The objections to reproductive cloning on grounds of human dignity are presented as "extremist positions in the world". Although not stated explicitly, human dignity also seems to be understood as a rejection of genetic determinism, apparently in accord with Article 2(b) of the Universal Declaration on the Human Genome and Human Rights, which states that the inherent dignity of all members of the human family "makes it imperative not to reduce individuals to their genetic characteristics and to respect their uniqueness and diversity." In this sense, it is difficult to find fault in the creation of genetically identical human beings:

> There are extremist positions in the world that oppose cloning, claiming that it negates human dignity because the genetic constitution of the newborn is identical to that of the cell donor. These positions ignore the fact that a cloned child is no different from a naturally born genetic twin, and that behavioral science has proven that genetic twins are persons whose individual and unique dignity is preserved, despite their genetic identity. The idea that cloning might lead to the wholesale creation of 'Einsteins' or doubles of Marilyn Monroe, or any other dead or living person, is actually an illusion, because even intelligence, beauty and character traits are the result of education and environment no less than of genetic characteristics.

And while fears of mass production of genetically identical persons might justify a ban on non-medical cloning, it should not extend to the use of reproductive cloning in individual cases of special medical need:

> In addition, even if there is a consensus that justifies an absolute ban on the mass use of cloning for non-medical purposes (since it is important to preserve the genetic diversity that stems from natural sexual reproduction), that is not sufficient reason to prohibit forever even the thought of singular medical applications in special cases that have no other cure.

An appendix to the Paper specifies three possible applications of reproductive cloning for medical treatment. The first is treatment for otherwise untreatable infertile couples, that is couples suffering from "total" infertility, who do not want sperm or eggs from an anonymous donor (since there are halakhic concerns about possible incest between offspring of the same donor). The second is the prevention of the transmission of serious genetic disease, for example where both partners are the carriers of a recessive gene. The third is the deliberate birthing of a child that could be a donor of tissue (such as bone marrow) for treatment of an existing sibling suffering from a condition that requires transplantation of histologically compatible tissue.

Finally, there is a paragraph which seems to respond to the idea that human dignity is violated by the instrumentalization of human beings through their deliberate creation. Setting aside concerns that arise from genetic determinism, the deliberate duplication of a human being with therapeutic intention does not differ significantly from other forms of medically assisted reproduction and is not contrary per se to human dignity:

Similar arguments about the violation of human dignity or instrumentalization or the destruction of family values were raised when in vitro fertilization and test-tube babies first appeared. But after decades, the benefit that this research brought to those in need, and continues to bring thanks to the development of novel technologies, is not disputed.

6 Jewish Values

As is evident from the above, the government approach to reproductive cloning is informed by notions of the uniqueness of Israel's identity as a Jewish democracy. The value system that favors biomedical science is indeed highly influenced by the support of rabbinical authorities and *halakhic* bioethicists. Moreover, Jewish values are also culturally accepted in the biomedical context by secular Jews and by scientists and professionals. Several considerations explain why the *halakhic* authorities do not invoke any principled opposition to the use of cloning for medical treatment, including reproductive cloning for the treatment of infertile couples.

(a) *Healing and saving lives*: A major concern about reproductive cloning is about the dangers of "playing God" in tampering with the very essence of human nature. This is countered in the Jewish tradition by the value of *tikkun*, which refers to Man's relation to God's creation and to the human role in mending, repairing and improving the world, and correcting defects in nature. This value, for example, explains the tradition of male circumcision. A baby is born. Is it not complete? Why do we cut the foreskin? The answer is that God's work is not complete, and "even man requires *tikkun*". Similarly, the Jewish tradition seems to have a cultural affinity with medicine, and there is a therapeutic imperative for the prevention of suffering – "thou shalt surely heal" (*rafo yerape'*), which is particularly compelling for saving lives (*pikuah nefesh*). Hence, the positive attitude to scientific progress in general, and to medical science in particular.

(b) *Reproduction*: The first of the 613 commandments derived from the Old Testament is to be fruitful and multiply (*pru urvu*), and this is a strong religious imperative. Celibacy is not a practice, and male genetic continuity is important. On the other hand, the curse of female barrenness, which afflicted the wives of the Patriarchs, is a consistent and profound theme in the bible and in Jewish tradition. When Rachel found that she bore Jacob no children, she said to him: "Give me sons, or else I die." (Genesis 30:1). Infertility is regarded as a cause of great suffering, to the extent that treatment of infertility is considered to be "life saving" by many contemporary Israelis. The social acceptance of the centrality of family and reproduction is extremely high, and we shall return to this theme below.

(c) *The status of the embryo*: In the cloning debate, the term "human dignity" seems at times to be invoked (some say "hijacked") by conservative opponents as a camouflage for objecting to any form of research in human embryos, by asserting the moral status of the embryo and even the human rights of the fertilized egg. As a consequence, Ruth Macklin has commented that human dignity is a useless concept, a hopelessly vague restatement of other more precise notions, and a mere slogan that adds nothing of substance to an understanding of the topic at hand (Macklin 2003). In the Jewish tradition, however, the value of embryonic forms of life depends on the stage of development. The word for embryo and fetus is the same, and in one Talmudic text, it is referred to as "just water" (*maya b'alma*) until 40 days of development (*Yebamot* 96b) (at which stage according to Aristotle there is ensoulment). Hence, there is no principled religious objection to research in early human embryos, that is, in fertilized eggs. Moreover, the therapeutic principle of the healing imperative applies even in late stage abortion, so that the value of embryonic life is regarded as secondary to the value of the life of an existing human being. Thus, one of the classic sources on abortion in the Mishnah relates to the case of a laboring pregnant woman whose life is at stake, and describes graphically that the fetus may be extracted limb by limb, "because her life precedes its" (*Ohalot* 7:6).

(d) *Dignity as moral agency and difference*: In the Jewish worldview, biological determinism is rejected by the human capacity for choice and moral agency. Moreover, human beings are not biologically determined but develop as the product of environment – family, community and education, all highly esteemed cultural values. A text from the Mishnah says: "When a human being makes many coins in the same mint, they all come out the same. God makes every person in the same image – His image – and each is different" (*Sanhedrin* 4:5). In other words, as asserted by Jonathan Sacks, our dignity is rooted in our very difference, in our unique irreplaceability: "Our very dignity as persons is rooted in the fact that none of us – not even genetically identical twins – is exactly like any other. Therefore none of us is replaceable, substitutable, a mere instance of a type. That is what makes us persons, not merely organisms or machines" (Sacks 2003: 47).

7 Secular Culture

It is important to understand that religious values as such are not necessarily accepted as binding or even guiding in Israeli society. There have always been points of friction between religion and state that create political tension, most notably in the area of marriage and divorce. Israel's constitutional identity as a Jewish democracy embodies the tension between religion and secular humanism. But in the area of biomedicine traditional Jewish values are accepted in most cases by society at large, because they reflect and resonate with other cultural themes and values. The bioethics discourse in Israel is largely dominated by scientists and rabbinical authorities, and social scientists are not usually included in the public debate around the making of policy. But they have studied the culture of medicine and reproduction, and their observations are enlightening.

Israel is a country that enjoys advanced medicine, health technology and scientific research. It has a public health infrastructure and a national health insurance system that guarantees the right to a comprehensive basket of basic services. At the end of life, rigorous care and intensive treatment are the norm. At the beginning of life, Israel boasts the world record in reproductive medicine – it has by far the largest number of IVF clinics per capita, as well as treatment cycles, almost all of which are undergone with the help of public funding. Israeli doctors have been pioneers at the forefront of infertility research and also in the production of stem cell lines. Israeli culture is highly receptive to all forms of technology, and particularly with respect to reproductive medicine.

(a) *Science and technology*: The idea of a technologised society had been part of the Zionist vision from its start. Theodor Herzl describes in *Altneuland* (1902) – a utopian novel of the new society that he envisioned as the Jewish state – how technical achievements were introduced in every field. "The accumulated experience of all the advanced nations of the world was there to be used by the settlers who streamed into the country from every corner of the globe. And the professionals, the graduates of the universities, of the technical and agricultural high-schools, who came here from civilized countries, were well equipped with scientific knowledge." It was this youth who "brought the greatest blessing to Palestine, by its technological application of the latest scientific discoveries." (Herzl 1960: 101–102).

Zionism also envisaged a "new Jew", socially constructed as the embodiment of health and muscular strength, as opposed to the pale and puny Jew of the Diaspora. Barbara Prainsack describes how this new identity is also associated with a fearlessness that replaces the image of the Diaspora Jew, which is associated with persecution. "The absence of fear in the *sabra* does not lead to ruthless and unscrupulous behavior, but to a well-balanced mix of rationalism and morality. Nobody is allowed to question this morality any longer .... The fearlessness of the *sabra* is balanced by the correcting and steering force of Jewish religion and morals." (Prainsack 2003: 156–160). These cultural themes explain the moral pride taken in the pioneering of new frontiers in medical science, as well as the openness of Israeli culture in general towards progress and technology.

While cloning raised fears of playing God in other countries, this did not happen in Israel, and the parliamentary debate on cloning went almost unnoticed by the public and the media. The only front-page headline came as a result of the Advisory Committee report that disclosed its decision one year earlier, to approve cloning human embryos for scientific research. The newspaper article[27] quoted the Knesset Ombudsman for Future Generations as calling the decision "scandalous", because the Ministry of Health and the Committee had "failed to inform the public or to act with transparency". But the matter ended there. Cloning was no different from other subjects that trigger lively and critical public discussion in other countries and are virtually non-issues in Israel (such as, for example, genetically modified food). These matters do not gain much attention from the media, and do not raise concern among the public.

Yael Hashiloni-Dolev says it is impossible to understand the cultural logic behind the attitude to medical genetics in Israel without taking into consideration the society's scientific mentality. "Whereas other post-industrial societies are characterized by a pervasive discourse of risk, this is almost completely absent in Israel, and the public is generally trustful of science and 'progress.'" She suggests that this complacence about bio-technology can be attributed to the special relation of humans to God in Judaism and the idea of *tikkun*, of improving the world, to alleviate suffering. (Hashiloni-Dolev 2006: 139–140).

The positive attitude towards the potential of medical science to alleviate suffering also explains the widespread acceptance of prenatal genetic diagnosis, even among disability organizations (Raz 2004). Indeed, the problematic history of medical genetics and eugenics, with its fatal consequences for European Jewry, seem not to have any impact on Israel's pro-science culture:

> Jewish tradition supports the prevention of life with disability, especially prior to conception, and contemporary Israeli medical genetics is not haunted by the negative history of eugenics.... Israeli-Jewish culture does not perceive the technological manipulation of life either as 'playing God' or as threatening to human dignity or rights, since the prevention of life with disability is not seen as endangering human dignity, but rather as preventing suffering and improving on God's creation. In general, advanced medical technologies are understood to serve the common good and not to pose risks. (Hashiloni-Dolev 2006: 143–144)

(b) *Fertility*: The Holocaust might not cast a shadow on attitudes toward medical research and genetics, but on the other hand it plays an important role in Israel's pronatalist culture, which is related to the historical and geo-political context of the Jewish state. Israel arose out of the genocide of six million Jews in the Holocaust, at the hands of the Nazis. As early as the 1940s, when the news of the Holocaust in Europe first reached the Jewish community in Palestine, David Ben-Gurion spoke of a "demographic duty" to the nation and warned that Jewish existence was at stake. In 1967, the Government adopted a resolution to establish a Center for Demography, so as "to act systematically to realize a demographic policy, directed to create an atmosphere which encourages birth, taking into consideration that it is vital to the future of the Jewish people." (Hazleton 1978: 58–61) Israel's population today is around six million, and that includes over a million non-Jews. It is surrounded by Arab nations whose populations outnumber its by far. Demography is also an express issue as regards the Israeli-Palestinian conflict and is seen as a threat to the continuing existence of a Jewish majority inside the state.

Yet, Israel is also a liberal democracy, and a pronatalist government policy would not be effective if it did not comport with other accepted values – in this case, the centrality of family life. That is to say, that Israel has a national culture of fertility, and in particular a culture of reproductive medical technology. "For Israeli Jews," writes Susan Kahn, "the imperative to reproduce has deep political and historical roots as well. Some feel they must have children to counterbalance what they believe to be a demographic threat represented by Palestinian and Arab birthrates. Others believe they must produce soldiers to defend the fledgling state. Some feel pressure to have children in order to replace the six million Jews killed in the Holocaust. Many Jews simply have traditional notions of family life that are very child-centered." (Kahn 2000: 3) The threat of losing a child in war or in a terrorist attack may also have an influence on the desire to have many children (Hashiloni-Dolev 2006: 130).

The family in Israel is a central social institution, and the suffering associated with childlessness is a reverberating cultural theme. According to *halakha* (Jewish law), which governs the laws of marriage and divorce for Jewish couples in Israel, a man has the right to divorce his wife if the marriage has failed to produce children over ten years. Although this may be more in the nature of law in the books, rather than actually enforced in practice, it carries symbolic significance. Voluntary childlessness is virtually non-existent in Israel, including among secular and educated women. And involuntary childlessness – that is, infertility – is experienced as suffering. "The barren woman is an archetype of suffering in the Israeli/Jewish imagination. From the childlessness of the matriarchs in the Book of Genesis, about which every Israeli schoolchild learns from the age of six, to the later biblical image of Hannah weeping over her inability to have children, Israelis learn that barrenness is a tragic fate for a woman. The childless woman is to be pitied and prayed for; her suffering is the quintessential form of female suffering and her joy at childbirth the quintessential form of female joy." (Kahn 2000: 32)

Parenthood is regarded as an innate deep natural desire for genetic continuity, particularly for women. Infertility carries stigma, and is experienced as feminine deficiency and existential failure. The cultural importance of reproduction is reflected in the high rates of utilization of medically assisted reproduction, and the right to public funding for almost unlimited cycles of infertility treatment. Indeed, the accessibility of publicly funded medical services makes it virtually mandatory for infertile women to seek medical intervention, thus reinforcing the motherhood mandate and labeling women who forego treatment as deviant (Haelyon 2006: 177). These themes even find expression in Israel's jurisprudence. On more than one occasion the Supreme Court struck down restrictions on access to assisted reproduction, as undue infringements on reproductive liberty. And in a landmark case, concerning a dispute over frozen embryos, it recognized a positive "right to parenthood".[28]

8 Constitutional Law

Israel is defined constitutionally as a Jewish and democratic state. The legal equivalent of a bill of rights is Basic Law: Human Dignity and Liberty, enacted in 1992 [hereinafter – the Basic Law].[29] Its purpose is "to protect human dignity and liberty in order to anchor . . . the values of the State of Israel as a Jewish and democratic state". In other words, human dignity is a basic constitutional value. The debate on human cloning has not come before courts of law in Israel, but it is interesting, nonetheless, to consider briefly relevant themes in the jurisprudence.

Israel's legal tradition is rooted in a liberal jurisprudence, which places much importance on political and civil rights to non-interference by the government. The protection of individual liberty, as a constraint upon the power of the state, is a pillar principle of its constitutional law. This means that there must be weighty reasons to interfere in personal liberty, and it is not the function of the law to enforce morality. Gut feelings of common revulsion from certain conduct (the sentiments of "the man on the Clapham omnibus"),[30] deep felt as they may be, are not sufficient justification to restrict the behavior of consenting adults. There has to be a concrete harm, which results or is highly probable to result from the exercise of one's liberty, for the state to intervene. Many liberty interests are implicated in the debate on cloning, genetic medicine and biotechnology: on the one hand, there are the rights of participants in medical research. On the other hand, there are freedom of science; patient autonomy, privacy and choice; freedom of reproduction; the right to found a family; the right to parenthood; freedom of transaction and of contract; freedom of occupation; and rights to intellectual property.

Liberty, in the sense of autonomy and privacy, also informs Israel's law on abortion. Technically abortion is a crime (for the abortionist), but the criminal statute also provides for legal abortion, with the approval of a medical committee and on certain specified grounds.[31] In actual fact, abortion is accessible both legally and illegally, and only in extremely rare cases of injury to the woman have criminal charges been brought. In the only case that came before the Supreme Court, the judgment dismissed the claim of a putative father to have a right of standing before the medical committee, and recognized that the woman had a constitutional right to privacy that was over-riding.[32]

Israel's abortion law also reflects cultural attitudes towards the embryo. The statutory grounds for legal abortion include "a defect in the embryo", without any qualification as to the severity of the disability. Furthermore, no time limitations are set, although requests for late-pregnancy abortions are channeled to a central medical committee under administrative rules. As regards the legal status of the embryo, Israel's law is clear: "All persons are competent to [be the subjects of] rights and duties from the moment of birth until death."[33]

Israel's legal culture also includes a liberal value of human dignity as a norm of constitutional human rights law in Israel, even before the enactment of the Basic Law. The value signifies a secular, rational and humanistic conception of humankind, which is universal in its embrace. It indicates the inherent worth of human beings as such, their existential equality and inalienable right to freedom. It defines the ontological status of human beings, their unique position in the universe and distinction from other forms of life, and their endowment with attributes of reason and conscience (Arieli 2002: 1, 3–9).

The Basic Law was taken to have merely proclaimed an existing socio-legal reality, created by the case law of the Supreme Court in line with the notion of human dignity in the international human rights regime. However, the enactment of the Basic Law led to a dramatic development in the legal culture, referred to as a constitutional revolution. In the process, human dignity became a kind of super-norm and a catch-all for lawyers.

In the case law that preceded the Basic Law, human dignity had been invoked a handful of times, essentially to protect vulnerable individuals from degrading and inhuman treatment, in connection with the rights of prisoners and detainees. The enactment of the Basic Law resulted in a radical increase in the frequency with which the Supreme Court refers to "human dignity" and in the scope of its meaning. The term is mentioned in two substantive provisions (together with the right to life and to the body), as both a negative and a positive right.[34] Drawing upon this, the Court found individual rights in various contexts that bear upon the essence of individual personality, including, for example, the right to parenthood, the right to know the identity of one's parents, the right of an adult to be adopted by a family with whom he or she has a special relationship, and the right to determine the inscription on a tombstone. In some instances the Court ruled that certain aspects of the right to equality may be derived from human dignity, to the extent that discrimination is degrading, even though the Basic Law had omitted any express mention of equality for political reasons (Kretzmer 2002). In more recent developments, the protection of human dignity in the Basic Law is emerging as a legal source for claiming economic and social rights, in the sense of guaranteeing the right to minimal conditions of existence in dignity.

The vagueness and open-endedness of the concept of "human dignity" (*kvod ha-adam*) is compounded by the fact that the Hebrew term for "dignity" (*kavod*) has multiple meanings. Orit Kamir has pointed out how the ambiguities represent distinct and even incohesive value systems, which are characteristic of the tension between a Jewish and a democratic state (Kamir 2002). For example, *kavod* as such can be translated accurately as dignity, but it is also the only word in the Hebrew language for "honor" and "respect", as in the commandment to honor thy father and thy mother, or to behave with respect for the dead. *Kavod* also has engendered meanings. As "honor" it connotes patriarchal constructs of the honor of the family and the merit of female modesty. On the other hand, *kavod* is also the only word in Hebrew for the male attribute of "glory". This is the glory of the warrior and hero, and of authority, power, rank and position. In the Old Testament *kavod* is ascribed first and foremost to God – as in *melekh hakavod*, the King of Glory. Human dignity in the sense of Man being created in God's image (Genesis 1:27) – as invoked in the cloning debate – appears to be related to this meaning. It is the divine spark in human beings that makes each and every one of us unique. Kamir says that this understanding of *kavod* is more related to glory than to dignity, and that it does not necessarily comport with democratic notions of individual rights. Interestingly enough, the examples she gives are of life, death and reproduction:

> Glory ... implies a rabbinical, religious, Jewish ideology, which attributes Man's glory to his heavenly creation in the divine image of God. According to this worldview, Man's glory is the source not only of certain human rights, but also of Man's duties to his creator. As 'glory,' the word *kavod* does not entail such rights as to end a pregnancy and to die at will; rather, it implies a person's duties to live and multiply. (Kamir 2002: 235)

Indeed, the debate on cloning in Israel has been framed in terms of Jewish religious values, rather than secular humanist concerns that derive from a human rights ethic which is fundamental to democracy. But at the same time a human rights discourse which revolves around individual liberty, such as that of Israel's Supreme Court, may not be sufficient to explain the moral intuition that cloning should not be allowed as a matter of human dignity.

9 Critique

Human dignity is the foundation and unifying meta-principle of the international human rights regime, from which derive the fundamental principles of liberty, equality and justice and all the civil, political, economic, social and cultural rights. The preamble to the Universal Declaration of Human Rights, 1948 opens with the premise that "recognition of the inherent dignity . . . of all members of the human family is the foundation of freedom, justice and peace in the world." And Article 1 goes on to state: "All human beings are born free and equal in dignity and rights." Thus human dignity attaches to all persons of woman born, regardless of any distinction, including the circumstances of their birth and their genetic characteristics.[35]

It may be said that the recognition of human dignity was the most important lesson to humanity from the Nazi Holocaust (Cohn 1983). The historical events that led to the adoption of the Universal Declaration were the same as those that led to the establishment of the State of Israel. Yet Israel's position on reproductive cloning is at odds with an international consensus that it constitutes a violation of human dignity and should be banned. Thus, while human dignity is considered to be a universal value, the debate on cloning in Israel seems to suggest that this fundamental meta-principle of human rights is a cultural contingency and subject to different contextual interpretations. In this respect, Israel's position on cloning poses a challenge to articulate the reasons for the persistent and deep-set moral intuition that reproductive cloning should not be allowed, within a human rights tradition that confers human rights upon individuals only from the moment of birth. This will have to be left to another paper. In the present context some critical comments on Israel's position will suffice.

The above analysis indicates that the understanding of human dignity is limited to a rejection of notions of biological determinism, and does not address the deeper questions of interfering in human "nature". That is to say, the reasoning behind Israel's position might be sufficient to permit the creation of genetically identical individuals by means of embryo splitting, but it does not answer the myriad concerns that arise around cloning through nuclear transfer. The view of genetic determinism or non-determinism does not resolve the moral quandaries about "the dominion of man over man self-determining the conditions of future human life" (Ramsey 1970: 94). It does not assuage the concern that the deliberate control of an individual's genetic constitution, in place of the contingency of bi-sexual reproduction, will alter radically "the moral self-understanding of the species" (Habermas 2003). Nor does it answer the argument that cloning breaks the moral ground for inheritable genetic enhancement, for if one may determine the entire genome of a future child, what justification can there be to prohibit the selection of a specific genetic characteristic? (Annas et al. 2003)

Israel's positive attitude towards reproductive cloning should not be understood as a policy of total *laissez faire* in the area of biomedical research. It is qualified by a distinction between therapeutic and non-therapeutic purpose. Thus, it does not imply approval of the instrumental use of cloning out of a purely narcissistic desire for self-generation, nor for the mass production of genetically identical individuals.

Nor does it confer any legitimacy on positive genetic enhancement, or on eugenic ideas of producing a superior human being, race or species. However, while the difference between therapeutic and non-therapeutic uses of medicine is, indeed, crucial to distinguishing negative from positive eugenics, it can be very fuzzy, and is influenced by social attitudes towards what constitutes illness and merits medical intervention, as well as by economic forces.

Therapeutic utility, or the potential of such, is the general value underlying the ethics of medical research in human beings. But the foreseen therapeutic benefit must be balanced against the risks. Because of the current dangers of reproductive human cloning, Israel has adopted a moratorium. Yet, within these terms of reasoning, one might well ask whether research in human cloning can ever be considered safe enough. Even if there are possible therapeutic applications to reproductive cloning in singular cases of medical need, the benefit of such must be weighed by considering their added value in relation to existing therapeutic alternatives, and the conclusion should be that the benefit is marginal or negligible. Moreover, the danger of mishaps constitutes a crucial moral problem (Ramsey 1970: 81). These involve, of course, the risks to the offspring, but also the risks to the women, both of which cannot be confined to the period of gestation. We are just now beginning to appreciate the long-term effects on the health of the children of in vitro fertilization and related interventions. And it is fairly clear from the emerging practice of global trafficking in eggs, that the development of cloning technology would present a moral hazard to the human rights and dignity of women, as the source of the eggs.

Proponents of the government's position in the internal debate suggested that any negative impact could be dealt with through legal regulation, rather than prohibition. However, this is quite cynical, for while Israel boasts of being a leader in medical reproduction, it is also highly under-regulated. Artificial insemination, in vitro fertilization, sex selection, and medical research are all regulated under either administrative rules or secondary legislation. Embryo research is not specifically regulated in any way, and the recommendation of the cloning law's Advisory Committee to regulate egg donations for research was also not taken up. Parliament has addressed only one subject of medical reproduction in primary legislation – the subject of cloning – and even so, with minimal content and in the form of a moratorium, which is a highly unusual method of legislation. The moratorium is supposed to encourage public debate, but this does not occur of itself and the government does nothing to involve the general public.

The ambivalence of Israel's position on reproductive cloning – against now in no uncertain words, but in favor for the future; to be prohibited, but only for the time being – remains a puzzle. One wonders, why so much fuss to maintain a position that is so contrary to that of the rest of the world? The faith in science and progress seems to be exaggerated. The expectations from the outcome of research seem to be inflated. What can explain this? Perhaps there is a sub-text of economic stakes and market forces at play.

There is no doubt that the relative laxity of Jewish religious values regarding research in early embryos places Israel in a unique position to be at the forefront of competition in repro-genetic research and the global bio-industry. There is an oblique reference to this in the Position Paper, which mentions in passing that a permanent ban on reproductive cloning would "obstruct investments in this research". Moreover, while biological determinism is rejected, there is an undercurrent of market determinism. During the discussion of the MoH Committee, Professor Bolek Goldman, chair of the National Helsinki Committee, says: "I personally have no doubt that in two, three or five years there will be [human cloning] . . .. Technologies cannot be stopped. And what cannot be done overtly will be done under the table. Therefore I think [it is good] that our law is the most liberal. It gives all the options, it keeps all the options."[36]

The official language of scientific and reproductive freedom is consistent with a market ideology. Also social receptivity to new technology – be it the latest range of mobile phones or the most novel medical intervention – is a reflection of a consumer market culture, in which individuals purportedly exercise autonomy and choice. The technology is market-driven, and the market is motivated by self-interest and profit rather than by concern for the future of humanity, or for the universal enjoyment of human rights, such as the right to health or the right to share in the benefits of scientific progress. The notion of human dignity as a constraint on the freedom of individuals to pursue scientific research or to exercise reproductive choice, cannot rely solely upon a jurisprudence of liberty. In order to correct market failures, it needs to be complemented with a jurisprudence of justice, which asks: Who stands to benefit? And at whose expense?[37] Such a jurisprudence might also go beyond the self-centered individualism of liberty to include responsibility as a self-imposed restraint on the exercise of power, with a view of empathy to others who are affected by our choices.

10 Postscript

Justification for the exercise of restraint, precaution and humility with respect to the further development of genetic medicine can also be found in Jewish culture. When Adam and Eve eat the forbidden fruit of the tree "in the middle of the Garden of Eden" (the tree of knowledge), their eyes are opened to know both good and evil. God expels them so as to prevent them from eating also from the tree of life, and living for eternity. This teaches us two things. The first is that the quest for knowledge opens us to both good and evil – science is a positive value, but it can be abused. The second is that human mortality sets a limit to the power of knowledge, including medical science.

There is another story in the Jewish tradition that is relevant to the cloning debate. This is the story of the *Golem*, which has evolved in various versions over the course of many centuries.[38] The original form of the story, referred to in the Talmud, tells us about the creative power of the righteous man. Through one's mastery of knowledge of the structure of the world, one may imitate the creative power of God and form a Golem out of earth, although the creative power is limited, because the Golem is not fully human. It is notable that the knowledge pertains to magical transformations of the Hebrew alphabet without any practical purpose. Once it is created it returns to dust. In later versions, however, the Golem acquires practical functions to serve its creator. The most famous version of the story is the legend of the Golem of Prague from the 17th century. By this time, the servant – an automaton or mechanical man – becomes dangerous. The technology gets out of hand and is a threat to its own creator. Frankenstein, written by Mary Shelley in the 19th century is another variation on the theme.

What I find most interesting in this tradition is that the first warnings of danger were not related to the fear of an overwhelmingly powerful creature becoming autonomous and turning against its human creator. Rather the concern was about the tension that the creative process arouses in the human creator. The problem is that playing God creates a conflict in the internal life of the creator. I understand this to mean that the threat of the Golem, or the clone, is not external but internal. It is not that our discoveries and creations actually get out of hand, but that we are morally troubled by the power of our own minds and have need for internal restraints upon our own actions, so that we cause no harm. In other words, from the place of empowerment that is bestowed by the liberal discourse of human rights, we move into the sphere of responsibility and empathy, which is respect for the human dignity of others.

Notes

[1] I should disclose from the outset that I was personally involved in portions of this debate, and that I have done my best to describe its content in an objective, unbiased and nonjudgmental manner.

[2] See, e.g., Paul Ramsey, *Fabricated Man – The Ethics of Genetic Control* (Yale University Press, 1970).

[3] "Helsinki" committees are appointed under Public Health (Medical Experiments in Human Beings) Regulations, 1980, to review the ethics and approve bio-medical research in hospitals in Israel, in accordance with the Declaration of Helsinki on Ethical Principles for Medical Research Involving Human Subjects. The National Committee was appointed to review all genetic research proposals. Subsequently it was vested with statutory advisory functions under the 1999 cloning law and the 2000 genetic data law.

[4] UN Doc. A/RES/59/280, 59th session, 23 March 2005.

[5] UN Press Release GA/10333, 8 March 2005.

[6] Statement by Ady Schonmann to the United Nations, Sixth Committee Working Group for the Elaboration of a Mandate to a Convention Against the Reproductive Cloning of Human Beings, 24 September 2002, Appendix 9 to the Activity Report of the National Helsinki Committee for Genetic Research in Humans (December 2003) [hereinafter – the 2003 Advisory Committee Report].

[7] This appears to be a reference to Article 2(b) of the Universal Declaration on the Human Genome and Human Rights, which states: "[Human] dignity makes it imperative not to reduce individuals to their genetic characteristics and to respect their uniqueness and diversity."

[8] Knesset Science and Technology Committee, Protocol of Meeting of 8 December 2003 – statement of Ady Schonmann.

[9] The Prohibition of Genetic Intervention (Human Cloning and Genetic Manipulation of Reproductive Cells) Law, 5759-1999, *Sefer HaChukkim* 5759, p. 47 (hereinafter – the Law).

[10] *Hatza'ot Chok* 5758 (1998), p. 482

[11] Section 1 of the Law.

[12] For the sake of clarity, the description and discussion of the Law in this paper relates only to the subject of cloning, and not to the genetic manipulation of reproductive cells.

[13] Section 2 of the Law.
[14] Mirah Hibner-Harel and Talia Edry, Opinion on Experiments to Create Embryonic Stem Cells, 22 August 2002, Appendix 6 to the 2003 Advisory Committee Report (note 5 above).
[15] The Prohibition of Genetic Intervention (Human Cloning and Genetic Manipulation of Reproductive Cells) (Amendment) Bill, 5764-2003, *Hatza'ot Chok* 5764 (2003), p. 290 [hereinafter – the Bill].
[16] Section 1 of the Bill.
[17] 2003 Advisory Committee Report (note 5 above), pp. 6–7. Note that the recommendation to regulate egg donations has not been implemented.
[18] Commentary to Section 2 of the Bill.
[19] The Prohibition of Genetic Intervention (Human Cloning and Genetic Manipulation of Reproductive Cells) (Amendment) Law, 5764-2004, *Sefer HaChukkim* 5764, p. 340 [hereinafter – the Amended Law].
[20] Commentary to Sction 3 of the Bill.
[21] By decision of the Committee itself, the protocols are publicly accessible on the Ministry of Health website – http://www.health.gov.il/pages/default.asp?pageid=1172& parentid=10& catid=6& maincat=1.
[22] Per Professor Michel Revel, Ministry of Health Committee on the Status of the Human Embryo, Protocol of Meeting, April 14, 2003.
[23] Per Professor Asa Kasher, *ibid.*
[24] Per Professor Avraham Steinberg, *ibid.*
[25] Undated position paper submitted and signed by the heads of three public committees – the National Helsinki Committee [i.e., the statutory Advisory Committee] (Prof. Bolislav Goldman), the bioethics advisory committee of the Israel Academy of Sciences (Prof. Michel Revel), and the internal Ministry of Health committee on the status of the human embryo (Dr. Shraga Blazer) – as well as by the chief scientist of the Ministry of Health (Prof. Rami Rahamimov), for the March 1, 2004 meeting of the Knesset Science and Technology Committee.
[26] Knesset Science and Technology Committee, Protocol of Meeting, February 1, 2004, p. 12.
[27] Tamara Traubmann, "Ministry gives nod to human cloning", *Haaretz*, 5 March 2004.
[28] FH 2401/95 *Nahmani vs Nahmani* 50 P.D. (4) 661.The question facing the court was how to balance the relative weight of rights to motherhood and to non-fatherhood. After a long process of litigation, the Supreme Court finally ruled by a majority, that the woman's right to become a mother imposed a correlative duty on the father to cooperate and refrain from obstructing its realization. This is in contrast to rulings in other jurisdictions that require such decisions to be consensual.
[29] The protection of individual human rights was well established in a long line of previous judicial decisions, which had the binding effect of precedent. The rights included in Basic Law: Human Dignity and Liberty, and its companion Basic Law: Freedom of Occupation, are those upon which political consensus could be gained, and the list is not conclusive. Several rights that were recognized in the case law were omitted. These include the right to equality, as well as freedom of conscience, freedom of expression, freedom of religion and conscience, and freedom of the press. These rights and freedoms continue to be protected as part of Israel's constitutional jurisprudence.
[30] This is a reference to the Hart-Devlin debate on the use of criminal law to enforce sexual morality, such as the aversion to homosexuality.
[31] The Criminal Law, 1977, Sections 312–321.
[32] CA 413/80 *Anonymous vs Anonymous* P.D. 35(3) 57.
[33] Section 1 of the Legal Competence and Guardianship Law, 1962.
[34] Section 2 of Basic Law: Human Dignity and Liberty states in the negative: "There shall be no violation of the life, body or the dignity of any person as such". While Section 4 states in the positive: "All persons are entitled to protection of their life, body and dignity".
[35] Universal Declaration on the Human Genome and Human Rights, Article 2: "Everyone has a right to respect for their dignity and for their rights regardless of their genetic characteristics."
[36] Per Professor Bolek Goldman, Ministry of Health Committee on the Status of the Human Embryo, Protocol of Meeting, April 14, 2003.
[37] For a preliminary exploration of this theme, see my earlier article: Shalev, Carmel. Human Cloning and Human Rights: A Commentary, *Health and Human Rights* 6, 2002, 137–151.
[38] For a more expansive elaboration of this legend, cf. Shalev, Carmel. "Clones and Golems". *Ethics and Law in Biological Research.* C.M. Mazzoni (ed). Martinus Nijhoff Publishers, 2002, 187–192.

References

Annas, George J., Andrews, Lori B. and Isasi, Rosario M. "Protecting the Endangered Human: Toward an International Treaty Prohibiting Cloning and Inheritable Alterations." *American Journal of Law & Medicine*, 28, 151–178, 2003.

Arieli, Yehoshua. "On the Necessary and Sufficient Conditions for the Emergence of the Doctrine of the Dignity of Man and His Rights." In *The Concept of Human Dignity in Human Rights Discourse*. David Kretzmer and Eckart Klein (eds.). The Hague; London: Kluwer Law International, 1–17, 2002.

Cohn, Haim. "On the Meaning of Human Dignity." *Israel Yearbook of Human Rights*, Faculty of Law, Tel Aviv University 13, 226–251, 1983.

Habermas, Jurgen. *The Future of Human Nature*. Polity Press, 2003.

Haelyon, Hilla. " 'Longing for a Child' Perceptions of Motherhood among Israeli-Jewish Women Undergoing *in vitro* Fertilization Treatments." *Nashim – Journal of Jewish Women's Studies & Gender Issues*, 12, 177–202, 2006.

Hashiloni-Dolev, Yael. "Reproductive Genetics in the Israeli-Jewish Context." *Nashim – Journal of Jewish Women's Studies & Gender Issues*, 12, 129–150, 2006.

Hazleton, Lesley. *Israeli Women – The Reality Behind the Myth*. Idanim, Jerusalem, 1978 (in Hebrew).

Herzl, Theodor. *Altneuland*. Paula Arnold, trans. Haifa Publishing Company, Haifa, Israel, 1960.

Kahn, Susan M. *Reproducing Jews – A Cultural Account of Assisted Conception in Israel*. Duke University Press, Durham and London, 2000.

Kamir, Orit. "Honor and Dignity Cultures: The Case of *Kavod* and *Kvod Ha-Adam* in Israeli Society and Law." In *The Concept of Human Dignity in Human Rights Discourse*. David Kretzmer and Eckart Klein (eds.). The Hague, London: Kluwer Law International, 231–262, 2002.

Kretzmer, David. "Human Dignity in Israeli Jurisprudence." In *The Concept of Human Dignity in Human Rights Discourse*. David Kretzmer and Eckart Klein (eds.). The Hague: Kluwer Law International, 161–175, 2002.

Macklin, Ruth. "Dignity is a Useless Concept." *British Medical Journal*, 327, 1419–20, 2003.

Prainsack, Barbara. *The Politics of Life: "Negotiating Life" in Israel – Embryonic Stem Cells and Human Cloning* (doctoral dissertation). University of Vienna: December 2003.

Ramsey, Paul. *Fabricated Man – The Ethics of Genetic Control*. Yale University Press, New Haven and London, 1970.

Raz, Aviad. " 'Important to Test, Important to Support': Attitudes toward Disability Rights and Prenatal Diagnosis among Leaders of Support Groups for Genetic Disorders in Israel." *Social Science & Medicine*, 59, 1857–66, 2004.

Sacks, Jonathan. *The Dignity of Difference*. rev. ed. London and New York: Continuum, 2003.

Shalev, Carmel. "Human Cloning and Human Rights: A Commentary." *Health and Human Rights*, 6, 137–151, 2002.

Shalev, Carmel. "Clones and Golems." In *Ethics and Law in Biological Research*. Cosimo Marco Mazzoni (ed.). Martinus Nijhoff Publishers, The Hague/London/New York, 187–192, 2002.

Conceiving of Human Life

Values of Preservation vs Values of Change

Boris Yudin

In this Chapter I am going to trace back some developments in mutual relations between culture, on one side, and scientific and technological advances, on the other side. To my mind, these observations could help us to understand some aspects of current debates on goals, possibilities and limitations of extensive use of biological and medical sciences for the sake of preserving, restoring, prolonging, reconstructing or even constructing anew individual human existence. I want to emphasize the tremendous importance of different moral or, more fundamentally, value positions, for our perception of arising biomedical possibilities and, even more, for directing scientific and technological developments in this field.

To begin with, with regard to difference in perceiving life in general, not just human life, I would like to point out that Darwin's conception of the origin of species through natural selection has not only had a profound influence on the scientific understanding of life. It turned to be a source of, or support for, different versions of naturalistic ethics as well. A very interesting trait of these developments is that there are striking intercultural differences in the acceptance and promulgation of these different versions of naturalistic ethics. First of all, it is interesting that the dissemination of Darwinism in Russia was often associated with a strong rejection of the ideas of "struggle for existence" and "survival of the fittest".

Take, for example, Prince Peter Kropotkin (1842–1921), Russian philosopher, anarchist, geographer, natural historian, who was one of the most eager proponents of Darwinism in Russia. P. Kropotkin developed his own conception of evolution which goes not so much through struggle for existence, but through mutual aid; he acknowledged the presence of the struggle for existence only in the form of the extremely severe "struggle for existence which most species of animals have to carry on against an inclement Nature" (Kropotkin 1902). Yet his views presented only one of many expressions of the same ideas, which were widespread among Russian zoologists, botanists and biologists in general, as well as among the general public. Many Russian biologists were strong opponents of ideas of the prevalence of the ethos of struggle and at the same time supporters of ideas of harmony in interrelations between not only biological organisms, but first of all between humans.[1]

B. Yudin
e-mail: byudin@yandex.ru

M. Düwell et al. (eds.), *The Contingent Nature of Life*,

In his "Mutual Aid: a Factor of Evolution" (1902) Kropotkin himself referred to "the well-known Russian zoologist, Professor Kessler, the then Dean of the St. Petersburg University".

According to Kropotkin, Kessler "struck me as throwing a new light on the whole subject. Kessler's idea was, that besides the law of Mutual Struggle there is in Nature the law of Mutual Aid, which, for the success of the struggle for life, and especially for the progressive evolution of the species, is far more important than the law of mutual contest." These views of Karl Kessler (1815–1881) were first presented in a lecture "On the Law of Mutual Aid" (1880), which was delivered at a Russian Congress of Naturalists in January 1880. Kropotkin also mentioned some other Russian zoologists who had gathered a lot of evidence of mutual aid in relations between animals, especially between birds.

Some of Kropotkin's arguments turn to be essential for my subsequent deliberations. For instance, he refers to characteristics of Russian wild nature, especially in the most remote and severe parts of Russia, such as Eastern Siberia and Transbaikalia, which allow an observer to grasp the genuine importance of mutual aid and social instincts (sociability) for struggle for survival in such environmental conditions. This argument was used by Kropotkin to substantiate not just his own views on the subject, but conclusions drawn by many Russian zoologists in general. Competition and interspecies struggle may be more suitable for affluent conditions, whereas cooperation and mutual aid are necessary in the less favourable environment characteristic for many areas of Russia. These considerations were interpreted as evidence of the (evolutionary) priority of mutual aid.

Another line of argument in Kropotkin's writings refers to an understanding of interspecies relations in animals as a model for explaining interrelations between humans. These considerations are of great importance not only for naturalistic ethics in general, which became so popular after Darwin, but also for a specific version of naturalistic ethics that Kropotkin himself had developed. Yet in other cases arguments which have been borrowed from a description of social interrelationships are used as a possible means of construing explanations of evolution in animals. For instance, referring to a study of French philosopher and sociologist, adherent of evolutionary theory A. Espinas (Espinas 1877), he remarks:

> Espinas devoted his main attention to such animal societies (ants, bees) as are established upon a physiological division of labour, and though his work is full of admirable hints in all possible directions, it was written at a time when the evolution of human societies could not yet be treated with the knowledge we now possess.
>
> (Kropotkin 1902)

We can see that in this observation Kropotkin in fact perceives developments in (and even notions from) science of human society as a prerequisite for the understanding and explanation of phenomena of animal behaviour.

Another essential aspect of Kropotkin's argumentation is that he distinguishes between two different approaches to studying living nature: "As soon as we study animals – *not in laboratories* and museums only, *but in the forest and the prairie, in the steppe and the mountains* – we at once perceive . . ." (ibid., emphasis mine).

Kropotkin here draws a sharp distinction between two positions: one of them is that of a *researcher* who gains knowledge through experiments in the laboratory and, consequently, interferes with nature; another is the position of a *naturalist* (or natural historian), who spends his time in expeditions and gains new knowledge through pure, non-interferential observations of nature. It is clear that Kropotkin prefers the second position. The same value is manifested when Kropotkin refers to the authority of a prominent naturalist, I. Goethe.

The importance of the Mutual Aid factor – "if its generality could only be demonstrated" – did not escape the naturalist's genius so manifest in Goethe. When Eckermann once told Goethe – it was in 1827 – that two little wren-fledglings, which had run away from him, were found by him next day in the nest of robin redbreasts (Rothkehlchen), who fed the little ones together with their own youngsters, Goethe grew quite excited about this fact. He saw a confirmation of his pantheistic views in this anecdote, and said: – "If it be true that this feeding of a stranger goes through all Nature as something having the character of a general law – then many an enigma would be solved." He returned to this matter on the next day, and most earnestly entreated Eckermann (who was, as is known, a zoologist) to make a special study of the subject, adding that he would surely come "to quite invaluable treasuries of results" (ibid.).

Strictly speaking, Kropotkin's reasoning in this case is incorrect, because Goethe's observation concerned aid not to an individual of the same species, but to a stranger. Nevertheless, this example is important for Kropotkin, because it demonstrates the "naturalness" of such generous behaviour in the animal world.

It is worth mentioning that Darwin took both positions in his conception of evolution. On the one hand, when he made observations during his travel on Beagle ship. On the other hand he had from the very beginning of his studies taken a mode of activity of selectionists as a pattern for grasping the genuine meaning of variability. In other words, his initial intuitions came back as interventions into living beings, in the manner of researchers conducting experiments.

Bearing in mind the main topic of this article – values of preservation vs. values of change – it makes sense to shortly explain Kropotkin's views on the evolution of social institutions. He discusses phenomena of possible "parasitic growth" of some Mutual Aid institutions and the revolt of individuals against these institutions, which become a "hindrance to progress". This revolt can take two different forms:

> Part of those who rose up strove to purify the old institutions, or to work out a higher form of commonwealth, based upon the same Mutual Aid principles; they tried, for instance, to introduce the principle of 'compensation,' instead of the lex talionis, and, later on, the pardon of offences or a still higher ideal of equality before the human conscience, in lieu of 'compensation,' according to class-value. But at the very same time, another portion of the same individual rebels endeavoured to break down the protective institutions of mutual support, with no other intention but to increase their own wealth and their own powers. In this three-cornered contest, between the two classes of revolting individuals and the supporters of the status quo, lies the real tragedy of history
>
> (ibid.).

So, even progress of social institutions can be carried out by those who are inspired by values of preservation!

The same values expressed in another form were also predominant in the thinking of a rather original Russian religious thinker and philosopher of that time, Nickolay Fedorov (1829–1903), who in his "Philosophy of Common Cause" (1906; 1913) posed before science and, even more, before humankind in general, the overall goal of not just preserving lives of all living humans, but to resurrect, to revive all those who had died. In this case we can speak even about over-preservation.

It is worth to also mention the position of one of the most famous Russian scientists of that time, botanist Kliment Timiryazev (1843–1920), who made a lot to propagate Darwinism in Russia. In particularly, Timiryazev prepared one of many Russian translations of "Origin of Species". In an introductory article to his translation he made this characteristic remark:

> all . . . complex aggregates of mutual relations between living beings, as well as with the environment, Darwin, allegorically and for short, called struggle for existence. It seems that nothing else brings so much harm to his teaching as this metaphor, use of which he would have been able to avoid, could he have foreseen the conclusions which would be drawn from it.
>
> (Timiryazev 1896).

Later, in the next decade, Timiryazev wrote: "I call the expression 'struggle for existence' an unfortunate one . . . It is far from necessary, as becomes evident from the fact that I was able to deliver the whole course of Darwinism ('Historical Method in Biology'), never mentioning the word 'struggle'" (Timiryazev 1938: 31).

Incidentally, Ya. Gall notes: "Seemingly, the article by N.G. Chernyshevsky (1888), in which Darwin's teaching was subjected to sharp criticism due to numerous attempts to use it from the part of Social Darwinians, had left a strong impression on Timiryazev" (Gall 1976: 17). I should point out here that ideas of the Russian Social Democratic thinker, writer Nickolay Chernyshevsky (1828–1889) were extremely influential at that time. And even he, who once called upon Russia "to take up the axe" to fight for better society, disagreed with ideas of struggle as a constituent of social interrelations (see Chernyshevsky 1888).

Now we will discuss some, perhaps rather limited, but nevertheless meaningful, correlations between, on the one side, ideas of intraspecific struggle (competition), experimental research (as a source of directed interventions which are carried out under artificially constructed conditions of laboratories) and change, and, on the other side, ideas of mutual aid (cooperation), observation (as non-interventional activity) and pre- (or con-) servation. It seems possible to maintain at least some degree of affinity of intuitions and/or intentions underlying each of these sets. The next suggestion will be that the grounds of such affinity can be found at the level of values.

We shall now try to distinguish two different value orientations in the relation of humans to nature, including living nature, and finally, to human life and human nature. One of these orientations stresses *values of preservation*, be it preservation of life on Earth or preservation of human life, health, rights, dignity, autonomy, etc. It underscores the need to preserve, to protect the surrounding order of things, which can be easily and irreversibly destroyed by our rash and unreasoned actions. These motives are particularly evident in tackling ecological problems arising in the course of biotechnological intrusions, such as the introduction of genetically modified organisms into the environment and the necessity of the protection of the environment.

Certainly, for the sake of preservation, we often need to make a lot of changes; yet all these changes are directed towards restoration of some impaired (presumably natural) conditions, states, structures, processes and functions.

According to another value orientation we can hold our interests and desires to be more important than the imperatives of the preservation of the nature around us. In this case nature is perceived, first of all, as a raw material to be transformed, more or less radically changed, on the basis of our designs and by means of our technologies in order to achieve our own goals. This means that nature is conceived as something devoid of intrinsic value and significance.

The opposition between these two value systems can be presented as an opposition between, on one side, the previously discussed stands of a naturalist as a (pure) observer of phenomena in the outer and inner world, and, on the other side, a researcher as someone who actively intervenes and produces changes in the world.

The first stand was vividly presented by I. Goethe, who urged that we should endeavour "to see things as they are". To be sure, contemporary philosophy of science would disregard such a position of a "pure observer" as overly naïve, because it does not take into account the constructive potency of our cognition and, even more, of our perception. Indeed, strictly speaking, such a stand cannot be termed "pure observation", because it presupposes some directedness of our interests and our values. Nevertheless, alongside its presumed weakness, it has also its own advantages.

According to such a position, we cognise nature in order to grasp its beauty, to admire its perfection or (in more modern versions) to find ways to save it. In other words, some kind of reverence for nature is presumed: nature has its own *raison d'être* and as such it deserves our respect regardless of our desires and intentions.

For the sake of clarification it is necessary to remark that research activity can be directed by such naturalistic aspirations. Yet research activity, first of all as it manifests itself in experimental research, contains this inner intention that today generates innumerable means for (sometimes drastically) changing nature around us as well as our own nature.

The second stand is very often perceived as the most adequate expression of the spirit of science as a research activity par excellence. One of the most influential proponents of this point of view was K. Marx, particularly in his famous 11th thesis on Feuerbach: "The philosophers have only interpreted the world, in various ways; the point is to change it" (Marx 1969: 15). In the first of these theses Marx criticized the naturalistic position (which in this context is synonymous with so-called "contemplative materialism") in these words:

> The chief defect of all hitherto existing materialism – that of Feuerbach included – is that the thing, reality, sensuousness, is conceived only in the form of the object or of contemplation, but not as sensuous human activity, practice, not subjectively
>
> (ibid.: 13).

To put it in another way, nature unfolds its truthfulness, its real meaning and its value not in itself, but only as a milieu of change through human activities.

The 11th thesis can be conceived in two different ways: either correct interpretation, explanation (and, consequently, understanding) of the world is a consequence, a by-product of our attempts to change the world, or in general the very creation of such interpretations is something non-obligatory and even superfluous for human activity. Our interventions can be effective even without any previous interpretation and understanding of phenomena, irrespective of whether such interpretation would be right or wrong.

It should be remarked here that Marx, in his writings after "Theses on Feuerbach", especially after 1848, was not as radical in his rejection of interpretative and explanatory functions of philosophy and science. Moreover, it was he who developed the notion of a "natural-historical approach" in relation to the social world. According to such an approach, the historical evolution of social structures and institutions can and must be presented as generated by something like natural laws. After all, it is these laws that determine human activity in its diverse forms, and only by relying on these laws can we succeed in our efforts to change the social world.

Nevertheless, in his 11th thesis Marx vividly expressed the essence of the position which asserts change of the world as a primary goal and consequently, value. According to an interpretation of Marx by P. Berger and T. Luckmann (1967), he made the most essential contribution to sociology of knowledge when he, in his "Economic and Philosophic Manuscripts of 1844", described interrelations between infrastructure (or basis, "Unterbau") and superstructure ("Überbau"). Infrastructure in this case is nothing but human activity, whereas superstructure is the world generated by such activity. It is worth to note that such kinds of cause-effect relations turn out to be valid not just with regard to the realm of human knowledge: the 11th thesis does not imply such limitations. Therefore, the construction (which may be, but does not have to be, a social one – the same, *mutatis mutandis*, can be said about physical and biological construction as well), of the world can be interpreted as a specific form of changing it. Needless to say, Marx understood changing the world as at the same time changing humans who transform the outer world. Yet this transformation of the transformer himself was thought as a mediated transformation.

As became clear in the twentieth century, especially during its last decades, as well as in the first year of the twenty-first century, the distinction of two value systems is relevant not just for nature around us, but for our own, *human* nature as well. From the beginning of the twentieth century a variety of projects to transform and improve humans began to appear. (Some reservations seem necessary here: there were cases in which the necessity to stop degradation of humans, in a definite sense – to preserve or defend the existing genetic pool – was proclaimed. Eugenic programs, including sterilization, with such goals were launched in for instance the United States.) So, at that time the main direction of interests, discussions and even actions had turned from biology in general to human biology.

In the first decades of the twentieth century Russia was strongly influenced by ideas of radically changing existing humans to form new ones, inspired by the new regime and substantiated by some interpretations of Marxism. In the 20s there were many ideas and even attempts to combine social and biological ways of improving humans (including, for instance, eugenics, the use of psychoanalysis, attempts to experiment with crossbreeding humans with apes, and so on).[2] Some Russian proponents of eugenics indicated the necessity of special programs of "social hygiene" to reach these goals. Yet in the end of the decade Soviet leaders had decided that it would be ideologically incorrect to use biological means for achieving this goal; only social means were acknowledged as permissible.

It is possible to give different explanations for this turn, but we have no opportunity here to discuss this complex and interesting issue. We can assume, however, that it is at least partly explainable by the Russian cultural traditions, which we described earlier. During the first year after the drastic changes that were experienced by the country in 1917, a sharp rejection of almost all previous traditions was extremely widespread. This meant that all kinds of changes were very much welcomed. Yet, at the end of the 20s, since about 1928, processes of returning to traditions had started – in this situation, a negative position with regard to radical biological interventions into humans gained a new impulse. Incidentally, for many decades this choice was a main obstacle for research in human genetics and attempts to use its possible achievements for therapeutic aims.

At about the same time, ideas of betterment of the population gained more and more influence in Germany. When the Nazis came to power, these ideas culminated in politics of "racial hygiene", including the (physical) elimination of various categories of the population, which were perceived as inferior and as carriers of genetic burden.

We can note that both Germany under the Nazis and the Soviet Union during Stalin's era tried to reach similar goals of forming new humans, but the means that were chosen to attain these goals were totally different. Nevertheless, in any case the issue at stake was not about changing the nature around us, but the nature of humans themselves.

The main accent in Germany was placed on biological measures. In the Soviet Union, on the contrary, there was at least a latent presumption that human biological nature should be preserved even in the course of enormous, radical changes of humans by social means. In other words, humans can be almost infinitely plastic in relation to social, educational influences, yet they are rather rigid with regard to interventions in their biology.

Note that ideas of change through education and upbringing were extended even to the realm of biology. In "creative Darwinism", which was developed by the grievously famous Soviet agrobiologist Trofim Lysenko (1898–1976) and his followers, intraspecific struggle was refuted for the sake of inheritance of character traits acquired in an organism's course of life, in particular due to the some kind of upbringing.

We can say that Russian culture in general is not so much inclined to borrow concepts for describing and understanding human capabilities and behaviour from biology and to use biology as a pattern for conceiving of social tasks. Even nowadays, when ideas of changing human nature through directed interventions in biological processes and mechanisms are becoming so widespread, it is possible to discern in Russia tendencies to improve humans mainly by social, psychological and pedagogical means.

An example can illustrate such tendencies. Some time ago one of my colleagues who is a psychologist told me that she had received an interesting proposal. She had been asked to take part in the preparation of specific training programs for children. The idea is that some wealthy Russians are interested in bringing up their children with particular personality traits.

There are many people in Russia who think that the previous Soviet system of education formed (with only rare exclusions) standardized types of personality – heavily dependent on social surroundings in his (her) attitudes and behaviour, trying to be indiscernible from others and easily subordinated to those endowed with any kind of power. Others, however, see a profound influence of traditional Russian culture in the prevalence of such types of personality.

In any case, we can state that the current system of education in Russia, despite all (even very essential) changes, to a large extent reproduces such personality types. Yet now some of the so called "new Russians", i.e. those who are very wealthy and very successful in their business, wish to raise their children with different personal traits – strongly goal-directed, oriented toward achievement and heavy struggle for getting essential results in their activity, having well-developed communicative and leader's abilities. All such traits, according to their understanding, are necessary for a person to be successful in future life. These parents were ready to pay substantial amounts of money for special educational courses, which would allow the development of such traits in their children. More than that, they are ready to provide financial support for the elaboration of psychological programs of training designed to form such young people with traits of personality that have been chosen in advance.

In this example, we can discern striking similarities with aspirations to choose personality traits for future offspring, which are so heavily debated in Western countries nowadays. Both cases represent the manifestation of a particularly technological approach to human life and human nature. Yet in the latter case interventions into genes are conceived as means for achieving the goal.

This example can be treated as characteristic for describing the main distinctive features of technological approach.

Firstly, it clearly shows an essentially technological way of not just doing but also grasping things, including such intimate things as the personality traits of one's own child. This technological way of perceiving and thinking about the world presupposes that if someone has a clearly defined goal (say, some personality traits) and the necessary quantity of resources (first and foremost – money), he (she) can reach this goal by hiring professionals or experts who are able to collect or create all needed means. In our case these means are thoroughly directed interventions into human personality and the processes of its development.

We can go even further: not just some traits of a child, but the child as such is perceived in similar situations as begotten, as "made" by parents not just in the genetic or usual psychological sense, but also in this technological sense. In other words, the child is treated as a kind of constructed and even re-constructed entity, as someone generated not so much by nature but mainly by implementation of a human design.

Secondly, such a technological approach clearly presupposes and even requires thoroughly elaborated systems of measurement through diagnostics. Indeed, in our example it is necessary to have both a preliminary diagnosis of the person's traits that are to be improved as well as diagnoses of subsequent stages on the way to the desired state.

It is evident that these diagnostic systems must be rather complex and multi-dimensional; they can be created only on the basis of developed categorizations, which allow systemization and classification of a huge variety of individual human persons. That means that those parents who want to get their child enhanced, in fact receive not just their own, unique child but some averaged product of technological manipulations.

Thirdly, this approach is based on an (latent) presumption according to which every human personality can be treated as nothing more than a collection of distinct traits. The possibility of systemic interconnections and interactions of these traits is not seriously taken into account. Nevertheless, due to these interconnections, such an "injection" of desired new traits can cause inconsistencies in the structure of personality with resulting frustrations. Similar considerations can be developed with regard to a systemic organization of links between personality and the social and cultural milieu in which the personality is included and formed. In other words, it is a real possibility that a personality, constructed or reconstructed by these psychosocial technologies, would meet quite serious difficulties due to his/her incongruities with prevailing social and cultural norms and values.

Fourthly, the example under discussion can be treated as one of the manifestations of the contemporary tendency to understand individual human life, or the individual human being, as something that is constructed, in this case – socially constructed. Due to this understanding it is possible to pose such goals as deliberate re-construction of an entity, which is "naturally" constructed in ordinary processes of social interactions, including processes of generating and changing meanings, which are necessary (and often decisive) parts of these interactions.

So, in our days a human being to an ever increasing extent becomes not only an object of scientific investigations, but a target for various technological influences as well. It seems that current bioethical debates on therapy vs enhancement of humans reflect, among other things, an opposition of these two sets of values.[3] Therapy in this case can be interpreted as a restoration (or preservation) of the existing human nature, whereas enhancement definitely implies its change.

A specific expression of this opposition can also be found in the realm of ethics of biomedical research. In its more traditional forms ethics of research stressed first of all *risks and burdens* for the participants. In every particular case, the involvement of humans in biomedical research is a risky endeavour that needs to be scientifically justified and ethically approved. A researcher has an obligation to guarantee a minimal or at least acceptable level of risks for a participant. The latter, in his/her turn, has a right to choose whether to become a participant in the research or not. This choice can be interpreted in the following way: the person in question decides whether to use ordinary, existing, approved methods of therapy, and to consequently preserve the current state of the art, or to promote search for new methods, hence, change.

Yet more modern versions of research ethics tend to stress the benefits a person can get by participating in research through progress in therapy. And some authors even talk about an *obligation* on the part of a person to agree to be a participant in research, in other words – an obligation to be personally involved in the promotion of change. As John Harris emphasizes: "where risks, dangers or inconvenience of research is minimal, and the research well founded and likely to be for the benefit of oneself or others, then there is some, perhaps very modest, moral obligation to participate." and "To fail to contribute to research is against the public interest and may harm others."[4] This argument is built on the premise that one's participation in research is for the overall welfare of the community, but it is also presupposed that this common benefit can be achieved exactly by the way of some changes imposed on the person or manipulations with data concerning the person.

To conclude this discussion of two distinct value positions in relation to human life it is necessary to draw attention to one problem, which arises when values of change become dominant in conceiving of human life. In case of changes imposed by us on the world around us we can turn – manifestly or latently – to wishes, interests and so on of humans as a reference point. It gives us an opportunity to make judgments on the desirability, permissibility or necessity of our changing influences. In such a situation a human personality, understood as a goal in him/herself, can be presented as "a measure of all things". This does not mean that in such a way we get a measure which is easily applicable to all situations; nevertheless, we have at least more or less reliable grounds for meaningful discussions of any particular case. In some sense this reference point makes it possible to speak of the unity of humankind as a whole.

Yet there is quite another situation that arises when the issue at stake is possibilities of changing humans themselves. Up to now, at least, we do not even have a hint of any commonly accepted measure to deal with different designs for technologically generated humans. The very possibility of the continued existence of humankind as a unity in this case does not seem to be certain.

Notes

[1] Cf. Gall 1976, especially Chapter 1. "The Problem of Struggle for Existence in Evolution Theory of the Twentieth Century."

[2] For more details see, for instance, Adams 1990: 153–216; Rossiianov 2000: 340–359; Yudin 1993: 83–99; Yudin 2004: 99–110.

[3] See, for instance, Kass 2003. See also the document prepared by the US President's Council on Bioethics: "Distinguishing Therapy and Enhancement. Staff Working Paper." (http://www.bioethics.gov/)

[4] Harris 2002: 128.

References

Adams, M. "Eugenics in Russia 1900–1940." In *The Wellborn Science: Eugenics in Germany, France, Brazil, and Russia*. New York: Oxford University Press, 153–216, 1990.
Berger, P. and Luckmann, T. *Social Construction of Reality: A Treatise on Sociology of Knowledge*. London: Penguin, 1967.
Chernyshevsky, N.G. "Origin of the Theory of Beneficience of the Struggle for Life." *Russian Thought*, 79–114, Sept. 1888.
Espinas, A. *Les Sociétés animales*. Paris, 1877.
Fedorov, N.F. *Philosophy of Common Cause*. (In Russian) Moscow: ACT, 2003.
Gall, Ya. M. *The Struggle for Existence as a Factor of Evolution*. (In Russian). Leningrad: Nayka Publishers, 1976.
Harris, J. "Research Ethics Committees – The Future." *Notizie di Politeia*, Anno XVIII, 67, 123–138, 2002.
Kass, L.R. "Ageless Bodies, Happy Souls: Biotechnology and the Pursuit of Perfection." *The New Atlantis*, 9–28, Spring 2003.
Kessler, K.Ph. "On the Law of Mutual Aid." (In Russian). In *Memoirs (Trudy) of the St. Petersburg Society of Naturalists*. Vol. XI, 124–136, 1880.
Kropotkin, P. *"Mutual Aid: A Factor of Evolution."* London: Heinemann, 1902. All citation from: http://dwardmac.pitzer.edu/Anarchist_Archives/kropotkin/mutaidcontents.html.
Marx, K. "Theses on Feuerbach." In *Marx/Engels Selected Works*. Vol. 1. Moscow: Progress Publishers, 15, 1969.
Rossiianov, K. "Gefährliche Beziehungen: Experimentelle Biologie und ihre Protektoren." In *In den Jungeln der Macht: Bildungsschichten unter totalitaeren Bedingungen. Ein Vergleich zwischen NS-Deutschland und der Sowjetunion unter Stalin*. D. Beyrau (ed.). Göttingen: Vandenhoeck und Ruprecht, 340–359, 2000.
Timiryazev, K. "Meaning of Turn in Present-day Natural Science, Made by Darwin." Introductory Article. (In Russian). In: Ch. Darwin. *Origin of Species*. Russian translation by Prof. K. Timiryazev. Saint-Petersburg: O.N. Popova Publisher, 1896.
Timiryazev, K. *Collected Works*. Vol. 5. Moscow, 31, 1938.
Yudin, B. "Russian Modernisations and Science." In *Development and Modernity*. L. Gule and O. Storebo (eds.). Bergen: Ariadne, 83–99, 1993.
Yudin, B. "Human Experimentation in Russia/the Soviet Union in the First Half of the 20th Century." In *Twentieth Century Ethics of Human Subject Research*. V. Roelcke and G. Maio (eds.). Stuttgart: Franz Steiner Verlag, 99–110, 2004.

Globalization and the Dynamic Role of Human Rights in Relation to a Common Perspective for Life Sciences

Carlos M. Romeo-Casabona

1 Globalization *versus* Multiculturalism: A New Challenge for the Ethical and Legal Debate on the Life Sciences

One is sure to note the growth and speed with which new discoveries and applications are being produced in the life sciences. This reflects the dynamism of the sector and the competitiveness that exists among different groups of researchers. One will also note the social perplexity that is being created by the novelties that are testing the foundations of long-established social perceptions and values. This perplexity also indicates that legal reasoning is often unable to offer efficient and/or calming answers.

Cultural peculiarities, especially the moral, religious and legal traditions of different countries, states and nations entail relevant differences when dealing with the juridical framework of biomedicine. Nevertheless, when these assessments are carried over to human genetics, which in turn does not always connect solidly to previously well-defined axiological principles, it can often lead to quick changes in perception and even to the import of mimetic solutions by the policy makers. Paradoxically, these cultural divergences have been coexisting in some countries: their subjects accept ideological pluralism. In some cases, in order to re-examine concrete traditional assessments, especially those related to the respect and protection which human life in its different forms deserves, this pluralism has met strong resistance by ideological or religious groups. Resistance has also been met when confronted with new phenomena or realities, as for example in relation to the establishing of the critical moment of the beginning and of the end of human life (i.e. the ethical and legal status of the in-vitro human embryo as well as the decision to terminate a vital medical treatment). The finding of points of agreement on the acceptance or not of some life sciences' novelties has been made more difficult due to such axiological divergences and ways of dealing with new situations.[1]

The researches which provide these novelties in the life sciences and specifically in human genetics and biotechnology frequently require neither additional

---

C.M. Romeo-Casabona
e-mail: cromeo@genomelaw.deusto.es

infrastructures nor exceptional material means, as these are not very costly or difficult to obtain. The qualification of the researcher in the specific field is the decisive factor. A small group of researchers can establish themselves backed with a minimum infrastructure and resources. This means that these activities theoretically could be undertaken in any country regardless of the local potential research capacity. That is to say, it is evident that globalization also occurs in the field of human genetics and biotechnology.

Globalization has specific consequences for criminal law. Criminal law is still based on the principle of territoriality, that is, the application of law exclusively to violations within the own state territory. This entails great limitations in the application of its legal norms beyond its political-legal frontiers. International or border-crossing criminal acts are not easily brought to court. Judges have only jurisdiction to apply the laws of their state within the limits in which it exercises its sovereignty.[2] Therefore they cannot persecute persons who have committed punishable acts outside the territory or those who are outside of it, unless some international agreement provides permission is specific situations. Such agreements in fact limit a country's autonomy, and precisely the exercise of sovereignty is little prone to waive its rights in criminal matters, as this field of law is that in which sovereignty is most significantly reflected. To be sure, there are laudable attempts to overcome these limitations (e.g. the International Criminal Court). However, it does not seem at present that human-genetic and biotechnology-related abuses, serious as they are, are going to be heard by the ICC, except perhaps in cases of reproductive human cloning in the (not immediate) future.

Another consideration to take into account is the trans-national collaboration among scientists of several countries which have different legal approaches in some matters including, as it happens, human embryonic stem-cell research. We should find a solution for this kind of research when performed according to local ethical and legal principles and rules.[3] Whose rules should be applied?

In this contribution I reflect on how to prevent undesirable kinds of globalization in the life sciences, particularly when some cultural approaches try to impose their values on other cultural traditions among people of different countries. And secondly I want to consider whether it is possible to share worldwide certain relevant ethical and legal perceptions, in particular those that have a bearing on human rights theory.

2 The Route Towards 'Establishing' Pre-legal Regulatory Norms for the Modern Life Sciences

Bioethics emerged from the cross-disciplinary nature of new methodological procedures for the research and applications on the life sciences. Bioethics, i.e. ethics applied to the life sciences, continues to be a fundamental task of the specialists in that discipline. The important issues confronting the biomedical sciences also demand a joint effort on the part of ethics and law, even though the latter has its own tasks and methods (Kemp 1998: 9ff.; Romeo-Casabona 2002: 30). The meeting point would be found in bioethics, as used in its first meaning as a cross-disciplinary proposal and as a presupposition for bio law (Kemp 1998: 11).

The debate is not the exclusive heritage of the researchers, of empirical scientists or of the experts on the human and social sciences (González de Cancino 1995: 53). The issues involved here concern society as a whole, all of its members. And there has to be room for plurality with the intent of looking for ways of integrating the different ideological and cultural perspectives (flexible bioethics) (Casado 2000: 21ff.). And, thirdly, it should be trans-national and international, bringing together the economic, cultural and ideological diversity throughout the world in order to harmonize national legislations on this matter.

It is important to emphasize the unifying role that the construction of human rights has had. This is a creation of Western culture and has been accepted, to a greater or lesser extent, by people of other cultures. Specifically along these lines, in relation to life sciences, there have been recent contributions from UNESCO, the Council of Europe and the United Nations which should stimulate that unifying process. In a sense, these legal instruments have contributed to the globalization of bioethics and will be further promoted through UNESCO's Universal Declaration on Bioethics and Human Rights (19 October 2005). In this respect it is noteworthy that many countries that neither have a cultural tradition in these matters nor are at the forefront of biomedical investigation, have readily adopted legal measures aimed at, for example, establishing 'informed consent' as a basic right of the patients, or have been quick to prohibit human reproductive cloning.[4]

In fact, International Law has encouraged a trans-cultural perspective in relation to the life sciences.[5] This trans-cultural perspective benefited from the concurrence of several factors:

(a) National laws lack clear and undisputed ethical and cultural references applicable to the new challenges raised by the life sciences. This means that certain values specifically related to human genetics and biotechnology (e.g. human rights, juridical goods) have gained universal recognition with greater ease than other, shall we say, more traditional rights,[6] (e.g. civil and political rights as related to social groups). Declarations on civil and political rights were formulated earlier, often not in tune with the cultural and ethical beliefs of other human communities.

(b) The initial rise and development of the 'Law of Human Genetics and of Biotechnology' has manifested itself generally as a 'soft law' (i.e. legal rules which are neither obligatory nor coercive, but rather advisory, so that disregarding them can not be penalized. The most noteworthy exception to this tendency is in regard to human cloning and to the deliberate release into the environment of genetically modified organisms (GMOs) where several offences have been identified in comparative law for which punishment is, in general, very stiff. This has led to a discussion on whether criminal law provides more than a purely symbolic effect merely to calm citizens and to reassure them that something is being done by policymakers, instead of an effective method to prevent such crimes. However, and specially in the scope of International Law, legislation has slowly but surely moved towards a law characterized by rules of ascertainment, by its legal binding character, and backed by sanctions or other legal consequences.

(c) The concerns proper to the life sciences that constitute the object of International Law do not compromise state sovereignty. Moreover, it is also true that we are beginning to contemplate something beyond the current conception of the nation-state as a manifestation of globalization rather than internationalization.[7] The factors mentioned above have possibly helped the international development of some bioethical principles. In order to reach universal recognition lawgivers must find ways to fulfil this task, if possible by integrating these principles in human rights theory.

3 The Dynamic Role of Human Rights in Relation to Establishing a Common Perspective for the Life Sciences

It is already commonplace to emphasize the link, each time closer and more frequent, between the advances and applications of the biomedical sciences and the rights of the individuals. At times, such a link is presented as an intense dialectic tension, especially when individual rights are mutually in collision or when they collide with collective or supra-individual rights. Lawyers, but not only they, have made an effort to establish relations and hierarchies among them, to identify new rights and new holders of rights, and bring them into the purview of the development of human rights. This is an as yet unfinished task of reconstruction of the theory of human rights. Such reconstruction is not new. Successive earlier reconstructions have yielded so-called 'generations of human rights'. In relation to the life sciences, several generations of these human rights would be affected; in fact, a new generation of human rights would result this time, linked to the most recent contributions of science and technology, in particular genetics, biotechnologies and the new information and communication technologies.

This task must necessarily begin with various premises: first, to take into account the real situation of the life sciences' data, of the biotechnological discoveries and of their possible applications, actual as well as those which could be reasonably undertaken in the near or remoter future. All of which is the object of the legal analysis. The second premise consists in taking into account the assessments provided in the ethical debate, that is, if we do not want to run the risk of the law being blind. Finally, one must take into account the constructions from the perspective of human rights and of other principles or values, especially in international law, but also those emerging in constitutional law, that may have a bearing on biomedical science in general.

The most intense preoccupation in the last decades has been with establishing every possible legal and political mechanism to guarantee respect for human rights and has led to their inclusion in constitutional laws, thereby designating them as rules of the maximum normative rank.

We can agree that 'human rights' (a set of historically evolved articulations that specify the demands of dignity, liberty and human equality) are in full force and should be positively recognized by the legal systems at the national and international levels everywhere. They also form part of the fundamental rights in a certain number of states, which would be those human rights guaranteed by the positive legal system, in the majority of the cases in its constitutional regulations, and which usually are the object of resolute protection.

The theory of human rights has an importantly ethical background and seems to reflect undeniably universal values, as attested to by the international acceptance of human rights in international law. The universality of some of these values is unquestioned. However, the construction of human rights has met with criticism as well, especially to the degree that it concentrates on the individual, which is a main characteristic of Western culture. Critics will point to the collective world view of other cultures (as in the Far East generally, in some parts of Africa and among the indigenous populations of Central and South America), where obligations are held to be more fundamental than rights. Communal harmony rests on observance of the mutual obligations of the community members. It is from the perspective of the duties imposed by the community and the obligations to the community that one will attain the respect of its members.

Be that as it may, one must take advantage of the universal acceptance which human rights have been enjoying and must continue to prudently take them as a reference point to universally identify, accept and share a set of legally based ethical values. Furthermore, human rights are not static, nor do they aspire to create a closed universe. On the contrary, they are in continuous evolution, taking in new rights in relation to human needs, and therefore constitute a very valuable instrument for the shaping of new rights in the context of genetics and biotechnology.

The challenge of our time is to assure that an ethical globalization will emerge in a trans-cultural framework. This framework should welcome the universal acceptance of certain shared values and rights which can provide the answers demanded by the challenges of truly worldwide globalization. To this end, it is necessary in the next decades to find ways to mesh the two dimensions of principles and rights, that is, the individual and the collective, which should be constituted as an axiological instrument and instituted as charter of good fellowship.

Although the list of desiderata is endless, I limit myself to mention those that I consider essential for the scope of globalization. I leave aside references to certain civil and political rights, including some social rights, recognized by universal declarations or treaties. I also pass over considerations which more specifically constitute the core of rights related to human genetics, such as the human genome as being a common heritage of humanity,[8] the right to self-integrity, the right to one's genetic identity, the right to the protection of personal genetic data, the right not to be discriminated on grounds of genetic characteristics, etc.

As I see it, the following principles are minimally required to shape the contours of a basis for universal recognition of human rights: the principles of responsibility

(Jonas 1984: 153ff.), of solidarity, of justice – whatever this could mean – (Rawls 1971: 453ff.), equity, tolerance (Arthur Kaufmann),[9] non-discrimination and responsibility towards future generations.[10]

The starting point for these principles can be found in several reflections. The first is based on the well-known affirmation that the present time is characterized by both a *proliferation of means* and *confusion about ends*. German philosopher Hans Jonas stated that the human being has been capable of increasing his knowledge and his power; however, he has not worried with the same intensity about knowing the consequences of his power (Jonas 1984: 153ff.).

The twentieth century has witnessed an evolution in the *raison d'être* of science. Philosophers of science make clear that science no longer limits itself to rational explanation of the universe and everything that exists or might exist in it, including living matter. Although there is still a long way to go, science is moving towards increasingly accurate prediction of future events, and is able to interfere in them, to modify them, as in the case of living matter. Science has as it were moved beyond itself, as it no longer only seeks the truth but seeks to create truth, at the whim of the human being. But if the human being is a moral being, therefore capable of self-consciousness, of reflection on his own acts and their consequences, of experiencing the sense of freedom to decide and act in consequence, that is, if he is a responsible being, then the scientist must also be responsible. In possession of knowledge, aware of its power, capable of 'making' the world, the scientist must assume responsibility for the 'truth' so created.

Against this background, Jonas is led to say that traditional ethics is barely applicable in the actual circumstances, as our behaviours are going to transcend to the human beings that come after us. Apel, loyal to the thinking of Kant, proposes the construction of an 'ethics for the era of science', and this is a universal and urgent task (Da Cósta 2002: 45ff.).

Max Weber elaborated an 'ethics of responsibility', when he was trying to find a policy ethics aimed at the consequences.[11] In fact, this ethics evaluates actions by taking into account the consequences expected as possible or probable. It is essential to pay attention to the existing relation between means and ends and to the real situation in which the human action must develop. However, this ethics has little to say with respect to the valuation that the action in itself deserves.

It was Jonas (building on Weber) who developed a more convincing ethics of responsibility applied to technique and medicine. This ethics is especially applicable to biotechnologies, as an ethics of the future that binds all of us with the natural surroundings: 'Given that it is nothing less than the nature of man that enters within the scope of power of human intervention, caution will be our first moral mandate and the hypothetical thought our first task. To think of the consequences before acting is no more than common intelligence. In this special case, wisdom requires us to go further and to examine the eventual use of the capacities before they are completely ready for their use.' Jonas summarizes his thoughts with the following maxim on behaviour: 'act so that the effects of your action are compatible with the permanence of genuine human life' (Jonas 1984: 36). Jonas considers that the biotechnological revolution must take into account that, while mechanical errors are reversible, biogenetic errors are irreversible.

Jonas proposes a fundamental rule for the treatment of uncertainty: '*in dubio pro malo*, in case of doubt, the stakes are so high that we must set our eyes on the worst case scenario' (Jonas 1985: 49).

Usually, the ethics of responsibility has been contrasted with the ethics of conviction, though they are not considered irreconcilable: 'an ethics of responsibility that is not based in correct convictions must degenerate into a trivial ethics of success .... On the other hand, an ethics of conviction that is blind to the consequences is no longer based on a correct conviction' (Kaufmann 1999: 542ff.).

Arthur Kaufmann believes that the tolerance principle makes it possible for us to behave responsibly, but this is completed by the principle of responsibility, as the solution is not found in omitting the behaviour in the case where there is a risk that cannot be evaluated in advance. A morality whose first concern would be to avoid every error and conflict is not responsible morality. Tolerance has its root in the idea of freedom, its framework in a plural society and its motive is reasoning, not indulgence. These and other reflections lead Kaufmann to formulate tolerance as a categorical imperative: 'Behave in a manner such that the consequences of your actions are compatible with the greater possibility of avoiding or diminishing human misery' (Kaufmann 1997: 582).

These axiological principles and rights have the added value that they can be implemented in their individual as well as collective dimensions, with the effect that they are rights that individuals as well as human groups and communities are entitled to. I am not about to analyse the content of each of them, as they are well known, and at the same time so complex that to do them justice each deserves an entire monograph.

Important though they are, we must be aware that they are principles or rights that are not always shared by all cultures, and that even in Western culture some of them are barely making their way, at least as legal regulations. Nevertheless, the unstoppable extension and penetration of the phenomenon known as globalization requires us to find counterweights and balances against the very grave risks attending sweeping sources of power that cannot be controlled by individual states or even the international community. Human genetics and biotechnology constitute most attractive temptations to surpass any limit, any control, and human rights can be an instrument suited to moderate them or least be a step in that direction.

A brief reflection on the dignity of the human being should be added here. Among the various perspectives in which human dignity can be articulated, there is no doubt that the mainstream Western conception is closely associated with Kantian thought. There is no doubt that its acceptance has increased and its status is now almost that of a universal principle, even if not necessarily always seen as a fundamental right.[12] 'Human dignity' is considered a quality inherent to the human being, which is projected legally over specific fundamental rights. Its relevance as a limit and barrier against potential abuses from biomedicine in the human being is of first-rate importance (Cortina 2004: 29), as it carries with it the prohibition of utilizing the human being – any human being – as merely an instrument rather than as an end in itself.

Unfortunately, appeals to the dignity of the human being have been excessively frequent and abusive. 'Human dignity' is presented as a sweeping, authoritative argument against violations, but, brushing aside the many genuine advances in the biomedical sciences, the argument evinces little affinity with or insight in the specifics of the issues under discussion. Such argumentation precludes dialogue meant to facilitate a meeting of minds and finding points of consensus.

In spite of this unfortunate development I suggest that we should look deeper into the potential content of the concept of human dignity. I think that its contributions in the field of the life sciences can still be enriching, on condition that it is used with deliberation and loyalty to its true sense.

For example, we should explore other dimensions of human dignity than its personalistic slant. I am thinking of situations where genetic intervention is aimed not only at improvement but beyond that is designed to select or bring about specific traits or biological characteristics considered desirable from a subjective point of view (e.g. of the parents in respect to their actual or unborn children, of those concerned or of public powers). We should reflect on whether the personalistic conception of dignity is sufficient to contain the diverse assumptions associated with genetic interventions. These genetic interventions imply conflicts that go further than the individual dimension of the human being because sometimes they affect collective dimensions. From the perspective of a theoretical analysis we are permitted to hypothesize that genetic interventions can affect the human species.

Further thought should be given to whether human dignity could also have, at the same time, a supra-individual dimension for such situations. We might start from an objective dimension of assessments of human dignity and project these, as is the case in relation to other fundamental human rights. Some future measures of genetic intervention for achieving perfection or improvement (which if practised in the gametes before reproduction or in the zygote could be considered as eugenic practices) will at least potentially affect the human species or human ethnic groups. This would happen because the genetic make-up which characterizes specific species or specific groups would be modified and so involve the future generations. A supra-individual concept of human dignity (which need not exclude other concepts) may serve to mitigate or reject such practices, independent of the specific individuals who could be affected. It may also provide more clarifying approaches when one has to determine what is 'normal' and what is 'pathological'. And finally the process of world globalization that we are witnessing now is certain to require this type of elaboration in relation to human dignity. In this connection one can bear in mind UNESCO's statement that 'in full respect of the dignity of the human person and human rights, the human genome must be protected. Scientific and technological progress should not in any way impair or compromise the preservation of the human and other species.'[13]

4 International Law: Source for the Expansion of Human Rights in the Life Sciences

The preamble of the Universal Declaration on Human Rights (10 December 1948) already proclaimed the principles or values which impregnate all the articulated text: 'Liberty, justice and peace in the world have as their basis the acknowledgement of the intrinsic dignity and the equal and inalienable rights of all the members of the human family.'

The convergence of international law and national state laws, as mentioned earlier, is increasing becoming manifest especially in relation to human rights. It has been noted that in the immediate future this relation will intensify in the specific areas of the biomedical sciences. This observation is of enormous importance given that, for obvious reasons, up until recently international legal texts and national constitutional texts contained few or no explicit references to human rights as affected by current scientific progress, notwithstanding the fact that the situation is undergoing truly radical change.

Fortunately, we have other integrating reinforcements of the new or renovated rights, goods, or values prompted by applications of some results of the biomedical sciences:

(a) There are, first, express references in some constitutions to the Universal Declaration on Human Rights and other international conventions for the interpretation of fundamental rights. This legal statement is of the utmost importance together with the European Convention on Human Rights and Biomedicine, and other similar covenants or agreements that may be approved in the future.
(b) Secondly, the *general principles of law* are part of the internal legal regulations – but are also part of international law, with a double function: to integrate (as a reference for the hierarchy of norms) and to inform (as an interpretative source) the legal regulations. This way, law does not identify itself with statutory law exclusively, but also includes a series of principles orientated to criteria of material justice. Even though these principles are not explicitly collected in the law books, they do shape, together with custom and (in some legal systems) jurisprudence, the totality of the legal system. According to the extended criteria among internationalists, the international declarations can be integrated in the general principles of law, because they are not considered coercive legal rules in a strict sense. We see this to be that case with the Universal Declaration on Human Rights and with the Universal Declaration on the Human Genome and Human Rights.

As we have been able to verify, many of these principles, in turn, also have been taken up in international law and are of a paramount importance to achieve adequate focus in relation to the biomedical sciences. The interaction between international law and the internal (state) law is gaining intensity and depth.

5 Achievements Towards a Trans-Cultural Law on the Life Sciences and Human Genetics (Bio Law)

5.1 International Law

In the specific field of the life sciences, numerous human rights have been developing or have been identified, at times as part of the content of some fundamental rights proclaimed by the greater part of the modern constitutions (e.g. the right to life, to physical integrity, to ideological freedom, freedom of conscience, privacy, etc.). Other times, these rights have also been inspired by the Universal Declaration, by international agreements (in particular, the International Pact of Civil and Political Rights of 1996) or by regional agreements, as in the European case, the Convention on Human Rights and Fundamental Freedoms of 1950, and the American counterpart, the American Convention on Human Rights of 1969. The Convention of the Council of Europe on Human Rights and Biomedicine[14] has meant a qualitative step in the identification of new human rights or of, at least, the new perspectives that are offered in relation to scientific advances. This European Convention has already been incorporated to the internal laws of several countries – among them Spain – with binding and obligatory force for all the respective public powers. The Universal Declaration of UNESCO on the Human Genome and the Human Rights (11 November 1997), even though it lacks obligatory competence, is also an indisputable moral force to guide the nation states.

This commendable approach towards more collective perspectives of human rights can especially be noticed in the UNESCO declarations. They prove sensitive to the need to protect the interests of future generations (Declaration on the Responsibilities of the Present Generations Towards Future Generations, 1997) and to the need of solidarity among people (Universal Declaration of the Human Genome and Human Rights, 1997).

In this direction, UNESCO has taken the initiative for a Universal Declaration on Bioethics and Human Rights,[15] where relevant aspects of the life sciences and human rights are included.

5.2 The Constitutional Rights of the 'Bioethical' Citizen as a Response to the Biocrats

Constitutional law has great potential, both as a 'collector' of human rights that specifically are being affected by the biomedical sciences, and as an instrument to resolve the conflicts that emerge here. We can find numerous examples of conflict resolution in modern comparative constitutional law. Undoubtedly, the incipient examples that exist over the acknowledgement of some rights associated with the life sciences constitute a novelty for contemporaneous constitutional law.[16] Moreover, this process, which has been slow in its beginning, is logical. If the human rights associated with these matters have found their niche in international law, it is logical that some fundamental rights are taken up into modern constitutional law, inasmuch as they offer new perspectives for the protection of citizens. In this sense, there is no doubt about the influence that international law is exerting – and is certain to exert in the future – over the fledgling constitutional law of bioethics.

These rights gained acknowledgement and are now included in the political constitutions of some states, such as Switzerland,[17] Portugal,[18] Greece,[19] Ecuador[20] and Venezuela.[21]

Thus, the following text was introduced in the Constitution of the Swiss Confederation (1992): 'The genetic material of a person will not be analysed, registered or revealed without a previous consent, except when expressly authorized or imposed by law' (art. 24). In 1999 this provision was replaced by some clauses on the protection of health, the transplant, non-human gene technology and gene technology in the human field. These matters are reserved as legislative competence of the Confederation, without pre-judgement as to whether or not they should be designated as fundamental rights. In reference to gene technology in the human field, the constitution states:

> Reproductive medicine and gene technology in the human field. 1. All human beings must be protected against the abuses of reproductive medicine and gene technology. 2. The Confederation shall legislate the use of human reproductive and genetic material. It shall ensure the protection of human dignity, of the personality, and of the family, in particular it shall respect the following principles: a. All forms of cloning and interference with genetic material of gametes and human embryos is prohibited; b. non-human reproductive and genetic material may neither be introduced into nor combined with human reproductive material; c. Methods of medically assisted procreation may only be used when sterility or the danger of transmission of a serious illness cannot be avoided otherwise, but neither in order to induce certain characteristics in the child nor to conduct research. The fertilization of human ova outside a woman's body will only be permitted in the instances provided by law; and only the number of human ova that are capable of being immediately implanted into her will be developed outside a woman's body up until the stage of embryos; d. The donation of embryos and all forms of surrogate maternity are prohibited; e. No trade may be conducted with human reproductive material or with any product obtained from embryos; f. A person's genetic material may only be analysed, registered or disclosed with the consent of that person, or if a law provides so; g. Every person shall have access to the data concerning his or her ancestry.
>
> (art. 119)

This constitutional regulation is rather undefined with regard to its legal nature. It is also a very detailed and prohibitive regulation – it opens the door to purely retributive legislation – which is by all means excessive in many instances.

In contrast, the Portuguese Constitution introduces an important fundamental right: 'The law shall guarantee the personal dignity and the genetic identity of the human being, specifically in the creation, development and utilization of the technologies and in genetic experimentation.' (art. 26.3).

Also the Greek Constitution has some rules on genetic data and the protection against biomedical interventions: 'All persons shall enjoy full protection of their health and genetic identity. All persons shall be protected with regard to biomedical interventions as provided by law' (art. 5.5).

The Constitution of Ecuador states the following: 'Unfair use and application of human genetic material is prohibited (art. 23.2); The State will promote the scientific-technological advance in the field of health according to bioethical principles' (art. 44).

Finally, I quote the Constitution of the Bolivarian Republic of Venezuela: 'Genomes of live beings will be not submitted to any copyright and this matter will be regulated by law on grounds of bioethical principles' (art. 127).

Apart from the extent to which these statements are on target, one must recognize that these constitutional precepts constitute the first references at this highest national level related to the concepts of autonomy of the individual, to the genetic heritage and to a right to a genetic identity of the human being, as rights of the 'bioethical citizen'.[22] They comprise the core of a constitutional law of biomedicine that will be developing in the years to come as a barrier against the pressures by certain researchers and enterprises that do not recognize any slowdown in the progress of science and the concomitant economic benefits (the biocrats).

6 Difficulties and Defeats for a Trans-Cultural Law for the Life Sciences

Introduction of adequate international legal regulation of the life sciences has always depended on the extent of consensus that can be achieved in relation to the following: the scope of its contents, the different constructions of the human rights involved, and the limits, either permanent or temporary, that can be imposed on the scientific activities in the clinical applications or in any other related activity.

This is not an easy task. The initiative undertaken by the UN to seek approval of a universal Convention prohibiting reproductive human cloning and so-called therapeutic cloning is a case in point. Actually, human reproductive cloning is one issue – practically the only one – in which there is a measure of consensus: human reproductive cloning has met with broad rejection in the scientific community at large, by national authorities and international institutions. This has been the case with the UNESCO's Universal Declaration on the Human Genome and Human Rights (in particular art. 11) and the corresponding Protocol to the Convention on Human Rights and Biomedicine of the Council of Europe. In the supranational sphere, the future European Constitution has likewise provided an explicit prohibition on human reproductive cloning and there are numerous states that have enacted laws with the same purpose, at times even designating cloning to be a criminal offence. This favourable climate probably encouraged the corresponding authorities at the UN to take the correct initiative to approve a Convention aimed at universal prohibition of the cloning of human beings.

But even here, one can conceive of scenarios in which its illegitimacy is not so evident (e.g. if and when it were a really safe technique, it could be used to combat the infertility of a couple or to prevent the transmission of hereditary diseases to the descendants).[23] And this is where the trouble started. As the proposed convention was being drafted, representatives of some states advocated that its framework be broadened to include all types of human cloning, thereby also seeking the interdiction of so-called 'therapeutic cloning' for research purposes (perhaps more adequately termed 'research cloning'). In this way, there was a shift from the aim to regulate a matter on which there was a wide consensus, towards the inclusion of another matter that still today is the object of intense debates, as the bio scientists, the experts on the human and social sciences, the policy makers and the public are divided and the opposing positions seem irreconcilable, at least at the present time.

The consequence of this enlargement of the initial objectives was that the adoption of the Declaration of the UN on Human Cloning underwent a tortuous process before it was at last approved.[24] In fact, the document has been approved as a 'Declaration', rather than a 'Convention' – implying from the legal point of view a significant weakening of it as normative instrument. The approval was preceded by several failed attempts.

The vote count that ultimately, after many postponements, led to its approval is very revealing: there were 84 votes in favour, 34 against and 37 abstentions, 37 member states absent. As I see it, this vote clearly reflects the lack of a sufficiently broad consensus. In view of the relevance of the issues concerned it would have been advisable to continue with the negotiating efforts. It seems certain that restricting the document to the explicit prohibition of human reproductive cloning would have been the more suitable course of action. In that case it could have been left up to the states, for the time being, to adopt internal legal regulations in relation to other types of cloning, in accordance with their own criteria.

As it is, the prohibition of 'therapeutic' or non-reproductive cloning is facing irreconcilable cultural conceptions, and pressure on the part of scientific community (Graumann and Poltermann 2005: 209ff.). The miscarrying of this plural and conflicting situation has led to much criticism of this UN Declaration.[25] If the first criticism that can be made is that the Declaration is not carried by the desired broad consensus among the UN member states, the second observation is in reference to the fact that the approved text, in spite of having wanted to close all the possibilities to any type of human cloning, is very far from having achieved this. In fact, through the interpretation of legal norms, one can already notice that the prohibition of any type of cloning will only be so 'inasmuch as' (read: to the degree that) they are incompatible with human dignity and the protection of human life. This entails accepting that, at least hypothetically, there may be scenarios where such incompatibility would not exist.[26] The question will be to determine which scenarios will not be prohibited, and under what circumstances, and which scenarios will encounter interdiction.

A third point of criticism is that for other matters now included, although important in themselves, the Declaration does not seem to be the most appropriate legal instrument. On the one hand this blurrs the purposes of the regulation on human cloning and on the other, this seems to take detract from the prominence that these other points deserve. These matters should be regulated via specific and independent legal instruments. Examples of them as added to the text of the Declaration are the following: the obligation to take measures to prevent the exploitation of women in the application of the life sciences and the financing of medical research, including the life sciences, to take into account the pressing global issues such as HIV/AIDS, tuberculosis and malaria, which affect in particular the developing countries.

Finally, in relation to the obligations that this Declaration can entail for the states, it must be kept in mind that a Declaration does not have as strict a binding nature as a Treaty (or a Convention). Its status is regulative rather than prescriptive or obligatory. At best we can say that the principles of the Declaration *should* guide the acts of the legislators of the states.

I conclude that the United Nations probably has missed an opportunity to achieve the unanimous approval of a legal instrument, which in turn would have increased its prestige and its desirable leadership within the area of human rights related to the biomedical sciences. As it is, the confrontation between the states, where some sought to impose on others a specific position on human cloning as a whole in respect to which there was no pronouncement expected, as well as the foreseeable lack of efficacy of the approved Declaration, do not seem to have contributed to the credibility and unifying capacity that such an important Institution should have.

The dialectic tension generated by the ethical and legal debates on and status of the human embryo is also reflected in the difficult and self-contradictory equilibrium maintained in the Council of Europe Convention on Human Rights and Biomedicine. There is a ban on the creation of human embryos for use in experimentation, but, at the same time, there is (implicit) permission to use them for these ends in the allusion to the surplus frozen embryos coming from the techniques of assisted reproduction.[27]

7 Concluding Comment

Let me try to summarize the following conclusions:[28]

International Law has promoted a global perspective in relation to the life sciences' technologies. This global perspective has been favoured by the lack in national state laws of ethical and cultural reference points for clear and undisputed legislative response to the new challenges created by the biomedical technology. UNESCO's Universal Declaration on the Human Genome and Human Rights and the Universal Declaration on Bioethics and Human Rights, along with the European Convention on Human Rights and Biomedicine are significant contributions for the development of a Bio law on the international level.

Globalization does not mean an automatic sharing of cultural values and traditions, as there is a specific cultural and historical basis for the diversity of these in society. Nowadays we need to go further in order to reach a truly trans-cultural ethics and trans-cultural law in the field of human genetics and biotechnology, that is to say, to share a common worldwide threshold.

Some values (human rights, juridical goods) related to the human life sciences have achieved universal recognition with great ease, but at the same time these have not always found an adequate match with certain cultural and ethical conceptions by some non-Western human communities, as these have a holistic view of human social relations.

In relation to human genetics and biotechnology, we also find a 'soft' law and some 'soft' values. However, though inevitable in some cases, I feel that this soft approach to the implementation of law and values is undesirable, and at best only a temporary strategy, as it could give rise to the following consequences:

(a) The intrusion of several value contradictions in the national legislation, as can be seen in a disparity whereby there is simultaneously an intense protection in some legislations and an evident lack of protection in others. An example is the legal framework that is being established on the possibility of research using human in-vitro embryos.
(b) The introduction and institutionalization of a perspective of globalization that is not trans-culturally sustained, e.g. in relation to the blanket prohibition of human reproductive cloning in all the states that have legally undertaken such research activities.
(c) The recourse to a symbolic law, whereby the legislator is more worried about expressing a moral and social rejection of certain activities, meant more to combat social panic than to establish effective legal persecution procedures to tackle current (and future!) activities including the means available through Criminal Law.
'Human rights' continues to be the unavoidable reference point to capture the multiple challenges of biomedical technology. The achieved development of these rights is the result of an ethical construction that gives them conceptual support and axiological credibility. In the future, it will be necessary to look further into the following aspects:
(d) Human rights present an objective dimension that permits the protection of realities or situations through them, independent of the possibility of accepting the existence of a subject titleholder of a specific right.
(e) Human rights are not only individual rights, they have a collective dimension that must be implemented. This dimension serves to guarantee adequate protection of specific social groups and communities, without detracting from the individual dimension as a hypothetical side effect; and
(f) From this point of view, there should be far greater awareness of the tremendous importance of a collective perspective on human rights, i.e. on those aspects that guarantee coexistence. In order to develop them, we must take into account principles such as responsibility, solidarity, justice, equity, tolerance, non-discrimination and responsibility towards future generations.

Notes

[1] See further on this Romeo-Casabona (in press).
[2] The principle of personality (the application of criminal law to the citizens, even though the crime might have occurred outside the national boundaries) continues to present its own problems, as occurs in Germany in relation to the interventions with human embryos. Eser and Koch 2004: 37ff., 53ff.

[3] As a result of this concern see The Hinxton Group (2006), *An International Consortium on Stem Cells, Ethics and Law. Consensus Statement*: 'In countries with laws that restrict elements of human embryonic stem-cell (hESC) research but that do not expressly prohibit international collaborations, research institutions should neither discriminate against nor restrict the freedom of their investigators who want to travel to do work that is undertaken with scientific and ethical integrity.'
[4] E.g. Peru (Peruvian Penal Code, art. 324), Vietnam and China, even though China has been quick to authorize the so-called 'therapeutic' cloning.
[5] See further on this Romeo-Casabona (in press).
[6] In this way Teboul (2004: 481ff., 496ff.), has outlined that universal norms rather than international norms are needed in this field.
[7] See in this sense, Singer 2004: 8.
[8] In a symbolic dimension, as stated by UNESCO's *Universal Declaration on Human Genome and Human Rights* (art. 1).
[9] Kaufmann 1999: 321ff. See Spanish translation by L. Villar-Borda & A.M. Montoya (Kaufmann 1997). See also on the various approaches of this principle, Saada-Gendron 1999.
[10] In reference to some of these rights or principles, see Romeo-Casabona 2002: 32–34.
[11] Weber, in AAVV, *Historia del pensamiento*. Vol. VI: 74ff.
[12] Cf. The German Constitution of 1949 states human dignity as a truly fundamental right (see art. 1).
[13] Declaration of UNESCO (12 November 1997) on the 'Responsibilities of the present generations towards future generations', art. 6.
[14] Convention for the Protection of Human Rights and Dignity of the Human Being with regard to the Application of Biology and Medicine, 4 April 1997.
[15] UNESCO, *Universal Declaration on Bioethics and Human Rights*, Paris, 19 October 2005.
[16] For a similar perspective see Gros Espiell 2005: 143ff.
[17] As reviewed in 1992 (art. 24) and in 1999, although there are not presented as rights, rather as exclusive jurisdiction of the Confederation.
[18] According to a reform of the Portuguese Constitution in 1997.
[19] As modified in 2001.
[20] According to the Constitution of 1998.
[21] Constitution of the Bolivarian Republic of Venezuela (1999).
[22] On this word see Fagot-Largeault 1985, *passim*.
[23] Cf. Grupo de expertos sobre Bioética y Clonación 1999: 98ff.; Romeo-Casabona 1997: 21ff.
[24] United Nations Declaration on Human Cloning, approved by the General Assembly on 8 March 2005. According to it: [...] '(b) Member States are called upon to prohibit all forms of human cloning inasmuch as they are incompatible with human dignity and the protection of human life; (c) Member States are further called upon to adopt the measures necessary to prohibit the application of genetic engineering techniques that may be contrary to human dignity.'
[25] See Romeo-Casabona 2005: 15ff.; Bergel 2005: 49ff.
[26] This is specially clear in the Spanish ('. . . en la medida en que sean incompatibles con . . .') and French ('. . . dans la mesure où elles seraient incompatibles . . .') versions of the Declaration.
[27] See further on this C.M. Romeo-Casabona 2003: 557ff.
[28] As I have suggested in Romeo-Casabona (in press).

References

Bergel, S.D. "La Declaración de las Naciones Unidas sobre la Clonación de Seres Humanos del 08-03-05." In *Revista de Derecho y Genoma Humano/Law and the Human Genome Review*, 22, 49–56, 2005.

Casado, M. "Hacia una concepción flexible de la bioética." In M. Casado (ed.). *Estudios de bioética y derecho*. Valencia: Tirant lo Blanch, 2000.

Cortina, A. "Una ética transnacional de la corresponsabilidad." In V. Serrano (ed.). *Etica y globalización: cosmopolitismo, responsabilidad y diferencia en un mundo global*. Madrid: Biblioteca Nueva, 15–32, 2004.
Da Costa, R. *Ética do discurso e verdade em Apel*. Belo Horizonte: Del Rey, 2002.
Eser, A. and Koch, H.-G. "La investigación con células troncales embrionarias humanas: Fundamentos y limites penales." Translated from German by A. Urruela Mora. *Revista de Derecho y Genoma Humano/Law and the Human Genome Review* 2004, 20, 37–64, 2004.
Fagot-Largeault, A. *L'homme bio-éthique:Pour une déontologie de la recherche sur le vivant*. Paris: Maloine, 1985.
González de Cancino, E. *Los retos jurídicos de la genética*. Bogotá: Universidad Externado de Colombia, 1995.
Graumann, S. and Poltermann, A. "No End in Sight to Cloning Debate." *Revista de Derecho y Genoma Humano/Law and the Human Genome Review*, 22, 209–227, 2002.
Gros Espiell, H. *Ética, Bioética y Derecho*. Bogotá: Temis, 2005.
Grupo de expertos sobre Bioética y Clonación. In S. Calles (ed.). *Report About Cloning*. Madrid: Fundación Ciencias de la Salud, 1999.
Hassemer, W. "Kennzeichen und Krisen des modernes Strafrechts." *Zeitschrift für Rechtspolitik*, 379–384, 1992.
Jonas, H. *Das Prinzip Verantwortung: Versuch einer Ethik für die technologische Zivilisation*. Frankfurt a.M.: Suhrkamp, 1984.
Jonas, H. "Técnica, medicina y ética." (Original work: *Technik, Medizin und Ethik: Praxis des Prinzips Verantwortung*. Frankfurt a.M.: Suhrkamp, 1985).
Kaufmann, A. *Rechtsphilosophie*. 2nd ed. München: C.H. Beck, 1999.
Kemp, P. "The Bioethical Turn." In *Studies in Ethics and Law: From Ethics to Biolaw*. Copenhagen: Centre for Ethics and Law University of Copenhagen, 1998.
Rawls, J. *A Theory of Justice*. Cambridge (Mass.): Harvard University Press, 1971.
Romeo-Casabona, C.M. "Legal Limits on Research and Its Results? The Cloning Paradigm." *Law and the Human Genome Review*, 6, 21–42, 1997.

Romeo-Casabona, C.M. "Embryonic Stem-Cell Research and Therapy: The Need for a Common European Legal Framework." *Bioethics*, 16, 6, 557–567, 2003.
Romeo-Casabona, C.M. (ed.). *El principio de precaución*. Cátedra Interuniversitaria Fundación BBVA – Diputación Foral de Bizkaia de Derecho y Genoma Humano. Bilbao; Granada: Editorial Comares, 2004.
Romeo-Casabona, C.M. "Editorial." *Revista de Derecho y Genoma Humano/Law and the Human Genome Review*, 22, 15–26, 2005.
Romeo-Casabona, C.M. "Is It Possible a Transcultural Law for Human Genetics and Biotechnology?" In A. Fagot-Largeault & J.M. Torres (eds.). *The Influence of Genetics in Scientific and Philosophical Thinking*. Kluwer Academic Publishers, 2007, 193ff.
Saada-Gendron, J. *La tolerance*. Paris: Flammarion, 1999.
Singer, P. *One World: The Ethics of Globalization*. 2nd ed. New Haven & London: Yale University Press, 2004.
Teboul, G. "À propos du Droit International de la Bioéthique." In E*tudes offertes à Jacques Dupichot: Liber Amicorum*. Bruxelles: Bruylant, 481–500, 2004.
The Hinxton Group. *An International Consortium on Stem Cells, Ethics and Law: Consensus Statement*. Hinxton: Cambridge, 24 February 2006.
Weber, M. In AAVV, *Historia del pensamiento*. Vol. VI. In Astral, Barcelona (ed.), 1985.

International Library of Ethics, Law, and the New Medicine

1. L. Nordenfelt: *Action, Ability and Health.* Essays in the Philosophy of Action and Welfare. 2000 ISBN 0-7923-6206-3
2. J. Bergsma and D.C. Thomasma: *Autonomy and Clinical Medicine.* Renewing the Health Professional Relation with the Patient. 2000 ISBN 0-7923-6207-1
3. S. Rinken: *The AIDS Crisis and the Modern Self.* Biographical Self-Construction in the Awareness of Finitude. 2000 ISBN 0-7923-6371-X
4. M. Verweij: *Preventive Medicine Between Obligation and Aspiration.* 2000 ISBN 0-7923-6691-3
5. F. Svenaeus. *The Hermeneutics of Medicine and the Phenomenology of Health.* Steps Towards a Philosophy of Medical Practice. 2001 ISBN 0-7923-6757-X
6. D.M. Vukadinovich and S.L. Krinsky: *Ethics and Law in Modern Medicine.* Hypothetical Case Studies. 2001 ISBN 1-4020-0088-X
7. D.C. Thomasma, D.N. Weisstub and C. Hervé (eds.): *Personhood and Health Care.* 2001 ISBN 1-4020-0098-7
8. H. ten Have and B. Gordijn (eds.): *Bioethics in a European Perspective.* 2001 ISBN 1-4020-0126-6
9. P.-A. Tengland: *Mental Health.* A Philosophical Analysis. 2001 ISBN 1-4020-0179-7
10. D.N. Weisstub, D.C. Thomasma, S. Gauthier and G.F. Tomossy (eds.): *Aging: Culture, Health, and Social Change.* 2001 ISBN 1-4020-0180-0
11. D.N. Weisstub, D.C. Thomasma, S. Gauthier and G.F. Tomossy (eds.): *Aging: Caring for our Elders.* 2001 ISBN 1-4020-0181-9
12. D.N. Weisstub, D.C. Thomasma, S. Gauthier and G.F. Tomossy (eds.): *Aging: Decisions at the End of Life.* 2001 ISBN 1-4020-0182-7 (Set ISBN for vols. 10-12: 1-4020-0183-5)
13. M.J. Commers: *Determinants of Health: Theory, Understanding, Portrayal, Policy.* 2002 ISBN 1-4020-0809-0
14. I.N. Olver: *Is Death Ever Preferable to Life?* 2002 ISBN 1-4020-1029-X
15. C. Kopp: *The New Era of AIDS.* HIV and Medicine in Times of Transition. 2003 ISBN 1-4020-1048-6
16. R.L. Sturman: *Six Lives in Jerusalem.* End-of-Life Decisions in Jerusalem-Cultural, Medical, Ethical and Legal Considerations. 2003 ISBN 1-4020-1725-1
17. D.C. Wertz and J.C. Fletcher: *Genetics and Ethics in Global Perspective.* 2004 ISBN 1-4020-1768-5
18. J.B.R. Gaie: *The Ethics of Medical Involvement in Capital Punishment.* A Philosophical Discussion. 2004 ISBN 1-4020-1764-2
19. M. Boylan (ed.): *Public Health Policy and Ethics.* 2004 ISBN 1-4020-1762-6; Pb 1-4020-1763-4
20. R. Cohen-Almagor: *Euthanasia in the Netherlands.* The Policy and Practice of Mercy Killing. 2004 ISBN 1-4020-2250-6

International Library of Ethics, Law, and the New Medicine

21. D.C. Thomasma and D.N. Weisstub (eds.): *The Variables of Moral Capacity*. 2004 ISBN 1-4020-2551-3
22. D.R. Waring: *Medical Benefit and the Human Lottery*. An Egalitarian Approach. 2004 ISBN 1-4020-2970-5
23. P. McCullagh: *Conscious in a Vegetative State? A Critique of the PVS Concept*. 2004 ISBN 1-4020-2629-3
24. L. Romanucci-Ross and L.R. Tancredi: *When Law and Medicine Meet: A Cultural View*. 2004 ISBN 1-4020-2756-7
25. G.P. Smith II: *The Christian Religion and Biotechnology*. A Search for Principled Decision-making. 2005 ISBN 1-4020-3146-7
26. C. Viafora (ed.): *Clinical Bioethics*. A Search for the Foundations. 2005 ISBN 1-4020-3592-6
27. B. Bennett and G.F. Tomossy: *Globalization and Health*. Challenges for health law and bioethics. 2005 ISBN 1-4020-4195-0
28. C. Rehmann-Sutter, M. Düwell and D. Mieth (eds.): *Bioethics in Cultural Contexts*. Reflections on Methods and Finitude. 2006 ISBN 1-4020-4240-X
29. S.E. Sytsma, Ph.D.: *Ethics and Intersex*. 2006 ISBN 1-4020-4313-9
30. M. Betta (ed.): *The Moral, Social, and Commercial Imperatives of Genetic Testing and Screening*. The Australian Case. 2006 ISBN 1-4020-4618-9
31. D. Atighetchi: *Islamic Bioethics: Problems and Perspectives*. 2006 ISBN 1-4020-4961-7
32. V. Rispler-Chaim: *Disability in Islamic Law*. 2006 ISBN 1-4020-5052-6
33. Y. Denier: *Efficiency, Justice and Care*. Philosophical Reflections on Scarcity in Health Care. 2007 ISBN 978-1-4020-5213-2
34. Y. Hashiloni-Dolev: *A Life (Un)Worthy of Living*. Reproductive Genetics in Israel and Germany. 2007 ISBN 978-1-4020-5217-0
35. M.A. Roberts and D.T. Wasserman (eds.): *Harming Future Persons*. Ethics, Genetics and the Non-identity Problem. 2008 ISBN: 978-1-4020-5696-3
36. D.N. Weisstub and G. Diaz-Pintos (eds.): *Autonomy and Human Rights in Health Care*. An International Perspective. 2007 ISBN 978-1-4020-5840-0
37. V. Launis and J. Räikkä (eds.): *Genetic Democracy*. Philosophical Perspectives. 2007 ISBN: 978-1-4020-6205-6
38. D. Birnbacher and E. Dahl (eds.): *Giving Death a Helping Hand*. Physician-assisted Suicide and Public Policy. An International Perspective. 2008 ISBN: 978-1-4020-6495-1
39. M. Düwell, C. Rehmann-Sutter and D. Mieth (eds.): *The Contingent Nature of Life*. Bioethics and Limits of Human Existence. 2008 ISBN 978-1-4020-6762-4

MIX
Papier aus verantwortungsvollen Quellen
Paper from responsible sources
FSC® C105338

If you have any concerns about our products,
you can contact us on
ProductSafety@springernature.com

In case Publisher is established outside the EU,
the EU authorized representative is:
Springer Nature Customer Service Center GmbH
Europaplatz 3, 69115 Heidelberg, Germany

Printed by Libri Plureos GmbH
in Hamburg, Germany